
DANCING THROUGH TIME

OXFORD STUDIES IN SOCIAL AND CULTURAL ANTHROPOLOGY

Oxford Studies in Social and Cultural Anthropology represents the work of authors, new and established, which will set the criteria of excellence in ethnographic description and innovation in analysis. The series serves as an essential source of information about the world and the discipline.

OTHER TITLES IN THE SERIES

Organizing Jainism in India and England
Marcus Banks

Sea Hunters of Indonesia
Fishers and Weavers of Lamalera
R. H. Barnes

Society and Exchange in Nias
Andrew Beatty

Nationalism and Hybridity in Mongolia
Uradyn E. Bulag

The Heat of the Hearth
The Process of Kinship in a Malay Fishing Community
Janet Carsten

Global Migrants, Local Lives: Travel and Transformation in Rural Bangladesh
Katy Gardner

Contested Hierarchies: A Collaborative Ethnography of Caste among the Newars of the Kathmandu Valley, Nepal
David N. Gellner and Declan Quigley

The Culture of Coincidence: Accident and Absolute Liability in Huli
Laurence Goldman

The Female Bridegroom: A Comparative Study of Life-Crisis Rituals in South India and Sri Lanka
Anthony Good

Of Mixed Blood: Kinship and History in Peruvian Amazonia
Peter Gow

The Archetypal Actions of Ritual: A Theory of Ritual Illustrated by the Jain Rite of Worship
Caroline Humphrey and James Laidlaw

Shamans and Elders
Experience, Knowledge, and Power among the Daur Mongols
Caroline Humphrey with Urgunge Onon

The People of the Alas Valley: A Study of an Ethnic Group of Northern Sumatra
Akifumi Iwabuchi

Nuer Prophets: A History of Prophecy from the Upper Nile in the Nineteenth and Twentieth Centuries
Douglas H. Johnson

Knowledge and Secrecy in an Aboriginal Religion: Yolngu of North-East Arnhem Land
Ian Keen

Riches and Renunciation: Religion, Economy, and Society among the Jains
James Laidlaw

The Interpretation of Caste
Declan Quigley

The Pastoral Continuum
The Marginalization of Tradition in East Africa
Paul Spencer

The Arabesk Debate: Music and Musicians in Modern Turkey
Martin Stokes

Inside the Cult: Religious Innovation and Transmission in Papua New Guinea
Harvey Whitehouse

DANCING THROUGH TIME

A SEPIK COSMOLOGY

BORUT TELBAN

CLARENDON PRESS · OXFORD
1998

Oxford University Press, Great Clarendon Street, Oxford OX2 6DP
Oxford New York
Athens Auckland Bangkok Bogotá Buenos Aires Calcutta
Cape Town Chennai Dar es Salaam Delhi Florence Hong Kong Istanbul
Karachi Kuala Lumpur Madrid Melbourne Mexico City Mumbai
Nairobi Paris São Paolo Singapore Taipei Tokyo Toronto Warsaw
and associated companies in
Berlin Ibadan

Oxford is a registered trade mark of Oxford University Press

Published in the United States
by Oxford University Press Inc., New York

First published 1998

British Library Cataloguing in Publication Data
Data available

Library of Congress Cataloging-in-Publication Data
Telban, Borut.
Dancing through time: a Sepik cosmology / Borut Telban.
(Oxford studies in social and cultural anthropology)
Includes bibliographical references (p.) and index.
1. Ambonwari (Papua New Guinea people)—Ethnic identity.
2. Philosophy, Ambonwari. 3. Ambonwari (Papua New Guinea people)—Social life and customs. 4. Cosmology, Pacific Island—Papua New Guinea—East Sepik Province. 5. Time—Social aspects—Papua New Guinea—East Sepik Province. 6. East Sepik Province (Papua New Guinea)—Social life and customs. I. Title. II. Series.
DU740.42.T45 1998 306′.09953—dc 21 98-4470
ISBN 0-19-823376-0

1 3 5 7 9 10 8 6 4 2

Typeset by Best-set Typesetter Ltd., Hong Kong
Printed in Great Britain
on acid-free paper by
Biddles Ltd.
Guildford and King's Lynn.

for

VESNA

ACKNOWLEDGEMENTS

My research in Papua New Guinea was made possible by permissions gained from the Institute of Papua New Guinea Studies in Port Moresby and the East Sepik Provincial Government in Wewak. I thank them gratefully for this. Beginning in 1986, Lorraine and Emil Pavsic provided me with support whenever I was in Port Moresby. My debt to them is obvious. I am also grateful to Elizabeth Brouwer for her friendship and help, not only in Port Moresby when I had a viral infection of the lungs but also in Canberra and Durras North. My greatest debt of gratitude in Papua New Guinea, of course, is to everyone in Ambonwari village.

Anayo, asayo, ama kanggrayn kupanma, ama wara wasakɨnday. O, mayn sararin ama kaykan ('Father-o, mother-o, I am already a big woman, I am not a little girl. O, I would like to marry and live so'). I am confident that Aranggmay and Kambrayma, my 'sister's daughters', and Kumbranggawnja and Kanbonma, my 'brother's daughters', still sing this song in Ambonwari while they clean fish in a creek or rest after a filling dinner of sago pudding. I also know that young Yanggimay and Kambianma, my 'brother's' and my 'sister's daughters' respectively, would be quite grouchy, or would pretend to be so, if I asked them about their future marriage. Such teasing would, on the other hand, quite please Saramambi, Simbɨnda, and Sanggɨrmari, my little 'nephew' and 'brother's sons', who would, in the attentive posture which allowed them to run away quickly, help me to tease their sisters and cousins. I miss their chatter on my verandah, and their concern to bring me sago grubs or betel nut whenever they found some.

The first and only house I ever owned was built in Ambonwari, a fact which makes my attachment to this place even greater. I therefore thank Tony Simiwaria Andiyapi and his wife, Augustina Awsay, Godfried Yangus Saramambi and his wife, (my 'sister') Yuana Warima, Jacob Sanggɨnbi Abram and his late wife, Veronika Kaima, for looking after me with their friendship, food, and shelter. Elias Wapun Kandanggway was my first friend, who not only explained different Ambonwari customs but even caught and gave me a pig. It was he who proposed that I work on the language with Julian Kapyamari Yanggan, which proved to be a wise and fruitful decision. Awsay and Warima, Cecilia Wusɨr, Stella Wasakay, Marta Yambakɨnda, Joanita Kogay, Angela Akɨsay, Aleksa Wakarimay, and Rita Akunbay invited me to their houses whenever they had 'special' food, such as grubs or a large fish, and they were also prepared to instruct me about 'women's matters'. There is no space to list all of the men who told me their stories or offered explanations. I want to mention, however, those who sat constantly with me throughout my fieldwork. Elias and Julian were always there, and the work we accomplished together cannot be adequately recompensed. Two village healers, Tobias Yangi Akawi and Bonifacio Marmi Sakmari, instructed me on all sorts of matters and took me with them whenever they were asked to perform a healing ceremony in the village. But my greatest respect and gratitude is due to Bob Kanjik Anjapi, recognized by all as the most knowledgeable man in the village, who not only was my 'father' but spent time with me over the entire period of my fieldwork. I remember my exhaustion after a whole day of discussion and clarification when he quietly lit his cigarette, stood up, and remarked that we would continue next morning. This tempo lasted for months while we were transcribing and translating songs. We shared the anguish when he was accused of being a sorcerer, and again when war began in Yugoslavia. We regretted that I had not come to Ambonwari when he was younger so that we could have shared our youth. Thank you, *anay amanan* ('my father').

I was invited to the Department of Anthropology in the Research School of Pacific Studies at The Australian National University by the late Professor Roger Keesing while I was still working on my M.Sc. in 1987. I also received the hospitality of another ANU professor, the late Anthony Forge, who invited me on several occasions to stay in his house and look after his birds, cats, and especially dogs. It was shocking and extremely sad for me that both men, whom I deeply admired and respected, and who were always supportive of my research, died during my Ph.D. course.

This book is based on doctoral research undertaken at the Department of Anthropology, Research School of Pacific and Asian Studies, The Australian National University. Without naming them all, I would like to thank many friends, fellow students, and colleagues in Australia. Bill Foley advised me to go to Ambonwari. Jadran Mimica, who provided hospitality whenever I went to Sydney, was never short of suggestions about what to read and how to be more critical. Discussions with Aletta Biersack were always stimulating. Within the Department of Anthropology in the RSPacS, Jim Fox as head of department skilfully provided all the field support I needed. Among fellow students I would single out Sophie Creighton for kindness and care, Richard Eves for shared intellectual interests, and Rozanna Lilley for her joy and love, and not least, for correcting the English of my very first draft. I thank all participants of the thesis-writing seminar, in particular Arlette Ottino who directed it. Gilbert Lewis, Nancy McDowell, and the late Alfred Gell offered productive critiques of my thesis and encouraged me to publish it. The penultimate version additionally benefited from the illuminating comments of two Oxford University Press reviewers.

The results and writing of this book were influenced by three scholars with whom I have shared not only an interest in Papua New Guinea but also deep and long-term friendships. Chris Ballard has been a true 'mate' ever since we first met in the Eastern Highlands in 1987. I have never known anyone who searched with such diligence for anything published on Papua New Guinea and be prepared to share it with the same enthusiasm. Don Gardner won my greatest respect and admiration. Not only as a pool and snooker player but as an academic who combined challenging critique with generous moral support, Don proved to be indispensable as a friend and supervisor. Without his enormous help during the final stage of my thesis I doubt that I would have been able to submit it in time. Many thanks to Anne, Sarah, and Catherine Gardner, too, for tolerating all the interruptions to family life caused by my last-minute demands. Finally, I wish to thank the person who was not only my chief supervisor but was instrumental in bringing me back to ANU in 1990, a friend with whom I shared periods in Australia, England, and Papua New Guinea. Michael Young was someone who became part of my own life through our Sunday bush walks, and weekends in Durras North, his visit to Ambonwari and mine to Vanuatu, his shared knowledge and appreciation of the richness of English expression, his unfailing care and support. My enthusiasm for certain ideas often went beyond the realm of possibility and though Michael shared much of excitement, he also tried to keep me down to earth (even when Don and I went sky-diving). It is especially on account of these three men that I was so sad to leave Australia.

My warm appreciation goes to the various other people whose services were indispensable. Ann Buller looked after my finances while I was in New Guinea, Margaret Tyrie formatted my figures and tables, Kay Dancey from the cartography unit produced all the maps, and Barbara Holloway proofread my final draft. I am most grateful for their patience and consideration. It can be said that without Ria van de Zandt's care the computers and printers in the Department would be inoperable—and so would I.

I acknowledge the generous support of The Australian National University through its Scholarship Stipend, Tuition Fee Scholarship, and other funding. I also thank the Slovene Research Association for a grant before I travelled to Australia, and the Royal Anthropological Institute of Great Britain and Ireland for a Radcliffe-Brown Memorial Fund/Sutasoma Trust award which enabled me to complete my Ph.D. thesis after my ANU funding had expired. The Slovene Ministry of Science and Technology awarded me a small grant which also helped me survive during the final stage of my writing. The tenure of a Leach/RAI Fellowship in the Department of Social Anthropology at the University of Manchester during 1995–6 and my ongoing research fellowship at the Scientific Research Centre of the Slovene Academy of Sciences and Arts gave me the opportunity to rethink and improve the whole text. Chapter 3 was previously published in *Manchester Papers in Social Anthropology* No. 4. Some material in this book has appeared in my essay 'Body, Being and Identity in Ambonwari, Papua New Guinea', in *Common Worlds and Single Lives: Constituting Knowledge in Pacific Societies*, a book edited by Verena Keck (1998). I thank Berg Publishers of Oxford for permission to reproduce it.

Last but not least, I wish to thank my parents who provided loving support and understanding despite the turmoil and hard times they themselves were going through. I know that my brother, who is in many ways responsible for my wanting to become an anthropologist, will most appreciate the path that I have taken, and for his encouragement I am deeply grateful. Without naming them, I would also like to thank many Slovene friends who kept in contact with me all these years. And finally, one who has kept faith with me during my travels, long absences, and throughout our private wars; in remembrance of a glittering dew on the stalks of young grass, the footprint of a firefly, and the pale petal of a daffodil, I dedicate this book to her.

Kdo brez vesel lahko vesla
in brez vetra jadra?
Kdo lahko od prijatelja
loči se brez solza?

Jaz brez vesel lahko veslam
in brez vetra jadram,
a ne morem se—nikdar
ločiti od prijatelja.

CONTENTS

LIST OF MAPS

LIST OF FIGURES

LIST OF TABLES

LIST OF PLATES

NOTE ON ORTHOGRAPHY

My study relies heavily on the Karawari language, which is a member of the Lower Sepik Family of the large non-Austronesian group of languages in New Guinea called the Papuan language group. Fortunately, prior to my departure for Papua New Guinea, I was able to discuss the main aspects of Karawari with William Foley, who studied the neighbouring Yimas language, and even recorded some Karawari. Thanks to Foley's detailed research (1986; 1991), and the many similarities between Yimas and Karawari, my understanding of this Papuan language was greatly facilitated. Throughout the book Karawari terms are italicized while those in Tok Pisin are italicized and marked by (TP).

Phonetically, Karawari, like Yimas, is a four-vowel language /a/, /ɨ/, /i/, /u/, where /ɨ/ is the most often used. It is heard as /i/ in 'sir'. The mid vowels /o/ and /e/ are overshadowed by a single vowel /a/, though in some cases pronunciation remains unclear. Thus, the /aw/ or /wa/ combinations may be heard as /o/, for example, *mariawk/mariok* (talk, speech, discourse, story) or *mambaw/mambo* (crocodile). Sometimes /o/ and /a/ are so close that it is difficult to tell which one is actually used, e.g. *pariapo/pariapa* (front porch, verandah). It appears that the younger people more often pronounce /o/ than their elders. The /ɨ/ phoneme is often hard to distinguish from /i/ or /u/ and it is ambiguously phonemically present; sometimes it seems to be used to break up certain consonant clusters. On the other hand, the presence of /ɨ/ in many words justifies its retention as a separate phoneme. Quite often a semivowel /y/ interacts with a vowel /i/, as for example in the verb and noun *kay-/kai-* (habit, ritual; to be, to stay; canoe), or in *yimbian/imbian* (pig). In the initial position, however, the semivowel is far less distinct than in the terminal position, a fact which I have indicated in my orthography.

Among the consonants are two pairs of allophones, /r/ and /l/ and /t/ (sometimes /d/) and /s/. In Ambonwari the preference seems to be for /r/, e.g. *yamara* (bird of paradise) in the first pair, and for /s/, e.g. *sunggwi* (moon) in the second pair. In a few cases there is minimal distinction between the plosives /b/ and /p/. Whenever /g/ follows /n/, /ng/, it indicates an 'eng', nasal /n/ with /g/ inaudible. When /g/ is heard after such a nasal, I write it as /ngg/, as in *sunggwi* (moon).

Introduction

Preamble

When I was a boy in primary school a friend gave me a book for my birthday. The dedication was from Richard Bach's *Jonathan Livingston Seagull*:

> If our friendship depends on things like space and time, then when we finally overcome space and time, we've destroyed our own brotherhood! But overcome space, and all we have left is Here. Overcome time, and all we have left is Now. And in the middle of Here and Now, don't you think that we might see each other once or twice? (1970: 63)

Much as I liked this saying I was somewhat puzzled by it. Over the years, as I travelled through Europe, North and Central America, South-East Asia, and Australasia, I began to grasp those Heres and Nows, but I still wondered about that 'middle of Here and Now'.

Shortly after starting high school, I was lucky to befriend a bohemian film director with whom I used to exchange thoughts about the works of *fin-de-siècle* novelists which were a 'must' for that 'Here' and 'Now' of intellectuals. Influenced by him I 'departed on a long journey' through the many volumes of Proust's *Remembrance of Things Past*, which had been sensitively translated into the Slovene language. Among other things, Proust's description of the Narrator's great-aunt's room and his blaming human habit for his inability to familiarize himself with new surroundings made me wonder for a long time to come. Not that I had any such problems myself, nor was I particularly interested in 'regaining lost time', but I did begin to value my memories and my past despite my future-oriented daydreaming. I even understood why Proust thought of entitling his first volume *The Age of Names*.

While still studying pharmacy I journeyed for several months around south Mexico and Guatemala, and this experience was crucial for my future life. I spent time with Slovene ethnologists, whose lifestyle I admired compared to that of pharmacists. The two 'fields', in Bourdieu's sense, were far apart. Slovene ethnology was oriented towards folklore and national heritage and it was hard to find anyone who had conducted serious ethnographic fieldwork outside Europe. Since it was Latin America that fascinated several ethnologists of my acquaintance (among them my brother) I decided to take a different path—towards the East.

In 1985 I left Yugoslavia on a journey which still continues. I spent more than three years in Papua New Guinea, almost five in Australia, one in England, and the rest either travelling or in Slovenia. I conducted my first research in the Highlands of Papua New Guinea as a qualified pharmacist interested in ethnomedicine, and that is when I became truly fascinated by New Guinea's cultures. This led me to pursue the study of anthropology. I completed my M.Sc. at the University of Zagreb, where I became accepted as a pharmacist-turned-anthropologist. I then abandoned my previous occupation and enrolled in a Ph.D. course in anthropology at the Research School of Pacific Studies at The Australian National University.

My first period of fieldwork among people living in the scattered hamlets of the upper Yuat River gave me one criterion for choosing a second fieldsite: I wanted to live in a populous village where I would be able to make day-to-day observations. While I was trying to decide where to go, Jadran Mimica introduced me to Bill Foley, who enthusiastically proposed that I go to Ambonwari village in the Sepik. He provided me with a few cassettes of the Karawari language which he had recorded while carrying out linguistic research among neighbouring Yimas. The decision was made. Already fluent in Tok Pisin, and familiar with many general aspects of life in Papua New Guinea, I departed again, this time for the village in the middle of the mosquito-ridden sago forest. I lived in Ambonwari for eighteen months between September 1990 and March 1992.

I returned to Australia convinced that before I even began to draft my thesis I should write a Karawari grammar while the language was still fresh in my memory. And then, as if from nowhere, appeared the question of my identity, posed through its most explicit indicator: language. I remember once saying in a friend's house that I did not have a sentimental grasp of the English language, that I did not 'feel' it. After this moment I became more and more aware of, at times almost obsessed with, the interrelations and differences between those languages I had been using in recent years: my Slovene, my English, my Tok Pisin, my Karawari. I could swear obscenely in English without feeling I had broken any taboo. I was able to express certain things that I thought about only in English, some seemed better said in Tok Pisin, while I began to stutter in confusion when trying to say the same things in Slovene. I was perhaps fortunate never to master Karawari, a difficult Papuan language, and thus not to encounter similar trouble. I thought about Lévi-Strauss, who complained about the mistranslation and misinterpretation of some of his French terms. And I read about the multiple translations of and arguments concerning Heidegger's *Befindlichkeit* ('affectedness'). So I asked myself: how can I write in English about Ambonwari village where people speak Karawari, when my remoteness from both is obvious? This seemed to be an even more tangled thicket than the one which most anthropologists stumble through while interpreting other people. This book represents a compromise. My writing in these pages is an interpretation of an interpretation made by Ambonwari people about their own lives, written in my anthropological—and thus my learnt or adopted—language, English, which I interpret and understand in my own idiosyncratic way and in the ways of those with whom I have communicated during these years.

During my stay in Ambonwari I embarked on another journey—through phenomenology and existentialism. Reading Sartre while in the field led me to search for Ambonwari's own particular expressions, meanings, and ontological concepts. While compiling a dictionary of Karawari I was at first surprised that there was no term for time. As the relationship between temporality, historicity, and identity became apparent I decided that these concepts should become the main focus of my book. After returning to Canberra, I joined a postgraduate reading group which discussed Heidegger's *Being and Time*. Weekly meetings lasted for over a year, though by the end the group had shrunk to two participants. The works of Bourdieu, Cassirer, Ricoeur, and Schutz became my intimate 'consociates' during the period of my writing solitude. Though I never regretted these journeys, it was only towards the end of my course that I fully realized that writing about time *takes* time. On a short trip to England in 1993 I read Alfred Gell's *The Anthropology of Time,* which made me perceive all over again the complexity of the subject. I radically revised my writings, first following Michael Young's advice concerning the importance of the clarity of the text, and second following Don Gardner's enthusiasm for the Ambonwari's own notions, concepts, and practices, which I had somehow allowed to become obscured. This reworking sharpened the concept of *kay*, though at the expense of theoretical and comparative considerations of 'time'.

While I was living in Ambonwari, war broke out in Yugoslavia. Being a Slovene, my birthplace, in terms of the nation at large, suddenly became a foreign country. For people living there, as well as for many abroad, the importance of their identity suddenly loomed large. The history I had learnt in school was abruptly refuted and historians began to write new ones. I shall never forget the many discussions I had with my Ambonwari friends and the way none of us could comprehend the genocide that had begun in Yugoslavia. In addition to those who were killed, many people who had lived there all their lives were 'de-personified'. The tragedy of Boris Pasternak's novel *Doctor Zhivago*, in which, after the hero's death, his wife Lara disappears as a nameless number into one of the innumerable concentration camps, became a reality again. As Fraser remarks (1987: 179):

> making people into nonpersons is the socially invented and enforced version of Alzheimer's disease. It deprives a person of that single symbol that he could otherwise have retained for life and held up against the change of everything else. It steals from him his sense of human time and regresses him into the biotemporal umwelt of living but mindless organisms.

A few years ago I hurt the feelings of a person whom I have loved dearly through all these seasons of exile. She posed a heartbreaking question: 'What are we going to do with our past?' While my own process of identification was doing its very best in its 'Now', I was yearning for a 'secret treasure' (the kind that children make under a piece of glass in the earth) and that time which two of us had shared. I recognized the meaning of that 'middle of Here and Now' from Jonathan Seagull.

Then I realized that time for the completion of my thesis was running out. So my answer to friends who asked me to go out with them became 'Sorry, I don't have time.' At that moment, too, I knew exactly the answer to the famous question which puzzled Augustine: *Quid est enim tempus?* What then is time?

> Who can explain it easily and briefly? Who has comprehended it in thought so as to speak of it? . . . What then is time? If no one asks me about it, I know; if I am supposed to explain it to one who asks, I do not know; yet I say confidently that I know: if nothing were to pass away there would be no past time, and if nothing were coming there would be no time to come, and if nothing were to exist there would be no present time. (Augustine, *Confessions*; cf. Heidegger 1988: 229)

If anyone had asked me I would have known the answer in these terms: time was something that I did not have.

The Study

This is a study of cosmology, identity, and temporality among the Karawari-speakers of Ambonwari village in the East Sepik Province of Papua New Guinea. The published anthropological material on Karawari is sparse; no long-term research has been carried out amongst them, and the few publications that have appeared (Bühler 1961; Craig 1992; Haberland 1975; Tischner 1965) concentrate mainly on the area's most fascinating material objects—carved crocodiles. Yet the Sepik basin more generally has received substantial ethnographic coverage: see for example the 1984 issue of *Research in Melanesia* and *Sepik Heritage* (Lutkehaus et al. 1990), a large volume based on papers presented at a conference, sponsored by the Wenner-Gren Foundation, which took place in Basel in 1984. Not only have there been numerous researchers in this part of New Guinea, but some of the most influential monographs in the early ethnographic literature, such as Bateson's *Naven* (1958 [1936]), Mead's *Sex and Temperament in Three Primitive Societies* (1963 [1935]), and Whiting's *Becoming a Kwoma* (1941), also came out of this part of the world. Ambonwari have close cultural links with some of the neighbouring groups examined in such studies, and various publications have greatly facilitated my own work. First, Foley's research on the Yimas language (1986; 1991) enabled me to understand Karawari language better than I would otherwise have been able to do (see Chapter 2); second, similarities between the Ambonwari and the Iatmul (Bateson 1932; 1958; Schuster 1990; Stanek 1983; 1990; Wassmann 1990; 1991) allowed me to compare not only the kinship and naming systems but also other aspects of social life (see Chapters 3 and 4); third, despite the cultural differences between other Sepik peoples and the Ambonwari, studies by Forge,

Gewertz and Errington, Harrison, Lewis, McDowell, and Tuzin proved indispensable before, during, and after my fieldwork. Harrison's work on person, spirit, and understanding among the Manambu, for example, directly stimulated my own inquiries while I was still in the village (see Chapter 2), while McDowell's lucid examination of the *kamain* exchange partnership in Bun enabled me to compare it with a similar kind of relationship in Ambonwari (see Chapter 3), though regrettably only after I had already returned to Australia. The late Anthony Forge gave me copies of most of his publications, and despite the differences between the Abelam and the Ambonwari he made me think differently about the carved objects so important to the Ambonwari (see Chapter 6).

In a time when anthropology as a 'craft' is under scrutiny, and so many different kinds of cultural and textual studies take priority over fieldwork and plain ethnographic reportage, it was the above-mentioned ethnographies which enabled me to understand the Ambonwari's own notions better and to reflect upon and question many of the concepts I had acquired during my previous education. But despite the transformation in my life over the last decade there have been continuities. In particular, this book raises issues about identity and continuity that are relevant to those I have faced existentially, for its focus is the Ambonwari concept of *kay* (way, habit, manner; ritual; being). As I will argue, this concept articulates with mythical, historical, and personal temporalities and it is crucial to the Ambonwari's conceptualization of their present and future identity. *Kay* represents, among other things, habitual embodied practices, which, as Bourdieu would say, produce time in the very acts whereby they produce themselves.

Locating the Study Theoretically

My aim here is not to review the anthropological literature on time in different societies. Thanks to earlier reviews (e.g. Maltz 1968; Rigby 1983) and particularly to the recent ones by Gell (1992) and Munn (1992), this task has been fulfilled. Many societies, including Ambonwari, do not even have a term which can adequately translate the English term 'time', with its multiple and often measurement-oriented meanings. Their 'time' is not an abstract idea but an integral part of their life.[1] Moreover, in Ambonwari temporality is an indispensable part of people's identity: the ways activities are performed, the manner in which people behave, the ways the village functions, as well as most significant rituals, are all subsumed in the term *kay*, which refers to embodied processes, both collective and individual. The past as represented in *kay* is not just automatically reproduced, but requires an active process which is reflected

[1] This kind of conclusion was reached by e.g. Bohannan (1953: 262) for the Tiv; Evans-Pritchard (1974: 104–8) for the Nuer; Hoskins (1993: 58) for the Kodi of West Sumba; Howe (1981: 223) for the Balinese; Ohnuki-Tierney (1973: 285) for the Ainu; and Panoff (1969: 153) for the Maenge.

upon and guided by people's *wambung* ('insideness'). Schutz made it clear that '*the problem of meaning is a time problem*—not a problem of physical time, which is divisible and measurable, but a problem of historical time' (1972: 12; emphasis in original). Filled with physical events and having the nature of 'internal time consciousness' (Husserl 1964), historical time is always related to one's own life. 'It is within this duration that the meaning of a person's experience is constituted for him as he lives through the experience' (Schutz 1972: 12). Thinking of Ambonwari I agree with McDowell, who emphasizes that 'process is a key element in Bun worldview' and that 'Bun history is not only about the past but also the present and the future as well' (1985: 32–3, 36). These two views, so simple and so complex at the same time, are both characteristic of Ambonwari *imɨnggan kay*, 'the way of the village', of which *kupambɨn kay*, 'the way of the ancestors', is its past: a past projected into the future as the specific way of preserving the eternity of community. However, this projection of the past into the future is not automatic. As *kay* articulates with understanding, thoughts, feelings, and so on, people are confronted with an open future, which also means reflection on and reconstruction of their own past which makes their future.[2]

In order to foreground my presentation of *kay* as it unfolds in the book I would like to look briefly at some of the methods used by analysts in the comparison and differentiation of the human conceptualization of the past.

I begin with three oppositions used to contrast forms of temporality: qualitative versus quantitative; past versus history; eventful past versus chronological past. Despite the differences between them, these oppositions nevertheless have important characteristics in common. In all three paired oppositions there is one set of terms which can be numerically expressed or measured, and the second in each set is usually associated with modern Western societies and the former with pre-literate, pre-industrial or non-Western societies.

The notion of qualitative time is associated with Henri Bergson (1912), for whom time exists only in qualitative terms, its quantitative aspect being determined by space. Time appears to be a measurable magnitude, but it is not; whenever we try to measure time we replace it by space. Bergson saw the intensity of our conscious states as purely qualitative: 'pure duration is the form which the succession of our conscious states assumes when our ego lets itself *live*, when it refrains from separating its present state from its former states' (1912: 100). While duration might be a succession of qualitative changes melt-

[2] Relating mythology to history, Terence Turner argues that: 'history and myth are both primarily to be understood as modes of consciousness of the social present, expressed in terms of the relation of that present to its past (and future). As such, they form part of the sociocultural structures through which the present is produced, which always means in large part *reproduced*, as an instance of a pattern of the past. Social consciousness, in both its mythic and historical modes, is thus formed as an integral part of the process of social reproduction through which a society reflexively defines itself' (1988: 279).

ing together, such a duration would be 'pure heterogeneity'. For Bergson, the idea of a homogeneous and measurable time was simply an appearance, an illusion, a concept deriving only from the idea of space. He observed that there is a real space without duration, and that there is a real duration, but that it is heterogeneous. 'The comparison of these two realities gives rise to a symbolical representation of duration, derived from space' (Bergson 1912: 110). He called the link between duration and space 'simultaneity' and defined it as 'the intersection of time and space'.

While Bergson applied his theory to humanity in general, his notion of qualitative time found its way into the social sciences to illustrate the recollection of the past and the experience of time in many, mainly pre-literate, societies. Lucien Lévy-Bruhl, in his *Primitive Mentality*, concluded that 'primitive' societies see time as a qualitative entity which is felt as a quality rather than as an intellectualized homogeneous quantity, represented by a linear order of succession (Rigby 1983: 430–1). Though critical of his work, Leenhardt, a close friend of Lévi-Bruhl, found the qualitative aspect of temporality applicable to the Canaques of New Caledonia. Leenhardt believed that the notions of time and space for the Canaque were both discontinuous and qualitatively grasped. He asserted that the Melanesian notion of time is indistinguishable from the notion of period, and as such cannot be related to geometric measurement, and that therefore Melanesians cannot have history. An individual Canaque is enclosed in the spatio-mythic domain of his or her own existence defined by the social group, its totems, boundaries, altars, and names (1979 [1947]: 92). Leenhardt argued that the Melanesian is 'unaware of his body, incapable of classifying objectively', while space appears 'as a heterogeneous ensemble of places whose existence is felt by physical presence' (pp. 91, 106). Again, 'when the sensuous reaction to the resistance of the physical milieu is absent, space does not exist' (p. 47). Leenhardt's vision of Canaque person and lived myth is extremely stimulating despite its many shortcomings (such as his declarations that 'space does not exist' and that the Melanesian is 'unaware of his body'). One can also criticize his inconsistent use of myth, his confusion of analytic construct with empathetically felt reality, his stress on 'word' at the expense of 'deed', his blurred distinction between Canaque cosmology and his own philosophy (Clifford 1982; Crapanzano 1979; Young 1983: 13–16). Nevertheless, one of the virtues of his book is its indication of the direction which future studies of time, space, and the human body might take.

Evans-Pritchard (1974 [1940]: 94, 135) distinguished between ecological time (that is, representing the Nuer's relationship to their environment, conceptualized in terms of seasonal variations of social activities) and structural time (that is, representing people's relationships to one another in the social structure, conceptualized in terms of relations between political groups). He argued that both were socially determined. For Evans-Pritchard ecological time is dynamic, rhythmic, a process which is 'reflected' upon—in Munn's words—'through a lens of cultural concepts referring to activities (i.e. via the time-reckoning "system"), rather than experienced through immersion in the activities' (1992: 96). In opposition to ecological time, Evans-Pritchard found that the structural time of institutionalized relationships between political groups did not 'move'. It was persons and groups who 'moved' through the ideally constant structure: 'Time is not thus a continuum but a constant structural relation between two points, the first and last persons in a line of agnatic descent' (1974: 107–8). While ecological time was related to people's daily and seasonal activities in their social contexts, structural time was a 'model' which immobilized the relationships between, and arrangements of, groups in an illusory and idealized genealogical space-time. 'The space of Evans-Pritchard's structural time is thus not concrete and qualitative but geometrical and quantitative—rather like a social-structural version of Bergson's spatialized time' (Munn 1992: 97). The static nature of structural time so greatly influenced Lévi-Strauss that he used it not only to characterize a single society but a whole class of societies (Gell 1992: 23).

Lévi-Strauss introduced a distinction between societies he called 'cold'—those without history which thereby replicated without change—and societies he called 'hot'—those which internalized the historical process (1966: 233–4). The Lévi-Straussian dichotomy between those societies with and those without history entailed others: history/myth, dynamic/static, changing/unchanging. It is a series of oppositions that 'leads to the conclusion that native Amazonians attempt to freeze historical events into an ahistorical, atemporal, mythic order' (Hill and Wright 1988: 79).[3] But this is not the place to canvass opinion on Lévi-Strauss's contrast between synchrony and diachrony and I simply echo John Barnes (1971: 541) and Alfred Gell in saying that the dichotomy does not really refer to two distinct kinds or types of 'time' but depicts different kinds of temporal process and 'the contrast between models and reality, structure and events, essence and accident' (Gell 1992: 25).

A third dichotomy also attempts to characterize and explain differences in the ways of constituting the past in Western and non-Western societies. Errington (1974) follows Gellner (1964) in seeing history (and change) in the West as 'evolutionary', and he borrows Gellner's term 'episodic' history to characterize Karavarans.[4] The same distinction was used by McDowell in her discussion of the nature of episodic time in Bun. Evolutionary history and change emphasize the passage of time; successive generations are not interchangeable; there is a continuous accumulation of more and more history;

[3] Lévi-Strauss's dichotomy has attracted substantial critiques by Barnes (1971), Fabian (1983), Gell (1992: 23–9), Harrison (1990a: 73–6), McDowell (1985: 31–3), Rosaldo (1980), Sahlins (1985), Schieffelin and Gewertz (1985: 2–3), Wagner (1972), and, most importantly perhaps, by those writings on Amazonian and Andean societies in Hill (1988).

[4] 'History dissolves being into the never-ending sequence of becoming, in which no point is singled out but every point indicates the way to one farther back, so that regression into the past becomes *regressus in infinitum*' (Cassirer 1955: 106).

events are 'placed' in time (Gellner 1964: 2; cf. McDowell 1985: 28). In contrast 'an episodic view of change and the past is one of relative changelessness; changes occur only as a result of events of radical or monumental impact' (McDowell 1985: 29). Schieffelin takes a similar view when arguing that the perception of time for the Kaluli 'seems to be constructed around events, rather than events being primarily related to each other in time' (1976: 141).

These oppositions effectively set up a dichotomy between us and them, the West and the Rest. But we have to recognize that all people, everywhere, regardless of whether or not they render these oppositions explicit in standardized systems of time measurement, have at their disposal both aspects of any of them. What differs across peoples, however, is which aspect of any one of these oppositions people elaborate as an explicit element in their representation of themselves, and how precisely they do so (see Toren 1988). Therefore, in order to escape this 'sterile ballet' of symbols (in Barnes's felicitous phrase), in which qualitative and quantitative, episodic and evolutionary, past and history, reversible and irreversible time, synchrony and diachrony, discontinuity and continuity 'dance endlessly round each other until the audience decides to go home', we need to 'break down the dichotomies, establish continua and feed in more facts' (Barnes 1971: 545).

It was Bourdieu's concept of habitus and his emphasis on practices which guided my theoretical approach to the Ambonwari's 'way of the village'. He states that 'to restore to practice its practical truth, we must therefore reintroduce time into the theoretical representation of a practice which, being temporally structured, is intrinsically defined by its *tempo*' (1977: 8). Bourdieu advocates the study of people's practices, arguing that it is the strategies of interaction that give time its social efficacy (and vice versa). Manipulation of the 'tempo' of the action (by waiting, postponing, delaying, hurrying, surprising, 'giving', or 'witholding time') allows people to exploit limitless possibilities. As we shall see, in a somewhat different way, Ambonwari *kay* encompasses people's ways of doing things, their habits, manners, and so forth. It embodies what philosophers like Bergson and Russell called 'habit-memory' in contrast to personal and cognitive or 'true memory' (Connerton 1989: 23). However, as I shall be at pains to argue, it is the relationship between these two kinds of 'memory' which is most important for Ambonwari identity and its processual nature, which produces time as it produces itself.

Outline of the Book

It is my contention in this book that the concept of *kay* and those concepts with which it articulates closely, *wambung* ('insideness'), *mariawk* (speech, story), and *konggong* (path, marriage), are the means by which Ambonwari represent their life-world both as an enduring structure and as an ongoing process. Linking institutional and individual practice, *kay* temporalizes both. It is primarily grounded in embodied processes which it temporalizes in relation to both a past and a future that are mythic, closed, and necessary, yet historic, open, and contingent.

In Chapter 1 I introduce the village in its wider Karawari and Sepik context, discussing its subsistence pattern and such habitual activities as fishing and hunting. I show how *kay* is extended to and inseparable from those things which a person uses in these as in other activities. The tables in Appendix A provide the factual basis for this argument. I also examine food taboos, which show the openness and experiential nature of *kay*, and I discuss house-building, which exemplifies the strategic nature of social practice. In the section entitled 'Male and Female *Kay*' I argue that while men give identity to their own place, women form and preserve the relational dimension of group identity. I end this chapter with a discussion of the significance of *kay*, and relate it to some theoretical issues.

In Chapter 2 I look first at temporal expressions in the Karawari language and then discuss their concept of 'period', arguing that it is the distinctiveness of the process which defines temporal span. In the subsequent sections I examine those concepts which are most important in the Ambonwari's conceptualization of their collective and personal identity: first, *kay* itself associated with *konggong* (path; marriage) and *mariawk* (speech, discourse, thought; story, myth), and second, personal spirit and *wambung* ('insideness'), which represents knowing, understanding, and feeling. Though it is an aspect of *kay*, *wambung* is always able to transform *kay* through self-transformation. I end this chapter by emphasizing the processual nature of Ambonwari identity.

In Chapter 3 I discuss *kay* in relation to the key social institutions of clanship and lineage affiliation, the naming system, 'fathers' and 'mothers' of the village, *warɨmbar* ('exchange partnerships'), and *sɨmɨrinyar* ('ritual inaugurator') relationships.

In Chapter 4 I discuss the kinship system as an aspect of 'the way of the village', *imɨnggan kay*. Ambonwari 'Omaha' kin terminology seems to endorse two main characteristics of their society: the principle of equivalence of alternate generations (identification between a man and his paternal grandfather) and the concept of 'oneness' in which all generational differences are made to disappear. After elaborating Ambonwari marriage practices and one of the differentiating dimensions of *kay*, i.e. *konggong* (path, marriage), I look at formal kinship in its relationship to behaviour.

In Chapter 5 I present another important aspect of *imɨnggan kay* ('way of the village'), that is, *kupambɨn mariawk* (myth). It narrates the mythic-historical past which for the Ambonwari is the core of identity; it is part of lived experience and is thought and talked about. The myth of Moon and Sun suggests that Ambonwari once had two moieties. I focus on the first ancestor of the village, Kapi from the Crocodile-1 clan, and his myth of origin, which is also that of Ambonwari itself. I discuss the ways in which people identify with Kapi and how he thus becomes an element of collective identity. In the last part of this chapter I discuss the temporality of myth in its relation to Ambonwari identity. To show emplotment of the myth in terms of 'narrative' I append a significant story about a stone breast (Appendix B).

In Chapter 6 I first present the distinction between ordinary things and spirit-things. I argue that ordinary things are defined by their usefulness and are thus part of people's personal *kay*. Spirit-things, by contrast, are beings in themselves, embodied in carved objects with their own identity and temporality and are consequently part of *imɨnggan kay*. In a section about the differentiation of spirits, I argue that men manipulate the visibility of spirit-things to gain command over them and over their own existence. I propose that it is an elaborate concept of 'caring' which determines the crucial distinction between those spirits whom Ambonwari fear and those whom they do not. I discuss men's houses and the most important spirit-things, such as masks, flutes, posts, slit-drums, and spirit-crocodiles, in the second part of this chapter.

Finally, Chapter 7 is dedicated to rites and ceremonies, the practice of which is recognized either as mundane *kay* or, in the most significant rituals, as *kupambɨn kay* ('ancestral way'). The focus of ritual is *kay* itself: either its transformation or its presentation. I omit, at the people's own request, the description of the most important Ambonwari ceremony, male initiation ritual. However, because my main argument in this chapter concerns rituals that embody collective values concerning gender, social positions, and roles, the first menstruation ceremony and the all-night dance serve my purposes equally well. The latter is a celebration performed by people as frequently as possible. It unifies the heterogeneous elements of the village and integrates participants in a singular experience of collective identity. It re-embodies their shared past and establishes the grounds on which their future can be anticipated. This supremely popular—but in some sense ordinary—ceremony crystallizes so many dimensions of the complex notion of *kay* that its central element, dance, inspired the title of this book.

I conclude my book with a discussion of *kay*, *wambung*, and time.

1

Ambonwari: The Way of the Village

In September 1990 I left Wewak, the capital of the East Sepik Province in Papua New Guinea, and took a truck along the dirt road to Angoram, a well-known market town on the Sepik River. The larger part of the settlement near the river was built by non-local people from different language groups, some coming from the north but most from the swampy areas to the south. They bring fish and sago to sell at the market before they return to their homes with the goods they have purchased at the local trade stores. A large canoe, once used for the purpose of war, took me up the Sepik River. Heavy rain drenched the outboard operator, his son, and myself all day long. I tried to see, through the mist and rain, if anyone else was travelling in either direction. Except for islands of grass floating in the muddy river there was nothing. I was left to ponder my first trip to Papua New Guinea several years before, and the uncertainty of my arrival in Ambonwari. The people, of course, did not know that I was coming.

After passing the mouths of the Keram and Biwat (or Yuat) Rivers we might, if the waters had been high, have entered the Kangramai Creek and thence found a way through the flooded swamps to Karawari River. But as the waters were low we proceeded south up the Sepik. Immediately after landing in Timbunke, I was lucky to come across two canoes, one from Ambonwari and one from Imanmeri. Tony Simiwariya, who later became my *kupan*, 'elder brother', was willing to take me with him. We left the next morning and near Mindimbit, Iatmul village, entered Korosameri River. After a short time we arrived at the confluence of the Korosameri and Karawari Rivers and proceeded up the latter. The first Karawari-speaking village was Manjamai. Further south we entered the Konmei Creek just before reaching Amboin. Leaving Konmei village behind, we arrived at Ambonwari village (see Plates 1 and 2). The canoe trip from Angoram to Ambonwari usually takes between ten and fourteen hours, depending on the weight of the load and the power of the outboard motor.

The Village

Ambonwari people consider themselves to be an independent group in relation to their village membership, while at the same time acknowledging totemic relationships with communities beyond. They recognize their past and present relations (both amicable and hostile) with other Karawari-speaking people who

PLATE 1. Yanbonman, the main men's house in Ambonwari

PLATE 2. Ambonwari village as viewed from Maramun Creek

live in villages to the west and north, as well as Imanmeri (Arafundi language group) to the south, Kansimei (Biwat or Bun language group) to the south-east, and Yimas, south of Amboin, whose language and culture are the closest to Karawari. Karawari language (people say *anduk apianak*, 'our language', lit. 'our mouth') is classified as a member of the Lower Sepik Family which belongs to the large group of Papuan languages. A major characteristic of the Lower Sepik Family, as Foley (1986) has shown, is that all six languages of this family (Murik, Kopar, Angoram, Chambri, Yimas, and Karawari) are morphologically agglutinative, employing both prefixes and suffixes (see Map 1). Another common feature of Karawari nouns and verbs, as in Yimas and German, is their compounding or serialization (see Foley 1991: 3, 84–7). Because of this I often search for additional meanings by breaking up the words into their meaningful segments, as indeed the Ambonwari themselves do. About 2000 people speak Karawari.

There is a Subdistrict office in Amboin where the officer in charge is still known by the Tok Pisin term *kiap*. There is also a grass airstrip in Amboin where small planes bring tourists to the Karawari Lodge, and from which one can fly to Mt Hagen in the Western Highlands or Tari in the Southern Highlands. The whole area is under the administrative control of Angoram District Office.

The land of Karawari-speakers can be defined by the lower Karawari River (*Yimaria*), the lower Konmei Creek (*Kwanmar*), and the upper Kangramai Creek (*Kanggramai yinggun*), between longitude 143° 30′ and 143° 40′ E and latitude 4° 30′ and 4° 38′ S in the Sepik (*Ambiamai*) River basin area (see Map 2). Along with countless bush camps, of which new ones are continually being built and old ones left to decay, there are eight main villages where people speak Karawari. According to the Ambonwari people, small dialectical differences between some of these villages mean that the latter can be classified in four groups:

Konmei (*Kwanmar* or *Singgara*), Manjamai (*Masamai* or *Karɨnggupa*), Kundiman (*Kundiman*)
Meikerobi (*Maiprokin*), Kaiwaria (*Kaiwarian*), Kungriambun (*Kunggriambun* or *Warɨpanai*)
Masandenai (*Masɨndanggwan*)
Ambonwari (*Ambanwurin* or *Manjimakan*)

Ambonwari is probably the largest Karawari-speaking village, with over 400 people. The name 'Ambonwari' was given to the village by recruiters who passed through in the 1930s or 1940s searching for cheap labour. Ambonwari people call their place *Ambanwurin*. This is actually the name of the nearby sago forest where the original village stood long ago. A male villager is called *Ambanwurinyar* and a female *Ambanwurinma*; for two persons together there is

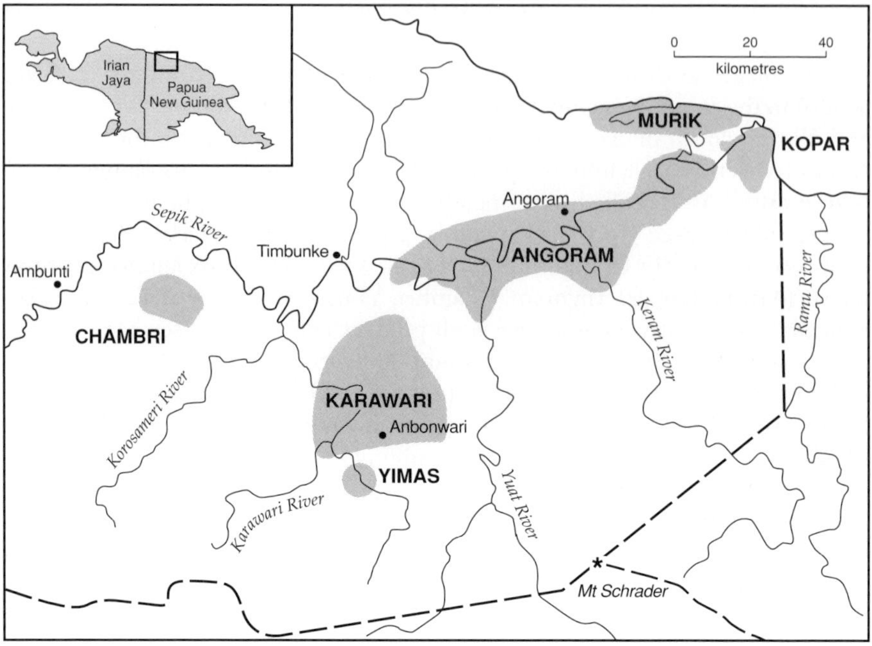

Map 1. The Lower Sepik Family

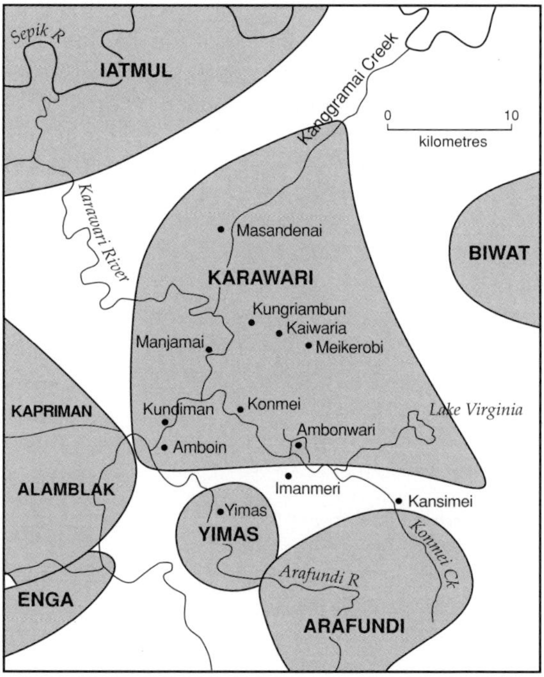

Map 2. Karawari and neighbouring languages

a single term, *Ambanwuringri*, irrespective of sex. Ambonwari people refer to themselves as *Ambanwurinmas* if men, *Ambanwurinkɨmbi* if children, and *Ambanwuringa* if women. If a collective term is needed for the people of the village (though such a concept is seldom used), the last, feminine term is employed.[1] The semantics of the terms *Ambanwuringa* and *Ambanwurinmas*, of which people are well aware, reveal how the Ambonwari posit a period of transition from chaotic homelessness into the familiar cultivated past. *Amban*, as a noun, signifies a dense jungle with lots of vines, through which the ancestors had to struggle. *Amban*, as a verb, means 'to plan' or 'to give birth'. The second part of the name, *wurin*, means 'at dark', 'at night'. Thus *Ambanwuringa* and *Ambanwurinmas* mean 'the people who plan at night', as the Ambonwari themselves explained, or 'the night people from a dense jungle'.[2]

People say that the village has the shape of the huge snake which lies beneath it. This *wakɨn* (snake) is the essence of Ambonwari and represents the *wambung* ('insideness') of the village; it is its protector and, if something goes wrong, its destroyer. The snake has to be firmly pressed down by all ancestral female spirits, who are represented on the round masks on men's houses and by the three 'mothers of the village', the wives of descendants of the first three ancestors, whose obligation is just to sit. They are excused from doing any work outside of the house (see Chapter 3). This snake can 'see' what people do and it communicates with them through its 'mouth' and 'ears' which are represented by the main men's (spirit) house in the centre of the village. Men refer to this snake only in the men's houses, and only then by euphemism: *imɨnggan saki*, 'the spirit of the village' or *kupan imɨnggabɨnan*, 'the big one of the village'. Its real name means 'strong' and is known only to a few old men, for it is the most awesome secret which is seldom whispered even in the men's house. Besides *imɨnggan saki*, people can refer to it indirectly by using the name of the other powerful spirit in Ambonwari, Ibrismari. This is the protector of the Crocodile-1 clan of the main men's house, and it presses the middle of the snake firmly inside the ground. The head of the snake lies 'down-river' (*masɨr*) and is anchored by the Bird of Paradise clan, in particular by the spirit of Ibrismari's son, Yawundaki. The tail of the snake lies 'upriver' (*pambɨn*) and is controlled by a spirit called Amburmari, and anchored by the Crocodile-2 clan. The people of these three clans are the descendants of the

[1] When men painted their bodies before going to fight they called themselves *wundumanga*, 'the spirits of dead women/ancestresses'. When discussing the periods of fighting with neighbouring villages, a man would often praise himself as being 'a mother of fighting' (in Tok Pisin *mi mama bilong pait*). I often heard people saying in Tok Pisin *draipela mama bilong kanu* for a large canoe. As I will show in Ch. 6, a canoe has a female character. As Harrison (1993: 31) points out, a canoe carries a person as a mother carries her child.

[2] When I discussed these issues with several people, someone commented on the pidginized name *Ambon-wari*, saying that the ancestors had 'worry' (*wori* in Tok Pisin) to settle down in a village. The second name, the use and meaning of which is not known to everyone, *Manjiman* or *Manjimakan*, comes from *manji* (a non-venomous snake which, people say, can jump into a canoe) and *makan* ('to hide'), thus likening their place to a hiding snake.

first ancestors, brothers and founders of the village. Other clans occupy the areas between these three, while the last to arrive are attached to the snake's 'tail'.[3]

Though the Ambonwari people recognize the existence of land and people beyond their boundaries, especially those neighbouring people who were their enemies in the past and those who share the same cosmogony (see Chapter 5), they are more directly concerned with their own relationships and their own environment. Ambonwari say that darkness covers the village every night like a huge *arɨng*—a mosquito bag of plaited sago shoots (see Mead 1963 [1935]: 179, 245; Gewertz 1983: 37–8), preserving and protecting the village, its people, and 'the way of the village'. The village is surrounded by swamps, land, hills, creeks, spirits, all of them with their own distinctive names. As is true for many people, names make their world, with which they feel familiar and in which they find their own identity. This environment is not something that is just there in the sense of 'nature'. It is something which makes sense because it is in a permanent relationship with people who use it. A sago palm gives flour and grubs, other trees and palms are used for building houses, canoes, and tools, animals are a source of food, and rivers provide fish and means of transport.

At first, it may seem as if animals and plants, especially when anthropomorphized, are entities which occupy the environment in the same way as humans and spirits do. But then one realizes that only humans and spirits can make a village (*imɨngga*), a place which 'stands' (*imɨng-*), and is built by humans or spirits to live in. A tree also stands (*yuwan mɨn imɨnggan*) but it does not build for itself; it simply stands. A bird (*ming*) makes a nest (*andɨm*) from things gathered from earth (*andi*), and a bush spider (*apiak*) makes its web (*apiakɨs sɨndi*, lit. 'urine of a spider'), but *imɨnggan kay*, the way of arising/standing up ('the way of the village') is the prerogative of humans and spirits. What distinguishes Ambonwari people from spirits is their own way of being, i.e. the way they 'stand'. This 'way' (*kay*) is for the Ambonwari the crucial term in the language of identity. However, it is a term which signifies much more than just 'way', as we shall see. The Ambonwari refer to their collective identity as *imɨnggan kay*, 'the way of the village'. This represents the habitual practices of daily life, collective ceremonies, and the entire valued past as an embodied collective history which 'functions at every moment as a *matrix of perception, appreciation, and actions*' (Bourdieu 1977: 83). Although much of the rest of the book is devoted to establishing this claim, here I shall merely begin the task by examining some village practices with a view to giving some ethnographic content to the notion of *kay* before proceeding to a linguistic discussion in the next chapter.

[3] There are 12 totemic clans and 35 patrilineages (including 6 with no present members) in Ambonwari. Each clan has at least one men's house. Residence is patri-virilocal (see Ch. 3).

Ways of Finding and Using Food

Ambonwari people live in wetlands surrounded by hills and forest to the south-east and sago swamps in all other directions. The land where the village stands is in fact an island cut off from the mainland by Konmei (*Kwanmar*), *Maramun*, and *Ibris* Creeks. While sago is a staple, hunting, fishing, and gathering are the most important sources of meat, fish, greens, mushrooms, and other products of the environment. People also practise subsistence agriculture. Gardens up and down the Konmei Creek are often used as bush camps where people (usually nuclear or extended families) stay for a week or more in privacy, away from village life.

Melanesian people value food not just as something necessary for life, but as a tool for the creation, maintenance, and manipulation of social relationships; with its 'sociopolitical value' representing the 'ethic of the social system', food becomes 'the measure of all things' (Young 1971: 146; Schieffelin 1976: 46–52, 63).[4]

Food is not only the main preoccupation of Ambonwari people but also a major component of rituals, myths, and the relationships between people. Their ancestors moved from the hills to the wetlands because of the plenitude of sago; the first ancestor, Kapi, made a fool of a bush-spirit who, though living in a sago forest, did not know how to prepare baked sago; Kapi brought with him the stone breast which is kept in the village and is shown to those children who refuse to eat (for a story about this breast, see Appendix B). The most important totem-spirits in the village are carved crocodiles. The carvings on their backs represent the food which is forbidden during the initiation period. Whenever something goes wrong (sickness, bad luck in hunting) or whenever there is a major success (when many pigs are killed) food is supposed to be brought into the men's house for spirits. The large number of food taboos is one of the main characteristics of every important transition: initiation, first menstruation, pregnancy, and the period after birth. Many ways of hunting and fishing can succeed only if food prohibitions are obeyed. Whenever a man builds his first house, kills his first pig, carves his first canoe, or whenever a child finishes six years of school, the event is celebrated with a feast (a *kurang* ceremony). In everyday life food is exchanged continually between kin. Food is used for settling village disputes, and on many more occasions.

Life in Ambonwari cannot be imagined without sago, fish, sago grubs, and coconuts. The staple is sago extracted from different types of edible sago palm (*Metroxylon spp.*). Sago palms are distinguished according to the size of their thorns, with the one with the shortest thorns being most valued for its flour. The way of processing sago is an activity which is habitualized and acquired

[4] Heeding the advice given to Jonathan Seagull: 'If you must study, then study food, and how to get it. This flying business is all very well, but you can't eat a glide, you know. Don't you forget that the reason you fly is to eat' (Bach 1970: 15).

throughout childhood. Once a month a whole family, sometimes accompanied by additional members, goes to the forest to process sago. Late in the afternoon, after finishing half the work, they return to the village. Their task is usually completed at the end of the next day. The flour obtained from one palm usually lasts for a month. The sequence of activities is as follows: getting up early, going to the location of the palm which has been already chosen, cutting it down (men's work), stripping off the bark (men's work), removing the leaf stalks (men's work), making a container of them (women's work), pulverising the pith (men's work), carrying the pith in baskets to the container (women's work), washing the pith (women's work), putting the wet flour into the baskets (women's work), and finally carrying it home.[5] The 'way' of making sago includes the proper use of tools such as an axe, wooden spades, a blunt adze, a coconut shell tightly attached to a stick with a rope, a container, and baskets. The tools define the activity as much as the activity itself defines them. This applies to both men and women, and to the tools they use in their own distinctive ways. The sequence of sago-processing activities is well known to everyone in the village, as is the actual method of performing each sequence. Thus, we see here an instance of a more general characteristic of *kay*, namely, its inherent temporal nature: as I shall be at pains to emphasize, *kay* is not just a manner, nor even a practice (conceived as an abstract procedure), but an embodied process characterized by a specific temporality.

The Ambonwari have access to so much sago that sago flour is the foremost market product of the village; it is usually sold in Angoram. The preferred meal is *karis*, a pudding made of sago flour mixed with boiling water, combined with a little meat or fish and sprinkled with greens as garnish.[6] Dry sago pancakes, *sipi*, are second to *karis*. People take these pancakes on canoe trips or if they are going away for a whole day. They eat them as a snack whenever they feel hungry, and pieces are often carried in men's personal baskets.

Appendix A lists meat-obtaining methods in the following order: fishing in Table A1, hunting in Table A2, and other methods in Table A3. In addition to their descriptive value, these tables show that Ambonwari recognize and define these practices according to the tools they use. Such a 'classification' breaks down generalizing notions such as fishing and hunting into more precise units of the activities themselves. The tables show that many of these practices are differentiated in relation to the water level in the surrounding creeks as well as according to gender.

The main protein sources are fish and sago grubs. People fish with traps,

[5] Many previous researchers have discussed the production of sago and its importance for Sepik societies (Foley 1991: 12–13; Gewertz 1983: 21; Schindlbeck 1980; Townsend 1974; Tuzin 1992). On the culinary aspects of sago in Papua New Guinea, see May (1984: 52–5, 110, 170).

[6] Meals are supplemented with coconut (*urang*) and various types of edible greens: *tulip* (TP) (*munggum*) and *tulip* (TP) fruits (*munggumɨnɨnggi*), *aibika* (TP) (*kapɨs*), fern of the lake shores (*kanggu*), greens with white leaves (*mandia*), fern with short leaves (*manggrum*). Mushrooms (*konggrambanɨnggɨnma*) from decaying tree-trunks often garnish sago pudding.

PLATE 3. Women fishing with *sɨmbang* baskets

hooks, nets, arrows fitted with a bicycle inner tube that are used as a sort of spear-gun, ordinary spears, and bush knives, and occasionally by poisoning the creeks. All these fishing practices are acquired during childhood and become an important part of the common 'way of doing things'. Thus, people may refer to these practices as 'the way of catching fish' (*aria awur kay*) when they use hooks; 'the way of pulling out the fish' (*aria awkwiyar kay*) when they use fish traps; 'the way of putting fish inside' (*aria apɨr kay*) when they drag a hand-net or a basket through the water (see Plate 3); and 'the way of shooting fish' (*aria wandɨr kay*) when they use pronged spears or homemade arrow-guns. Every year there are short periods when the waters from different creeks and ditches commingle. This usually happens during the dry 'season' when backflows are produced by the Konmei Creek swollen by rains in the mountains. The muddy water from the sago bush makes the creeks turbid; it appears oily, with many whirlpools. Fish drift on the surface and are easy to catch by spearing. People refer to such periods as *pɨpɨrkwanmay* (in Table A1 marked with an asterisk), where *pɨpɨr* means a coconut-shell spoon used for serving sago pudding, and *kwanmay* a creek or stream. The expression thus refers to the fact that fish are as easy to catch as if one were using a coconut-shell spoon. When fishing with purchased hooks—traditional hooks made of sago thorns are not used any more—people use the following bait: sago grubs, large grubs, sago, small fish, chopped pieces of green or brown frogs, small white larvae from the sago palm, worms, grasshoppers, and insects. Fishing is practised by everyone—men, women, and children—though they use different methods. All kinds of small fish, each having a name of its own, are called by the general term *aria*. Among the individual larger fish species (which can be boiled, cooked in leaves, roasted in the fire, or smoked) are eels; an introduced species from South America, *Tilapia mosambia*, known by the Tok Pisin term *makau*; another recently introduced species known as *kol pis* (TP); and the native species *big maus* (TP), *pis palang* (TP), cat fish, and *pis nil* (TP). Large and small turtles quite often make a perfect meal. Crayfish, and very rarely crocodiles, are a supplement to the *maray mɨn amɨng*, 'food of the water'.[7]

Hunting is the domain of men, but a woman does not hesitate to throw a fishing spear or use a bush knife if she comes across an edible animal. The most important source of meat is wild pig, which is hunted in several different ways (see Tables A2 and A3 in Appendix A). Cassowaries and possums are valued, while wallabies are scarce and rarely killed. Large pythons, lizards, flying foxes, and bandicoots are additional sources of meat. The area is rich in birds. People prefer wild fowl, wild duck, wild pigeon, and the crested or crowned goura pigeon, which is especially prized for its tasty meat. Eggs of wild fowls are often eaten, while those of cassowaries and crocodiles, though relished, are rarely found. People also eat certain types of grasshopper and praying mantis, and other insects such as flying yellow *panjas* or black *yuwanɨnggɨnma*. Small lizards are the preserve of children, but they avoid different kinds of frog, as elders tell them that eating such creatures will make them soft. Frogs are eaten by men in the men's house, however. All these animals are grouped under the term 'food' (*amɨng*), and humans, when killed, once belonged to the same category. Like fishing, hunting practices also fall under the general aspect of *kay*, more precisely *kumɨndɨ kay*, 'the way of finding meat'.[8]

Everyone has to find meat in whichever way they can, but it is the mastery of different ways of performing the activities of fishing and hunting which become habitual and thus characteristic of an individual Ambonwari. It is the profi-

[7] Foley (1986: 113) would disappoint those who believe that all languages must have a verb which only refers to the act of eating. He says that the Yimas verb *am-* covers eating, drinking and smoking (tobacco) too. The same verb is used among Karawari covering these activities and in addition chewing (betel nut, for example) and kissing (touching with the lips). For chewing a plain betel nut or plain baked sago they use the verb *suwang-*, but for chewing betel nut together with piper betel and lime they use the verb *kɨpɨs-*.

[8] Not every man in the village is a hunter. Those who are usually keep tallies: a bundle of little staffs (*kunaypa*) of sago leaf stems which mark the number of the pigs they have killed. For a cassowary a man can use a shorter staff or tie a string in the middle of it. A man can use his deceased namesake's (father's father's for example), mother's brother's or his dancing partner's staffs and 'count down' (by putting them into a new bundle) to see when, if ever, he reaches the deceased man's number. Thus, Tobias and Bonifacio, who are in their 50s and hunt only with dogs, spears, and traps, have 167 tallies for pigs and 1 for cassowary, and 152 for pigs and 5 for cassowaries, respectively. Bonifacio says that he misses from his bundle 60 staffs. This would mean that in 30 years these two men killed on average 5–8 pigs each in a year. Elias, in his late 20s, hunts with a gun and a spear and has 78 sticks for pigs and 2 for cassowaries. Hunters may mark the number of successful killings on their spears. Thus Rowi, in his late 50s or early 60s, who hunts with dogs very successfully, has 90 notches on his spear.

ciency in using particular tools or weapons which brings recognition. Based on past experiences, it formulates someone's expectations about the future; the acquired *kay* always relates 'before' to 'after' and past to future. To put it in Ambonwari terms: a man who did not acquire in the past the practice of making a pig trap 'the way of pulling out' (*awkwiyar kay*) does not even try to make a trap, and thus has no hunting expectations in the future concerning this way. Many people are reluctant to admit that they have never really mastered particular ways of hunting. They find all sorts of excuses for not involving themselves in these activities. If they nevertheless do go hunting from time to time, their catch is invariably meagre. With the sympathetic agreement of family and friends, they blame crocodiles for chasing fish away, spirits for making them blind, thieves for stealing fish, husbands for not observing taboos, and wives for promiscuity. Less sympathetic people from other wards simply recognize that such an activity is not really that person's *kay*, because if it were there would be visible results and something to anticipate.

People inside and outside the village are differentiated according to the ways or manner they perform their activities. For example, the Ambonwari remark differences between themselves and neighbouring Imanmeri concerning canoe-making and fishing, arguing that because Imanmeri are hill people they do not make good canoes. The Ambonwari think that the final shaping of a canoe in particular is done better in their village; but this might simply be a case of local chauvinism. More obvious is the difference in fishing during the dry season.

PLATE 4. Yuana Warima weaving a *sɨmbang* fishing-basket

While Ambonwari women make fishing baskets (*sɨmbang*; see Plate 4), Imanmeri women make fishing nets in a round frame (*yuway*), though both are held in the hands while fishing. Ambonwari women who have good relationships with Imanmeri women can obtain a few of these nets, but they are not made in Ambonwari village and are rarely seen there.

Domestic animals include chickens (killed mainly for important visitors) and a few ducks; domestic pigs were never looked after in large numbers and no one had a pig during my stay. Dogs and cats are eaten sometimes, though only in secret: people are afraid that others will tease them. Mice and house rats are not normally eaten, though there are a few individuals who, when hungry, cannot resist roasting them in the fire and secretly consuming them.

Gardens provide red and yellow pandanus fruits, different types of banana, sweet potato, taro (which has no local term), sugarcane, papaya, pineapples, and watermelons. The food from gardens is regarded as being only supplementary, to help over short periods when there is not much else. Women and children also like to eat an edible type of white clay dried over a fire. Men plant their own tobacco. Betel nut is important both in ceremonies and in everyday interpersonal relationships.

Ambonwari women make their own salt (*isin*) which, besides its use in cooking, has symbolic significance. A woman makes fire from sago palm leaf stalks, the dry wood of the Malay apple tree, and dry coconut husks. When everything has burnt down and the ashes are cold, she sprays water over them and mixes them with a stick. She puts the moistened (not wet) ashes on a dry sago palm leaf stalk, carries it into the house, and puts it in a basket. Whenever she wants to use salt, she washes some ashes in a coconut shell (or a tin) with a hole partly plugged with a dry grass so that the yellowish water (*isin maray*, 'salt water') slowly drains into a container. The ashes are washed twice and then thrown away. For a family pot of food a woman usually needs two shells of washed ashes.

Symbolically, to drink salt water means that one is cleared of one's wrongs and that one has fullfilled one's obligations. In the past, when a man killed someone he was secluded for five days in a men's house, where he ate only unsalted food. After a celebratory ceremony in which he did not participate himself (but was represented by his namesake or 'exchange partner') he was able to eat sago pudding stirred with salt water. During the time of cannibalism a human skull was washed with salt water, and sago pudding was then made from the same water.

Prohibitions Concerning Food

The symbolic aspects of food become visible through its prohibitions before, during, or after particular activities and events. As the human body is defined by ways of doing things and as this notion is extended to the objects which a person uses, it is understandable that people do not touch those things which could have undesirable consequences for their bodies or the bodies of those with whom they are in frequent contact (a child, a wife, a husband). Likewise, people do not want others to interfere with their *kay*.

One can often see in the village that the owner of a coconut or betel palm has wrapped some leaves around its trunk to prohibit others from taking the nuts. Whenever people plant tobacco they leave a sign of prohibition on their plantation: 'they put inside a sign on the tobacco garden' (*mbu mambɨr yamba apɨr yakin wak*). Signs (*mambɨr*) are specific to every clan: for example, Wallabies have *sɨmbiyawi* greens, Crocodiles the leaves of a vine, and Cassowaries the leaves of *awanma*, 'female cassowary' tree. They put these signs inside the 'v' of a split stick which they thrust into the ground. To disregard these prohibitions would be to interfere with the *kay* of the owner.

Some prohibitions concern the totemic animals of clans. People from the Crocodile-1, Crocodile-2, and Bird of Paradise clans do not eat the large white bird *saun* because their first ancestors (three brothers) decorated their paddles with the feathers of this bird. Nor do people from these clans eat birds of paradise (though they can shoot them). People say that they would vomit blood and grow thin if they even smelt the cooked meat of these birds. People from the two Eagle clans are not supposed to eat eagles, which are rarely caught anyway. All other totemic groups, Cassowary, Pig, Eel, and Wallaby, eat their 'totems'.

The food prohibitions surrounding particular practices and particular situations differ between clans in the village and between individuals; the greatest taboo for some may not be a taboo at all for others. But all have taboos which have similar origins and share a common anticipation of success and well-being.

There are prohibitions related to fishing and hunting activities (see Appendix A). Ambonwari make fences or barricades on the flooded bank of the river in the same place as their ancestors did. People refer to prohibitions as 'very ancestral' (*pan kupambi*) or 'the way of grandparents' (*arkɨn kay*). By observing them they can expect a successful catch with the approval of spirits. For this reason, all those who are going to eat the catch from the traps placed in the barricades are not allowed to eat eel beforehand. People say that if they disobeyed this rule, which they inherited from their ancestors, the eel's fat which they had eaten would seal the entrance of the trap and fish would not go inside. If we recall that the body of a person is 'connected' to those things he or she uses (the traps in barricades in this case), then we can easily comprehend the taboos which people impose on themselves for a successful catch. People do not eat bandicoot and possum because the strong smell of their burning fur would chase fish away from the traps. Neighbouring Imanmeri do not eat *tulip* (TP) or *aibika* (TP) leaves when fencing the fish as they fear that these leaves would block the fish traps. Prohibitions also apply in the use of fish poison. Because of the direct contact between the body and the poison, the man who beats the roots of the poisonous vine is not supposed to have had sexual intercourse the previous night, to have had 'wet dreams', or to have a pregnant wife. Women cannot beat the vine if they are pregnant or menstruating.

There are some very general prohibitions concerning hunting. A man does not eat pig's snout and he does not eat a grasshopper-like insect called *wambunggrambang* which lives in decaying tree trunks near the water and can bite. People explain these two prohibitions in the following way: when a pig eats sago or when dogs chase a pig it makes a frightening noise with its snout. They say that if they ate its snout they would be afraid of being bitten by a pig when going hunting. For a similar reason, men do not eat the insect, as they fear that their chest (heart) would beat as fast as the insect's does if held in the hand. No wonder, then, that the term for this insect, *wambunggrambang*, is composed of two roots: *wambung* ('insideness', 'heart') and *kra-* ('to fight, to beat'). Women and young girls can eat them, but boys are usually advised by their fathers not to do so.

People say that the personal spirit of a man wanders around the bush before the hunt and finds the place where pigs are. The spirit later guides the hunter. Hunters in general do not eat the meat of the pig that they have killed with a spear, because this would make a man's spirit lazy. This would result in unsuccessful hunts in the future. Hunting with a spear brings the hunter and his victim close together. The hunter 'gives his strength to the pig', as Elias, one of the most successful hunters in the village, puts it. To eat such a pig would mean to eat his own strength, i.e. himself. This is a good example of how people understand *kay* as being extended not only to the things which a hunter uses (a spear, for example) but also to game or a human victim. Gilbert Lewis also noted that a Gnau man would not eat anything that he had shot: 'by shooting it he has put something of himself into it', that is, his blood (1980: 174). While among the Gnau this taboo is extended to the game caught by a shotgun, the Ambonwari emphasize that when using guns there is no direct contact between man and game, and because of this it is not part of a hunter's *kay*. Ambonwari men can therefore eat the game they have shot with a gun. The prohibitions surrounding hunting are very idiosyncratic, following previous experiences and the elders' instructions in each lineage. Whenever Elias builds a trap or a hide near the wild sago palm he has just cut down to attract pigs, he will not eat turtles or eels. These prohibitions remain until he uses that particular hunting spot. He says that he does not eat turtle because it has a strong shell. If he did eat it the skin of the pig would become as strong as the shell of the turtle and his spear would not penetrate its body. On the other hand, the eel is very fast; it jerks its head backwards and disappears in the water. Elias would not like any pig he was hunting to be as swift as the eel. Among those few who have mastered the art of trapping a pig is Rius, an old man. He says that whenever he builds a deadfall trap he does not eat pork, cassowary, possum, eel, turtle, birds, or one type of large fish—all those animals which remind him of humans. He says that if he did eat these animals the trap could fail.

The use of mother's or wife's lineage land carries with it prohibitions attached to that land. For example, the land called Samɨnggumun belongs to the Crocodile-1 clan, which is Elias's maternal group. Members of this clan do not eat chickens or eggs because the placentas of their two famous ancestors, Moon and Sun, were buried on this land (see Chapter 5). Women married into the clan are allowed to eat them but feel ashamed to do so in front of their children, who are not allowed to eat them, so they prefer not to. The people from the Crocodile-1 clan also do not eat the cassowary's eggs from this area. So Elias would not eat these things while on their land, otherwise, so he says, he would be sleepy all the time. He is not allowed to eat cassowary, as in this bush it is like a human. If he ate it he would feel tired and sleepy during the dance, because 'I do not know how to protect my skin'. To eat a lizard would cause his skin to be heavy and he would be unable to climb coconut and betel palms.

As one might expect, the greatest number of prohibitions accompanies the periods when the *kay* of persons, and of those in close contact with them, is in transition. One such instance is the period of pregnancy. When pregnant, a woman is expected to refrain from eating frogs, turtles, crocodiles, cassowaries, lizards, snakes, possums, flying foxes, the large fish *mɨndaun*, crayfish and eggs. Here the future baby's body, with its habits, manners, and behaviour, depends mainly on its mother (its father's activities become important for a baby after its birth). People say that it would not be desirable for a baby to follow the habits or assume the characteristics of animals that its mother had eaten. If a pregnant woman ate frogs, a baby might have a big mouth, a huge belly, and short arms; if she ate a large frog called *yarɨma*, the baby would walk around with legs astride; eating turtles, crocodiles, or lizards would induce the baby to urinate all the time; eating cassowary would cause a baby to turn its head as a cassowary does; if a woman ate a possum, her baby would have small arms and its huge eyes would twinkle all the time; eating a possum would also prevent a baby from sleeping at night; snakes have no legs and eating one would cause a baby to sit all the time and never begin walking; eating flying foxes would cause a child to be restless at night; to eat a *mɨndaun* fish would give a baby an ugly face and huge eyes; crayfish would induce the baby's eyes to stick out of his head; the eggs of domestic or wild fowl would cause a baby to wake up whenever fowls began crowing. A pregnant woman does not eat pandanus which has been nibbled by parrots, because parrots have sharp beaks and they can harm a baby. She cannot eat Malay apples, *laulau* (TP), otherwise her baby will get boils, abscesses, and sores. Also *laulau* fruits, *nggrɨmɨnɨnggi*, are hairy, and a baby would get scabies or the skin disease *tinea imbricata*. She does not eat the yellow liquid from the squeezed nests of sweat bees or the baby could get boils. A pregnant woman does not eat mushrooms, as she does not want her baby's ears to protrude from its head like mushrooms. Likewise, she does not eat any type of wild yam because if she did so the child's fingers might look like the excrescences of wild yams.

After this list of prohibitions one might wonder what a pregnant woman does eat! She can eat sago and everything from the garden, and pork, fish, sago grubs, the meat of wallabies (so that a child will be able to move around easily and find food). All the prohibitions mentioned above end after a woman gives birth.[9]

If one looks at the taboos on consumption, where every single prohibited food carries its own symbolic meaning, one might surmise that their purpose is to protect and preserve the body of a person. First of all, one does not eat one's own body, and because people have their own distinctive habits (*kay*) they also have many distinctive taboos. Another aspect of prohibitions refers to the interrelationship between one's body, the food that one eats, the things which one uses in a particular activity, and the activity itself. As all of them are part of a person's *kay*, he or she has to consider the effects of one upon the other (e.g. I do not hunt the morning after sleeping with my wife). In a period of transition, *kay* which is in a state of becoming is vulnerable to the influence of the habits and manners of some other body; thus, those things which could have a negative influence have to be avoided. Likewise, in any activity with a dubious outcome, such as hunting or fishing with traps, the practice is in process until completed—whether the outcome is successful or not. The fact that taboos differ between individuals and groups, and that new taboos might be tried out before becoming habitual, also suggests the openness and experiential nature of *kay*. Taboos represent a strategy needed to bring an activity successfully from a beginning to an end. Other strategies involve timing itself. In many cases, especially when other people are involved, the successful completion of a project is not simply a matter of completing a sequence of steps, but is also dependent upon proper pacing. I now turn to an example which will make this evident.

House-Building

When I arrived in Ambonwari I lived for the first few months in the house of Tony Simiwariya, who had become my 'elder brother'. Tony told me that his house had been built in 1987 from old house posts and old sago leaves covering the roof. It had now begun to fall apart. The roof leaked during heavy rain, there were holes in the walls, and the posts were cracking. Tony frequently said that it was time for him to build a new house, but in the meantime he had chosen six men to build my house. They were all closely related to Tony: classificatory brothers; real and classificatory sisters' husbands; classificatory

[9] Margaret Mead noted that among the Mundugumor (Biwat), taboos usually began after the birth of the child, though a pregnant woman also had to observe some prohibitions (McDowell 1991: 122–3; see also Kulick 1992: 96–7 for Gapun).

sister's husband's son. Tony acted as a supervisor. I told the workers that I would not provide any food for the work itself but would pay them 400 kina (the kina being the Papua New Guinea unit of currency which during my fieldwork was almost equivalent to one US dollar) for the whole house and would organize a party at the end. They agreed and the work progressed quickly, as everyone in the village was enthusiastic about the idea of having a 'white man' living among them. Also, as people said, my house was far smaller than those they make for themselves and was thus much easier to build. Whenever there was a need for female help the workers' wives, along with those of Tony and my clansmen, also worked on my house. It was completed in just over two months. On 15 December, while I was still living with Tony, he went upriver with his sister's husband, Godfried Yangus, and two of his wife's brother's adult sons, David, and James, in a large motorized canoe. They returned rowing four tree trunks: one for Godfried, one for David, and two for Tony. In the evening, when people had returned from their own daily occupations (fishing, sago-processing, gathering, and so on), the men from the Bird of Paradise and Crocodile-1 clans, with the help of Tony's sisters' husbands and sons, dragged the new posts out of the creek.

On 24 December I moved into my house. Tony, his children, and many others immediately offered to stay with me, as they were concerned about me living alone. I explained that because I would work on their language inside my house, and because different people would come and tell me their 'stories', I would prefer to be alone. On 12 January the topmost horizontal beam holding the roof of Tony's house cracked loudly in the middle of the night. As it threatened to collapse, Tony took his wife and their six children to the house of Joachim, his classificatory's sister's husband. Tony explained why did he not simply move next door into a house of one of his 'brothers': 'The eldest brother does not go and live with his younger brothers' (see Chapter 4). During the following four months Tony slept alone in his old house, while his family inhabited it only during the day. Whenever there was a gathering Tony always mentioned that his house had broken, that he lived in a 'bad' house. Working during the evenings, with the help of the same people as before, he succeeded in putting the posts for his new house into the holes that he himself had dug. Other events such as a death, a trip to Angoram, work at the school, and helping others on their houses interrupted the work on his own house. In addition, Tony did not have enough meat for the workers, as neither he nor anyone else from the Bird of Paradise clan was a successful hunter. Tony always mentioned that he was a master of sago grub-gathering but that the waters were still too high for this.

At the beginning of May Tony decided to go to Kansimai village, for his friends there would catch a pig for him. But five days later he returned from Kansimai without one. As the reason for the unsuccessful hunt he blamed people's gossip about his hunting in a forest that belonged to other Ambonwari clans. The next day he began, alone, to gather sago leaves for the roof, saying: 'No meat, no help!' A few days later, Joachim, who had been with his family fishing on Karawari River for weeks, sent a message to Tony to come and fetch him by motorized canoe. They returned with bags full of smoked fish. Finally, having something to put on sago pudding for his workers, Tony asked several men to come next day and help him collect sago leaves. Though they promised, nobody came. Tony was angry, especially with the men from his own clan. A man's closest 'brothers' are often reluctant to help him, particularly during the periods when they do not have enough food for themselves, because in any large-scale work the hosts (into which category the brothers fall) give food both during and after the work for people to take to their houses, but they do not eat themselves. The closest 'brothers' (but not more remote ones) have to refuse any food, thereby showing that they too are hosts. Tony talked to himself: 'What's wrong with you people? Do you have your houses covered with metal sheets like white men? There will be a time when you will need something.' Then speaking to me: 'What's wrong with these young men?' He recalled how a few days ago Adam (a man from the Hornbill clan) had put food and betel nut into the men's house, thereby asking the young men to help him carry seven thin tree trunks as he wanted to change the floor in his house. But they refused to eat: nobody touched the food. They explained that because the waters were low they wanted to go fishing.

Tony and his wife went processing sago and returned with new bundles of leaves. He had calmed down, and he told me that he had also cut many wild sago palms so that they would have enough grubs for those who helped them sewing sago leaves into thatch shingles. His five-year-old daughter protested: 'They will just finish all of our sago grubs. We can sew sago leaves ourselves without them and then we can eat all the grubs.' At the beginning of June, however, large amounts of sago grubs attracted helpers to Tony's house. The men erected the beams and the frame of the house while women sewed those sago leaves that Tony had collected himself. At the end of June, eight young men arrived from Meikerobi saying that they were the members of a youth group which offers its labour for cash. I proposed to pay them on Tony's behalf. In one day these eight men collected forty-one bundles of sago leaves. The next day they knocked down Tony's old house, making sure that he would still be able to use the well-preserved pieces. On the third day they placed the old poles and sticks on the roof of the new house, since Tony had been unable to cut new ones himself. He brought many thin logs which would support the floor of the house. At the beginning of July, 'sisters' and wives of the men of the Bird of Paradise clan helped Tony with sewing the sago thatch while men began to fix them on the roof. During the following days Tony and I cut and carried a heavy palm bark and, using also the bark from the old house, covered more than half the floor inside the new house, the part which by then had a roof. The progress on Tony's house during these weeks was obviously too fast, for people had begun to feel that they were neglecting their own tasks. For this reason Tony decided not to ask for further help, and his family lived in the unfinished house for several more months. Only then did he decide to fix the remainder of the roof, the walls, and the floor. Later, when his 'brothers' began to build their own houses, Tony did not offer to help them.

Every clan has its own area or ward in the village where its members, being from different lineages, can build their houses. Each ward is divided into smaller areas within which each lineage, the smallest named group in Ambonwari, builds its houses. What I want to point out here is the relationship between house-building and time. To build a house, especially of a size which will reflect the person's capabilities and supportive relationships, is hard work, in terms not only of physical activity but also of managing relationships. Though the owner of a house can do some work himself, he nevertheless needs a good deal of help over a longer period of time. This implies that he has to conceive a strategy which will keep satisfied not only those who help but also those who do not. This strategy has to be well prepared and has to take account of the strategies of others. As Biersack has described it: 'Cooperation is the paradigmatic arena of sociality and morality. Reciprocity is achieved by "doing-giving": sharing "work"' (1990: 69). House-building has its beginning and end: the activity between these two points is experienced through the rhythm and tempo of house-building. A man and his family who participate in the house-building *kay* experience time according to the specificity of practice (including the relationships involved).

If a man is in a good relationship with another of his own clan he will help him. The men from mother's and sisters' husbands' lineages from other clans give a hand and instruct the younger, inexperienced house-builders. A man who builds his house needs assistance in successive stages: first in bringing house posts to the village and placing them in their holes; second, when horizontal posts have to be lashed to those in the ground; third, when sago leaves are cut, collected in the forest, and brought to the village (all these tasks are men's work); fourth, when sago leaves are sewn into thatch shingles (women's work: see Plate 5) and then tied to the roof (men's work); fifth, when palms are cut and stripped of their bark for covering the house floor (men's work). In all these activities a man and his wife need assistance from other men and women. To get this help they need enough food (meat in particular) to feed those who assist them. Not only the women of the house and its lineage but also those from other lineages of the same clan help in providing and preparing food for the workers. Finally, a man has to make walls from the stems of sago leaves.

In the above example I showed not only how Tony found it hard to find enough food to provision those who helped but also how he did not try to finish the work when it was almost done. The latter behaviour was characteristic also of those who were successful hunters and who had enough meat to find workers. It became obvious that when men build their houses they do not

PLATE 5. Stella Wasakay sewing sago leaves into thatch shingles

attempt to complete them as soon as possible. They continually delay the final completion of the house, though they may live in it for a long period once it has a roof. The tempo of this 'movement' of the work, a *larghetto*, so to speak, shows people the modesty of the owners, who do not make their capabilities offensively obvious. That is, by not demonstrating their skills, prestige, or authority, their capability to command support by providing plentiful food, they avoid causing offence. The house-building *kay* ensures, as a part of a process itself, the continuity of important social relationships which would not receive the appropriate emphasis if the house were completed quickly. In this way, either a real or an artificially imposed scarcity of goods demonstrated by a person's apparent inability to finish his house quickly results in a squandering of that very thing which exists in abundance: time (Bourdieu 1990: 117). With the postponement of the completion of a house, a man gains in terms of good relationships. People do not want a man to think that their help was not needed. They want him to ask them for help, provide food, and later remember and value them. They want him to feel obliged so that he will assist them when they are in need. They want to be part of his activity—and through this, part of his life. They want him to coordinate the rhythm of his activities with theirs so that he becomes part of their lives. This kind of manipulation of relationships, of course, can give rise to resentment (Telban 1993), which again differs (in terms of duration and intensity) from individual to individual. Tony's 'brothers', for instance, felt that they were insufficiently compensated for the work they did on his house. They also resented Tony's peremptory attitude. Such is village *kay*, the way of doing things.

Of course, if the delay in completing a house is too long a man might be called 'lazy', 'rubbish', or 'bad'. People may gossip about how he is unable to find food, how he is stingy, how he thinks only of himself and his family. The whole system depends on an acute awareness of the need to share. Today we collected all these grubs (or gathered mushrooms, or caught a big fish); tomorrow you will find or catch something similar and give it to us when we have nothing. This instantiates an ongoing tension between individual *kay* (eating your own food and thinking about your own work and about people of your own household) and the collective *kay* (an awareness of others and of your obligations towards them). When a house is finally completed, a year or so after building has begun, the feasting and dancing with the participation of guests celebrate the achievement of the harmony upon which *imɨnggan kay* depends.

Male and Female Kay

A woman, who is grammatically in a different nominal group from a man, is recognized (by both women and men) as one who 'sits well', who is a mother, who is a protector, whose main activities are conceptualized as focused around fishing, cooking, and children. Her ways of doing things define her body as a man's ways define his. Men are supposed to find money, though in the majority of households the woman keeps it, because a man is more susceptible to the persuasion of other men. While in the village, women's work is carried out inside the house, but men's work is mostly outside. Women are not supposed to sit like men. Women weave baskets, mats, fish traps inside or at the front of the house (see Plates 5 and 6), while men carve canoes, paddles, spears, or house posts, masks, and crocodiles for a new men's house, either in the forest or in a *kawara*—a small house with no walls and no bark floor—where people spend their leisure time protected from the sun or rain (it is also called *haus win* 'house wind' in Tok Pisin). Women gather greens and fish, while men cut sago palms for grubs, and hunt. They have complementary tasks in house building and processing sago. A woman paddles a canoe by sitting inside it, at the back, while a man stands with one of his feet on the edge, at the front. A woman's paddle is shorter and its blade heart- or leaf-shaped, while a man's paddle is longer and its blade is forked at the end. Every Ambonwari child grows up into this kind of gender differentiation.

People say that those children who eat everything and eat quickly will grow fast. 'Wet' food (boiled food, sago pudding) makes people fat; 'dry' food (smoked and roasted food, fried sago) makes people thin. Pork can be smoked as long as it stays soft inside, dry on the surface and 'wet' inside. When a baby

PLATE 6. Marta Yambakɨnda weaving a *marangapaw* carrying-basket

is born its parents give it sago pudding so that it will grow fast and have a large belly.[10] The rope which is fastened around the child's belly after birth is soon loosened, which shows the growth of a child. People say that if a child has a large belly, it will spread into its legs and chest and such a person will be big and strong.[11] People also explain that hot food is bad because it 'slows' a person, and cold food is good because it enlivens a person. Hot food means 'dry' while cold food means 'wet'. People often leave sago pudding to cool, saying that cold food has more water and cools a person down in hot sun. If you eat hot food, they say, your lips will crack.

Collectively recognized ways of doing things differentiate gendered persons and thus their bodies. The different habitual activities of each sex are acquired,

[10] A baby is conceived from female blood and male sperm. If the blood stays on the surface a baby will resemble its mother, if the sperm stays on the surface it will resemble its father. Augustina Awsay, the wife of Tony Simiwariya Andi, says that if a man sleeps with a woman 4–6 times during a short period of time she will give birth to a single baby. But if a man is careless and persistent and sleeps with a woman more often, she can give birth to twins because he can split the 'water' inside her (in South America splitting of the water is attributed to women: Lévi-Strauss 1992: 204). The same thing can happen if a woman, after sexual intercourse or during pregnancy, sleeps on her back.

[11] One of the explanations for the faster growth of girls is as follows. In the past a couple had twins, one of each sex. Their father took his son and sat him on the stool while the mother took a banana stem and slapped her daughter's knees with it. Because of this she grew faster than her brother. Since then, people say, only men use stools, which were previously owned by women. As *kay* expands a person's body into the things this person uses, a man becomes 'stuck' to a stool, which prevents him from growing fast.

and thus embodied, from early childhood through play and teasing. Children build shelters of leaves where girls pretend to cook sago pudding and serve their 'visitors'. A boy says to a girl, usually to his sister: 'I have a penis, it is easy for me to screw you. You do not have one, and others will screw you.' A little girl replies: 'They will screw you as well.' And a boy says: 'Not me. I have a penis.' Sometimes a little girl says *ama yanda yawi*, 'I (have) this vulva', and a boy says *ama kɨnda usɨng*, 'I (have) this penis'. A girl might say to her brother *apra kamban yungun*: 'Come, I will screw you'. And parents, usually the mother, tell their daughter that girls do not talk like this and that only boys can do so. When children ask their parents why boys have penises and girls do not, parents can say: 'You girls cook fish in the *kunguri* leaves [which were once used for a female apron] so you became women' or 'you girls cook and eat a small bat, *sakwima* [though in fact they do not, as no one eats it], and because of this you do not have penises.' But most often mothers say to their daughters: 'Before we did not have penises and vulvas. But then all the boys ate human faeces, so they got penises. But you girls, you refused to eat shit, so you have vulvas.' And a little boy would proudly say: 'I ate faeces, so I have a penis. You, my sisters, you did not want to eat faeces. That was your choice.' And a little boy tightens his belly, holds his penis in his hands and shows it to his sisters, asking them: 'You girls, do you want me to chase you?' When a mother holds her baby boy—under two years old—she stands him on his legs and tells him: 'You have a penis. You will screw women.'

As a part of embodiment, differences in sexuality are also perceived and recognized through the way things are done. Marilyn Strathern says: 'What differentiates men and women, then, is not the maleness or femaleness of their sexual organs but *what they do with them*' (1988: 128; emphasis in original). But this is only a half-truth. Sexual organs are part of the 'skin' and are visible. They should not be thought about in terms of a physical body, for which the Ambonwari do not have a separate term to direct us towards an explicit abstract concept. Rather, they should be understood in terms of an extension of being (*kay*) and of the accomplishment of being (*kay*) in the visible world. The sexual organs are there on a newborn child for its parents to recognize immediately its maleness or femaleness. The sex of a child is actually the first thing enquired about after a baby is born. Accordingly, because the child's sexuality is recognized by its sexual organs, the child is directed towards acquiring the practices which pertain to its sex including, later in life, sexual intercourse. Sexual organs are visible organs which are used in particular visible ways (*kay*).[12] They, along with the manner in which they are used, constitute the gendered *kay*. To deny this would be like denying that a paddle is not connected with the activity of

[12] In an episode of one of the many myths about the hero Mambi, he encounters a hermaphrodite who has a penis on the belly just above the vulva. Mambi recognizes the being as half-man and half-woman, and calls it a bush-spirit. The spirit has a daughter whom 'she' (that is how Ambonwari people refer to this being) has conceived by herself. The hero 'recognizes' the dangerous spirit and kills 'her'.

paddling a canoe, or that a spear is not related to the practice of hunting. I do not want to imply that objects and body parts are perceived as one and the same, but rather that the embodiment of sexual characteristics is already acquired at birth in the same way that a baby is conceived from female blood and male sperm as the direct consequence of female and male *kay*. *Kay* produces things and beings just as these also produce and allude to the particular *kay*.[13] The male and female *kay* are functionally attached (and acquired through *imɨnggan kay*) to the sexless existence (*kay*). It comes as no surprise that Ambonwari has two different verbs to express urination. A urinating man is referred to as *sɨndi mɨn pasan*, 'he flushed/released urine', while a urinating woman is referred to as *sɨndi mɨn awan*, 'she "boiled" urine'.[14]

The 'body' of a baby is not talked about from the perspective of how a baby perceives the world and its 'body', but how others see and perceive it. People simply say that a baby does not have its 'insideness' yet and that it acts according to the common *kay* of a baby. Only later in life does his or her 'body' become 'composed of relations . . . beyond itself' and 'appears as a result of people's actions' (Strathern 1988: 208). Thus, for example, during *kurang* ceremony men and women not only reverse their roles and actions but also decorate their skins according to the roles they enact. People's practices determine the appearance of their skins.

> If the body is thus personified, so too are its physical parts . . . [T]he potential distinction between male and female is crucial, and body parts are personified in being given a gender identity as one or the other. Thus we may apprehend the apparent paradox that much ritual attention is paid to sexual organs not because the organs sex the person, as it were, but because in her or his relations with others, the person sexes the organs. (Strathern 1988: 208)

In my view both aspects are patent: the appearance of skin and the process of *kay* are interrelated and influence each other in one and the same human being.

[13] One aspect of female *kay* (as others have argued for the Sepik and elsewhere in Papua New Guinea) is the ability to ensure reproduction (Errington and Gewertz 1987; Weiner 1977; 1982). This reproduction should not be seen solely as biological or automatic, but rather as 'a process that demands continuous work' (Weiner 1982: 56). Weiner argues for Trobrianders that every person is reconstituted, reconstructed, and recirculated over the generations, and that his or her identity is formed through others, through material objects, through networks of social relations, through associations with ancestors and spirits, knowledge, character, and talents.

[14] The verb *awa-* has multiple meanings, all relating to important aspects of people's lives. *Maray mi awan* ('boil the water') is used in ordinary life only for a plain water. Another verb (*sɨmbang-*) is used when something, such as sweet potato, is boiling in this water; yet another (*wunɨng-*) if something is boiling in salted water. There are other verbs for frying, cooking in ashes or fire, and baking. To understand better the verb *awa-* ('to boil') we might speculate on its possible connection with other terms (though I am unable to tell whether they are cognates or homonyms): *wi* ('night, dream, name'), *awi* ('fire'), *kyawi* ('daughter'), *yawi* ('vulva'), *awa* ('cassowary'). The same verb *awa-* means defecation, the act which Ambonwari men deny performing. To summarize, we could say that the acts of defecation and female urination are expressed according to heat and the way the practices are performed. In this way the whole procedure resembles the single activity performed by a cassowary which has a cloaca. For another New Guinea example of the intimate linguistic and symbolic connection between vulva, girl, and daughter, see Gell (1979: 45).

When Strathern (1988: 131) writes that 'if the body is composed of relations, if it shows the imprint of past encounters, then the relations are not in a state of stasis', she alludes to a conceptualization such as the Ambonwari have elaborated in the concept of *kay*.

When boys go through initiation they have to learn one of the most important aspects of male behaviour: to deny that they defecate and to avoid being seen by women in the act of defecation. The shame and prohibitions surrounding the subject of defecation appear in joking relations and in the 'dirty (bad) talk' people use when abusing one another. Nakedness and sexuality are far less shameful than the act of male defecation.[15]

If a child passes wind, *mɨndi mɨn pmbɨsɨrɨ*, 'he sprayed wind', or *mɨndi mɨn awan* (the same expression for defecating), others will laugh and he or she will be ashamed and walk away. While terms such as *mɨndi* ('faeces') or *mɨndɨr morum* ('smell of faeces') can be used to some extent, and *kapɨs* ('buttocks') is used to refer to the nakedness of a child, as in *kambra kapɨs* ('buttocks nothing'), phrases such as *pupung mamak* ('bad arse-hole'), *pupung akurupmbɨnauk* ('leaking arse-hole'), or *pupung urakɨk* ('stinking arse-hole') are considered terrible insults, *mariawk maman* ('bad talk'). They may be compared to today's Tok Pisin words such as *hul* ('hole'), *kan* ('cunt'), *fakin* ('fucking'), and to the Ambonwari expression *yarwambunggan* ('blood from the inside [of vulva]'). This kind of 'bad talk' brings shame to those who are insulted and can have serious consequences for interpersonal relationships.[16]

When a woman menstruates it is said that her spirit, *anggɨndarkwanma*, has copulated with the spirit of *Sunggwi*, Moon (a mythological boy, elder brother of Sun, also a boy). Thus, *sunggwi sanggwanma* ('the moon sees-it-woman') or *sunggwin kay mɨn kaykan* ('she stays in the manner of the moon') are the expressions used for a menstruating woman. She observes certain prohibitions. She does not cook nor fish; she does not touch objects used for fishing, such as fish traps, hooks, or spears. If she did touch these objects they would become contaminated. Such impurity could have negative effects on those who handled these same objects (and food). Elias associates menstruation with the moon in the following way:

[15] The Ambonwari do not take it for granted that people have always known how to perform some activities such as giving birth or defecating. Ambonwari women and men told me a story about how a white female cockatoo taught the women how to give birth (before, they say, not having vaginas, women gave birth through their mouths or their anuses). This is why Ambonwari women do not eat cockatoos. In a different category but in a similar way, the men did not know how to defecate but vomited their food. Only after a cassowary, irritated by human ignorance, had made a hole in a man's buttocks did he begin to defecate. However, the men's act of defecation is kept as a male secret, as is the story about the cassowary.

[16] People remember how in 1987, shortly after their marriage, Jarambi began publicly to insult her husband Clemence, calling him *akurupiyan* ('leaking arse-hole') and *yarwambunggan* ('blood from the inside'; also 'yolk', though because of its 'bad' meaning this word is not used very often). During this time she gave birth to a stillborn child. Then she fell sick and Clemence took her to Angoram and Wewak, where she died in hospital. People say that *imɨngan saki* (the spirit of the village) had punished her because of her vulgar behaviour. 'Bad talk' is often heard in family quarrels and can easily end in a fight between husband and wife, mother and daughter, or father and son.

When the moon becomes full, during dry times or when the night is clear during a period of rain, we can see that the moon has a yellow ring. The moon sleeps inside this ring, just like a 'new' woman. The moon is in the middle and the ring fences it. It is the mark of a woman's vulva. See my little sister, she cannot give birth yet, she does not have space for her baby. So the moon has to shoot her, she has to lose her blood. Once she loses blood she will have enough space for a baby. When she looses blood you will see how her skin becomes slack. Once her skin is slack, she will gain more space, she will be able to carry a baby. Once she has enough space a man will be able to sleep with her. When a man wants to take her, he can do so because the moon has already taken her. The moon removed the block inside her vagina and made the space for a penis and for a baby. A man cannot copulate with a young girl who is blocked [before her first menstruation] neither can she carry a child. When the moon takes a woman there is lots of work; she loses blood and she loses her wetness. She becomes skinny. She has to be taken care of. You can see that there are always other women around to help her and to look after her. In general this is women's secret. We men do not talk about these things in public. It is taboo.

Ambonwari men experience an event in which their behaviour resembles the behaviour of women during their menstruation: nocturnal emission. Ambonwari men say: I had sexual intercourse with a spirit of a dead woman (*ama wunduman aring iminggin*, lit. 'I stood with the skin of the spirit of a dead woman'). After such an occurrence a man is considered unclean and observes similar taboos to a woman during menstruation. He does not go hunting or fishing, he avoids contacts with his children, and he spends most of his time in a men's house. He cannot, for example, beat a *Derris* vine to poison fish in the creek, nor can he make *sumbawi* fish traps. For five days after the wet dream he is not supposed to have intercourse with his wife. Sometimes, according to their own testimony, a young man may ejaculate when conversing with a girl he finds desirable. He can tell her later what happened, and she may ask him why did he not tell her about his yearning. Men acknowledge that women take the initiative in sexual matters. Women usually begin with initiatory gifts, and are far more loquacious than their husbands or lovers.

From the above account it is obvious that from the perspective of Ambonwari's *iminggan kay* ('the way of the village') the roles and practices of women differ from those of men; such differences are habitual, commonsensical, and taken for granted. The trajectories of men and women begun in childhood then follow their respective gendered destinies. In this regard Ambonwari resemble Manambu from the Sepik River.

The villagers tend to speak of marriages as physical movements of women between wards; from the domestic area of their natal wards, to the domestic areas of their husband's wards. A man's life-cycle, on the other hand, involves a slow progression within his ward from the domestic area to the riverfront. It is men who give the subclan its residential identity and continuity, simply taking on an increasingly senior ritual status as they mature. Women, on the other hand, are dispersed 'laterally' at marriage;

the permanent ritual inferiors of men, they create the transverse ties which link the wards together. (Harrison 1990a: 31)

Thus, while men give identity to their own place it is the women who form and preserve the links between their own and their husbands' places; in this way women give places and groups the relational dimension of their identity. A woman moves in space between two places when she marries. She is the founder or maintainer of *konggong* (path, marriage) between two groups. Ambonwari people say that while men argue and fight it is women, because of their pivotal position in relational identity, who determine the place and the people with whom the men fight.[17] Men can initiate exchange relationships with other villages, thereby creating links with outsiders. But in Ambonwari, at least, even these are reactivations of links forged in the past through the exchange of women.

Thus, for both men and women *kay* is habitual, something that is done without much thought. Bourdieu's argument is apt here:

When the properties and movements of the body are socially qualified, the most fundamental social choices are naturalized and the body, with its properties and its movements is constituted as an analogical operator establishing all kinds of practical equivalences among the different divisions of the social world—divisions between the sexes, between the age groups and between the social classes—or, more precisely, among the meanings and values associated with the individuals occupying practically equivalent positions in the spaces defined by these divisions. In particular, there is every reason to think that the social determinations attached to a determinate position in the social space tend, through the relationship to one's own body, to shape the dispositions constituting social identity (ways of walking, speaking, etc.) and probably also the sexual dispositions themselves. (1990: 71)

I do not see the difference between male and female *kay* in terms of male domination over women, but rather in terms of the cosmology which sees male activities and 'duties' in terms of protection from both inside and outside threats. In similar vein, I believe that Ambonwari men do not deny their act of defecation because they want to dominate women, but because they want to gain and preserve the power (with 'closed bodies') identical to that possessed by spirits. This power allows them to protect the village and secure the existence of all. Being seen defecating by women is a gross interference with that aspect of village *kay* (not simply a male *kay*) which *kay* itself (custom in this case) regards as important for its own preservation. Men see this *kay* in terms of ancestral, traditional power without which not only men but the whole village and the whole Ambonwari cosmology could be destroyed. Of course, one can question all this and label men's behaviour as 'symbolic domination' of women,

[17] While it is no problem in the village for women to organize themselves into womens' groups (the activities of which are supported by the provincial government) or to play soccer on Sundays with the men and boys, there is strong disapproval of women who show an interest in leaving the village in search of work (see also McDowell 1991: 299).

following their self-assertive priority in political and social matters. As this is a part of *kupambɨn* (ancestral) and *imɨnggan kay*, however, both men and women feel comfortable with such a division of labour. What characterizes *imɨnggan kay* is that its existence is achieved only through the totality and cooperativeness of male and female *kay*.

> [T]he strategies of men and women are sufficiently different so that, in most cases, men and women are able to pursue their separate concerns without domination by, or even interference from, members of the other group . . . The realm of male activity is, thus, characterized by a considerable elaboration of strategy and political form as each man attempts to succeed at the expense of the others. In contrast, because women are able to achieve worth in a more reliable and less competitive fashion, their strategies are relatively straightforward and can be independently pursued in a less complex social context. (Errington and Gewertz 1987: 13)

I am not sure if Weiner would agree with this notion of a 'less complex social context'. Recalling Bateson's *Naven*, she argues that 'in Melanesia female/male symmetrical and complementary oppositions constitute elements basic to a conceptualization of the social order' (1977: 11). She goes on to say that 'control and power are exercised by both men and women, but they must be seen operating not merely within the "politics" of social relations. Power extends beyond the social to concepts concerning articulation with cosmic and transcendental phenomena' (p. 12). Ambonwari men and women, as Weiner argued for matrilineal Trobrianders, 'basically control differing aspects of generational time' (1977: 20). Women control regeneration of identity,

> the essence of person . . . that moves through unmarked cosmic time . . . Men control property, a resource contained within sociopolitical fields of action. The male domain of power and control is situated in historical time and space that, unlike unmarked time, are particularly meaningful in relating specifically named individuals over various generations . . . The symmetrical and complementary integration of these two aspects of time and space produces complex patterns of interaction between women and men, enacted in exchange events throughout their lives.

The Significance of Kay

Kay (habit, way, manner) relates to the whole realm of practice: from mundane food-gathering to the most important ritual. For the Ambonwari it includes both the collective and the idiosyncratic; moreover, it denotes not just the practice as such but also the manner in which it is carried out. Only through its way does a practice achieve its specificity. Furthermore, *kay* not only relates to the subject who performs the action but is also comprehended through the object or subject (as an extension of the body) on which or with which this activity is performed. Thus, the things used by people in their daily activities are inseparable from *kay*.

Bourdieu says: 'To control the moment, and especially the tempo, of practices, is to inscribe durably in the body, in the form of the rhythm of actions or words, a whole relationship to time, which is experienced as part of the person' (1990: 76). The way of sitting, the way of cooking, the way of eating, the way of arguing, or even the manner in which adolescent girls walk, or gestures of support (Lewis 1990), or techniques of fighting (Harrison 1993: 35, 38–40) are all embodied habits acquired and recognized by those who share the same *imɨnggan kay* (the way of the village). The body (*kay*), not just the visible 'skin', manifests itself through the way acts are performed.[18]

Ambonwari do not have a term for any kind of physical body removed from the totality of human existence. Terms are given to individual organs, bones, body fluids, and so forth with considerable anatomical accuracy. People refer to the 'body' in its active, behavioural state by *kay* and by *arɨm* (skin). Thus, it may be said that body has meaning for Ambonwari people only through the external appearance of skin and the way it is observed to act. We could say for the Ambonwari that their 'body is in the world as the heart is in the organism' (Merleau-Ponty 1962: 203). Body has to be understood through the ways (*kay*) of its actions, which is not to be distinguished from its presentation. Following such a conceptualization of the human body in Ambonwari, we can readily comprehend Leenhardt 's puzzle concerning the Canaque body when he says that it has neither an existence of its own nor a specific term to designate it (1979: 164). When Leenhardt provoked his old friend Boesoou by saying that foreigners had introduced the notion of spirit to the Canaque way of thinking, Boesoou objected: 'We have always acted in accord with the spirit. What you've brought us is the body'. Leenhardt's discussion shows that it is not a conflict between different perceptions but rather a conflict between different cosmologies: one with a notion of abstract physical body and the other in which such a notion does not exist. Likewise, a person's body in Ambonwari cannot be conceptually 'removed' (abstracted) from the person.

We might pause for a moment to consider *kay* as a habit. With their bodily spontaneity, habits are acquired and enduring practices; they show relatively stable ways of sensing, perceiving, thinking, and acting (Ricoeur 1966: 280). Acquired through frequent repetition, habits become intimate parts of our bodily selves. Habit conveys 'the sense of operativeness, of a continuously practised activity. It conveys the fact of exercise, the reinforcing effect of repeated acts . . . Postures and movements which are habit memories become

[18] It is difficult to find a word in Karawari which expresses the meaning of work, either as a noun or as a verb. The closest, perhaps, is the verb *sasɨ-*, 'to do', in the sense of making something: *yana yan sasɨra yandia kanggɨnɨng?* ('who made this?'). *Sasɨ-* is a compound of *sa-* ('hold, use') and *sɨ-* ('become, do'). Such combinations reflect a connection between a thing which one holds and uses in order to bring something into being. Thus things define practices as much as practices define things. Being a good workman has in Ambonwari an unambiguous meaning. What we would call 'a good workman/workwoman', Ambonwari call *pan amɨng kamarar/kamarma*, with the literal meaning of 'a man/woman who knows how to find food'.

sedimented into bodily conformation' (Connerton 1989: 93–4). Both Ricoeur and Connerton see habit as they see emotion, in terms of embodiment.[19] 'To reflect on habit always means to refer to the time of life, to the holds which a living being offers to time and the holds which, thanks to time, he acquires on his body and "through" it on things' (Ricoeur 1966: 280). We might conclude that the body which the Canaque acquired was a Christianized, de-temporalized, abstract notion of body, while their own, as in Ambonwari, kept the past inside it. This embodied past enabled it to act fittingly and effectively within its own intimate environment. Those who look at habitual practices from the perspective of de-temporalizing science do not see that practices are constructed in time, which gives them form in terms of an order of succession and thus their direction and meaning (Bourdieu 1990: 98). Through the embodied past the anticipation of a future is possible, and we can talk about a temporalized body.

In the Ambonwari case, as we have seen, habits locate *kay* in relation to a past. But what of the future? In fact, *kay* is temporally 'thickened', to use Husserl's term (1964), for it is constituted as a temporal horizon embracing a past—sedimented as habit—and a future inscribed in the projective aspect of practice. To illustrate this point ethnographically, I need only refer to the prohibition signs placed around valuable plants, such as tobacco and betel nut, as discussed above. Underlying this practice is the notion that preventing the realization of a person's project is *ipso facto* an infringement of that person. In Ambonwari a person is constituted not only by a definite past but equally by a projection of that person's being in their plans and projects. An even more striking illustration of the simultaneously 'retentive' and 'protentive' (Husserl 1964) nature of *kay* is afforded by certain aspects of mortuary practices. Upon the death of a child, especially a first-born son, a man destroys all belongings associated with child's *kay*: tools, growing plants, even the house which the father occupies but which the boy would have inherited. With the extinction of the boy's *kay*, all those things protentively linked to it must also perish.

In their descriptions of traditional cultures anthropologists have, not surprisingly, stressed the past. More recently some anthropologists have sought to stress the future as being at the heart of a people's cultural vision (Biersack 1991: 232). The perspective of Ambonwari, however, would suggest that there is no dichotomy here; it is in the nature of practice itself to unify temporality within a single horizon.

Having introduced the crucial concept of *kay* in this chapter, we need now to look at other concepts which articulate with it in framing the Ambonwari person as a totality.

[19] 'Habit is a knowledge and a remembering in the hands and in the body; and in the cultivation of habit it is our body which understands' (Connerton 1989: 95). In similar vein, Kleinman, following Foucault (1979), argues that 'the body mediates social structure and cultural meanings, making them a part of our physiology' (1986: 195).

2

Person, Time, and Identity

Mi ngok. Kwasɨkain ama kuriyakɨr. Ama sayn auraur ama kuriyakɨr. 'You wait. I will come now (I just have to do something) but I will hurry and come.' I was sitting in a large motorized canoe waiting for my Ambonwari friend to do what he had to do, expecting that we would leave in a few minutes. The time went by, and I rolled another cigarette while children were cheerfully jumping around teasing each other. After two hours I had to stretch my legs. Women brought a few more things for our trip: a cluster of betel nuts and dry coconuts, a sleeping mat, some smoked fish. A young man asked me if he could join us on our trip. I agreed. It seemed that people were deliberately delaying our departure and, by not leaving immediately, we were showing them our compassion and care. Another half an hour went by before my friend returned saying that we could leave now. And we left. I had to realize, all over again, that it is unwise to be the first in a canoe and that 'now' means 'when we are all ready', including those who stay. It meant 'today'.

Though this example says nothing explicit about how the people of Ambonwari village perceive and experience time, it shows their attitudes towards and strategies for the management of time in their everyday practices. We are not dealing here with a 'time is money' attitude but rather with a notion that 'time is relationships'. My friend had to wait for money from those who required kerosene, soap, tobacco, salt, and other items of everyday use. After collecting all these small orders we were able to leave.

Temporal Expressions in Ambonwari

As has occurred in other Sepik societies and parts of the Pacific, the spread of Christianity and Westernization has brought the Julian-Gregorian calendar to Ambonwari. People have radios and they follow the events and days of the year, in many ways just as the citizens of the wider nation-state do. While there was never a mission station in or close to the village, there was always someone from the village who, after a short course in a Catholic centre in Timbunke, took over responsibility for Sunday prayers and the modest celebrations of annual Christian events. Sunday is respected as a holiday for villagers; it is a time when village courts are held and community projects are debated by those who are interested. It is also a time to visit, gossip, play, laugh, and sleep. This calendrical time-reckoning nowadays orders people's experience of changing 'seasons', and together they represent the way of reckoning time in

Ambonwari. Calendars and clock time represent 'modernization' in the sense of measurement by hours, weeks, months, and years. Because it regulates the presence and absence of children in the village school, it also (among other things) affects men's decisions to stage initiation rituals during the holidays. In daily life, however, this 'modernization' does not dictate activities, nor does it organize people's lives.

Before I begin my inquiry into the complex understanding of temporality in Ambonwari village, I will present some temporal expressions in Karawari language. The Ambonwari have no word for time or temporality, though, as I will show later, the particular periods covered by the term *pɨsɨnɨm* (span of time, limited duration, period, 'season') can in some ways be compared to the French word *temps*, as in *beaux temps*. *Pɨsɨnɨm*, which is always qualified in terms of what kind of period it is, serves Ambonwari people as both an abstract and a lived concept of temporality by which different processes can be compared according to their simultaneity, duration, and succession. They also have other complex and elaborated concepts which relate them to their past, which influence their present and future, and which are crucial for their self-awareness and thus for their identity as Ambonwari. These concepts are interrelated and, as we shall see, are temporal in themselves.[1]

Karawari language has twelve tenses expressed by verbal suffixes. Nine of them are reserved for the past, one for the present, and two for the future. Three tenses mark an event which is happening or has happened 'this' same day: present, immediate past (e.g. events just completed), and near past (events which happened this morning). Four tenses, of which one is used for future and three others for past, are further divided into two subgroups each, thus making eight tenses of the total twelve. This division is based on the time of the day when the event occurred or will occur, either in the morning or in the afternoon or during the night. There is an additional distinction in the remote past—that is, another tense is used only for the afternoon of two days ago (see Fig. 1).

The suffix *-kia* in Ambonwari is used only while describing an event occuring during the late afternoon and night (but see Foley 1991: 240 for alternative meanings in Yimas).[2] In Ambonwari, as in Yimas, 'a full day is seen to elapse from sunset to sunset, rather than from sunrise to sunrise as in English or Tok Pisin' (Foley 1991: 240; see also Mimica. 1988: 13 n.). Whether or not *-kia* has a near-future meaning, as it has in Yimas, I am uncertain. Thus *-kia* may be used in all tenses except those which mark the events of 'today' (present, immediate past, near past). In these tenses the suffixes are as follows:

[1] Lewis's observation is apt here: 'Concepts may sometimes be held, yet expressed in actions far more often and conspicuously than in words. They do not necessarily become concepts only because they are lexicalised . . . An exclusive concern with words may blinker us so that we fail to grasp important ideas that are not so easy to identify because they are not conveniently labelled' (1980: 72).

[2] The suffix *-kia* is mainly used together with tense morphemes in order to convey that an action was performed at night. Thus, though *-kia* is not strictly a tense morpheme in itself, I have included it for the purpose of easier comprehension in my discussion of Karawari tenses (cf. Foley 1991: 241–2).

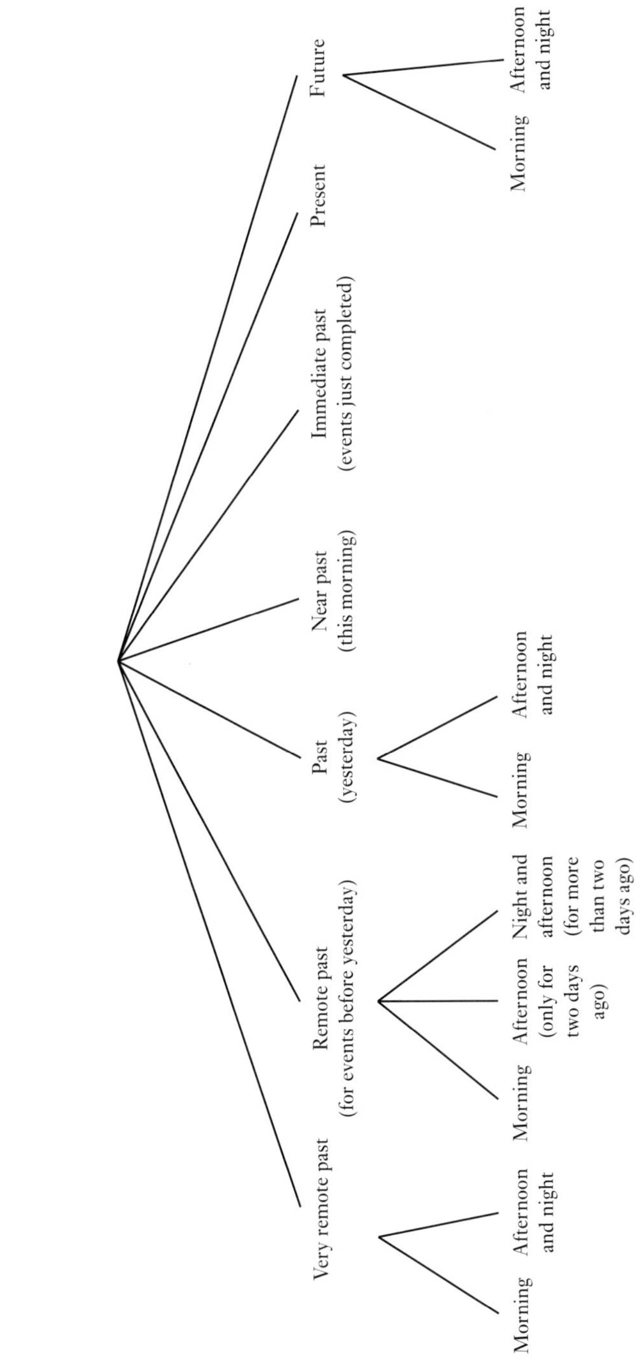

FIG. 1. Karawari tenses

present has *-kan* (or *-gan* when the verb root ends on nasals); the immediate past for the events which have just occurred is marked by the perfective suffix *-r* (or *-ndɨ* in the case of nasal endings); the near past takes the suffix *-nan*.

All other tenses have two 'day' and 'night' forms. The words describing events which occurred yesterday take the suffix *-kanan* for the day and *-kia-nan* for the night. The remote past from two days ago to the remembered 'lived' past is marked by the suffix *-n-dukun*. There is also a form which was presented to me as being used only as an 'afternoon' form for the day before yesterday, with the ending *-kia-kɨ-kanan*.

While collecting myths and legends I came across the tense in which Ambonwari, like Yimas, use the suffix *-k*. It appears to stand for the mythical juncture between the beginning of the world and the beginning of Ambonwari time. This juncture, I shall argue, represents one example of that class of periods that I shall call 'liminal' (see Chapter 5).[3] It is not unusual for a narrator while telling a legend to 'jump' from this 'irrealis' (marked by *-k*) into a remote past tense (marked by *-n-dukun*) as he 'jumps' from liminal events and situations to those characteristic of the post-transitional period of the familiar world. I never heard this form being used for the indefinite future among Ambonwari, as Foley reports for Yimas. I suggest that one reason for this might be that Yimas anticipate 'the end of the world' (certainly not a new theme in Papua New Guinea societies), that is, the end of cosmic human time and a return to chaos and homelessness. Foley speculates that this could mean that Yimas view time 'not as an open infinitely expandable line, but as a closed circle' (1991: 237).[4] Ambonwari express future simply by the suffix *-kɨr*.

What may readily be seen from Fig. 1 is that the Karawari language has a great variety of expressions for the past. Besides the main tenses, and the additional division of a day into day and night periods, Ambonwari also use inner operators, such as aspect (perfect and imperfect), to signify the continuation or repetition of an event or its limited duration. Because of its uncertainty, the few future tenses are augmented by different modalities, many of which are also used for the past.[5] It is beyond the scope of this book to discuss this any further.

[3] Foley says that this 'irrealis' suffix represents something between tense and aspect distinction, between real and unreal events (1991: 236).

[4] I would like to comment on a well-known notion concerning 'cyclical time'. The most notable exploration of time as cyclical 'eternal return' is the study by Eliade (1972 [1954]). On the relationship between cyclicity and linearity, see also Adam (1990), Barnes (1974), Farriss (1987), Howe (1981). Leach (1966), in his analysis of the myth of Cronus, proposed a zigzag conceptualization of time. While his starting-point was that certain phenomena of nature repeat themselves, and that life changes (ageing) are irreversible, he made a distinction between secular or 'profane' time (time goes forward) and 'sacred' time of rituals (social time has stopped, while sacred time is played in reverse!). As I shall argue, it is not cyclical, i.e. repetitive or periodical, *time* which is at issue in Ambonwari ritual, but cyclical or periodical *action*. Ritual secures the future by consolidating Ambonwari existence in terms of collective identity, which depends upon their relationships with ancestral spirits.

[5] As Foley puts it: 'Modality is an inflectional category relating to the intention of the actor and his potential or actual performance of the action' (1986: 152).

Before I explore the other ways of time-reckoning in Ambonwari, let me consider the interrogatative 'when?'. Quite often I had to use the sentence 'when will you come?' (*sɨndukumbar mi makuriyakɨr*), but I became aware of the full meaning of 'when?' (*sɨndukumbar*) quite late in my fieldwork. The word is composed of *sɨnduk* and *mbar*. *Sɨnduk-wa* is the interrogative meaning 'how much?' and 'how many?' (e.g. *sɨndukwa yarmas imana?*, 'how many men are in the men's house?') while *-mbar/mbas* is the suffix used in the sixth (seventh in Yimas) nominal group to express the plural. The word for period and duration, *pɨsɨnɨm*, which I discuss later, is a member of this nominal group. Thus *sɨndukumbar* ('when?') means 'how much of the period/duration?' or 'how many of the periods?'. This becomes even more obvious if we look at another way of asking 'when?', *wara pɨsɨnɨm*, lit. 'what period?' or 'what "times"?' (cf. Foley 1991: 115). This second way of asking 'when?' was seldom used in Ambonwari. The connection between 'when?' and 'how many?' possibly lies in the specific way of counting days. Ambonwari use a knotted *Cordyline* leaf to mark the days which will elapse before two people or groups are scheduled to meet. Once such a message is sent from one place to another the people in each place count the knots, which correspond to passing nights. Thus, 'when?' may refer to the number of knots on a *Cordyline* leaf used to coordinate activities.

The answers to the question 'when?' may be formulated in several ways. One of them is by using temporals. These may be placed in three sets: general temporals, parts of the day, and day-counters.

General temporals

first/at first	*yarpor*
just then	*yamba*
before/in the past	*kanggrayn*
long ago/long before	*kanggraynya kanggraynya*
after/later	*sambin*
another time	*ma-sambin*
not yet (wait)	*nggok*
only once	*mban-dukur*
again	*sayn*
once again (lit. 'and again')	*sayn-ok*
all the time (lit. 'again and again')	*saynyasayn*
sometimes (lit. 'not again and again')	*wara saynyasayn*
once (lit. 'on one knot')	*kɨnɨng mbanggɨna*
twice (lit. 'on two knots')	*kɨnɨnggri kripanya*
three times (lit. 'on three knots')	*kɨnɨnggi krianmawna*
sometimes, now and then (lit. 'on the other knots')	*ma-kɨnɨngg-inya*

Parts of the day

dawn, daybreak	*ariar wi*

morning	*sumbur*
during the day (sun)	*sɨmari-nya*
midday ('big sun')	*kupa sɨmari*
afternoon	*sɨnamana*
evening (lit. 'night is rising/becoming')	*wi mi sɨkan*
darkness, night	*wur*
night	*wi*
big night	*kupa wi*
Day-counters	
today	*kwasɨkain*
one day removed (yesterday/tomorrow)	*arin*
two days removed	*urasinggin*
three days removed	*kambarinjinggin*
four days removed	*mamaynjinggin*
five days removed	*sɨmbraynjinggin*

From day-counters it may be seen that Ambonwari count days from 'today', using the same terms for five days counted on the fingers of one hand for each direction, future or past.[6] The distinction between future and past is made by the choice of tense.

Terms used for time-reckoning, or rather their roots, are in Ambonwari, as in many other languages, often related to locationals. Thus, the term for 'before/in the past', *kanggrain*, has the same root as 'at the front', *kanggiambɨn*. However, many Ambonwari temporals are related to human practices, and so too are many locationals. From a cursory glance at the connection between these expressions we can see, for instance, that the term for 'not yet, later', *nggok*, is used in everyday conversation to mean 'wait', and it has the same word root as 'there', *nggun*. Those temporals which begin with the *sa-* ('to hold'; also 'to sit', with different suffixes) express a kind of duration and repetition, i.e. 'again' (*sayn*), or something that is still going to come or be, i.e. 'after, later' (*sambin*). Again, the same root, *sa-*, which characterizes the activity of 'holding on' or 'sitting' can be found in locationals, such as 'here' (*saman*). Thus, terms like again, later, after and here are all related to 'stay!' (*sa-ra*, lit. 'hold on') and reflect the activity of a person. I am confident that a closer look at the structure of temporals, locationals, and terms used for daily activities, would reveal even more connections.

The temporal order of events may already be represented in verbs (I do not mean tenses here) with the suffix *-bɨn*, which is attached to the other suffixes representing different tenses. This way of expression is continuously used in narration (see Appendix B) so that listeners easily comprehend the narrative sequence of events. The suffix *-bɨn* is used when two actions are performed one

[6] 'Five-day terminology' was reported also among Abelam (Kaberry 1941: 346) and Yimas (Foley 1991: 110).

after another. It stays the same in all persons and numbers, e.g. *ama kwasa-m-bɨn ama mariawk kusɨkan*, 'just after I stood up [also while standing up], I spoke'.

The nominalized verbs can take the suffix *-na* in the same way as the nouns in case relations. In this way people express simultaneity of events ('at the same time', 'while'). When this suffix *-na* is used with nouns, usually those designating tools, it means 'with' (e.g. *wandɨkakɨr-na* 'with an axe'), emphasizing the intimate relation between a thing and the person who uses it. In a similar way *-na* attached to a verb suggests the impossibility of separating an activity and the person who performs it, e.g. *kuriyara-t-na*, 'while (s)he walked around'; *amɨndɨ-nga-na*, 'while they (f) ate'; *amɨndɨ-mas-na*, 'while they (m) ate'. As practice, which I discussed in Chapter 1, incorporates a person's body, the logic of Ambonwari affixes discussed here—*sa-*, *-na*, etc.—underlines the inseparability of body, practice, and temporality. Thus, for example, a man cutting a tree with an axe is seen as a totality under the aspect of *kay*.

The Concept of Period

Anthropologists have used their empirical data to explore indigenous concepts of 'time-reckoning'. These studies not only cover the indigenous terms used for natural cycles (such as seasons, months, and days) but also look at the connection between recurring social activities or events and natural sequences. In Melanesia, as doubtless elsewhere, where the main preoccupation of daily life is centred around food-obtaining practices, people observe the flowering of plants and the growth of important crops as a guideline for their 'seasonal' time-reckoning. The Ambonwari, who monitor the flowering of grass and *Saccharum sp.*, would probably agree with the Kaluli of the Southern Highlands that 'the *dona* trees do not come into fruit because it is the season of *dona*; rather it is the season of *dona* because the *dona* trees have come into fruit' (Schieffelin 1976: 141). Panoff (1969: 156–7) observed the same use of flowering plants for the marking of time, lunar months in this case, among the Maenge of New Britain. On Simbo in the Solomon Islands, the stages of maturity of two species of canarium nut, important in exchange, are referred to in names of the months (Burman 1981). Gardening activities (Malinowski 1927: 210–11) or agricultural cycles of particular crops, such as yam among the Abelam (Kaberry 1941: 346) and Avatip (Harrison 1982a) or tobacco among the Umeda (Gell 1975: 163), are important for the way people reckon time. Ambonwari also relate a particular period ('season') to their activities and domestic work as well as to the food typical of that period.[7]

[7] None of these 'seasons' relates to yams, which they do not plant. Ambonwari do not even have a word for the domesticated yam, which in many other parts of the Sepik is the main crop (Harrison 1982a; Lea 1964; Roscoe 1989; Tuzin 1972). Harrison, for example, states that 'the ritual identification of yam cultivation with the age-grade system is the basis of a scheme for what Leach calls the symbolic representation of time' (1982a: 141).

In Ambonwari, the concept of *pɨsɨnɨm* (period) refers to the continuation of sameness which has its beginning and its end. It refers to a limited duration of a particular happening. It is most explicitly recognized in 'seasons'. However, *pɨsɨnɨm* refers not only to environmental seasons but to any definite processes, such as period of hunger, which affect people's *kay*.[8]

While for an external observer it seems obvious that there are generally two 'seasons' in a year: *kangɨm* or *kangbɨn pɨsɨnɨm* (dry 'season') and *maray kupay(nya)* or *maray-mɨn pɨsɨnɨm* (high tide, wet 'season'), the Ambonwari do not recognize only this kind of dichotomy. They distinguish four periods which follow one another. *Yuwayn maray(m)*, 'water of the flowering *pitpit*' (*Saccharum sp.*), is a period from December to May (the wet 'season') when the *pitpit* (TP) starts to blossom and the buds open sometime in March. This period may also be referred to as *manambɨn pɨsɨnɨm,* 'period of rain'. At its height the water may flood the village so that people must use canoes when visiting from house to house. When the *pitpit* loses its flowers in the middle of May, the dry 'season' (*yuwayn kangɨm*, lit. 'dryness of the flowering *pitpit*') begins. This lasts from May to the middle of September. August is the height of the dry 'season', when people are unable to use their canoes in the nearby creeks and so tie them at the bank of their main river and walk home. Along with these two main 'seasons', there are two shorter, transitional ones. *Kambɨs maray(m)*, 'water of the flowering grass', with the rain in September and October, is a period during which people make new canoes, as they are able to float tree trunks close to the village with the first turbulent, muddy flow of the fast-rising river. As the height of the water fluctuates during this time, and because of the effects of this on fishing, it is considered to be a period of hunger. This is followed by *kambis kangɨm*, 'dryness of the flowering grass', in November. Along with these four 'seasons', an additional one, *ambrɨma*, occurs when the water is neither high nor low, but fluctuates somewhere in between. During *ambrɨma*, which may occur at any time during a year and cause fishing to be unproductive, people complain of eating only plain sago. Ambonwari also refer to other quite regular periods, such as period of *supo* (*Pandanus* fruit) which is eaten at the end of the dry 'season', period of *ambing* (breadfruit) enjoyed from September to November, or period of different kinds of Malay apple fruits. The *tavan* (TP) tree and mango bear fruit at the beginning of the wet 'season', which is also rich in betel nuts (*payn pɨsɨnɨm*, period of betel nut). When rubber trees begin to lose their leaves at some point in June this is also a sign of the dry 'season'. People also talk about *sunggwin pɨsɨnɨm,* 'period of moon', and *sɨmarin pɨsɨnɨm,* 'period of sun'.

At first glance one might assume that *pɨsɨnɨm* refers to calendrical seasons. But whenever I use the term 'season' as a translation of Ambonwari *pɨsɨnɨm*, I

[8] A good example is the past volcanic eruptions in New Guinea, which shed ashes over the land. Many people remember this event (period) as a 'time of darkness'. Researchers, however, have difficulty placing the event in chronological terms (see Blong 1981; 1982; Lacey 1990; Mai 1981).

do not refer to anything related to a natural cycle construed in calendrical terms. 'Season' is not something determined by dates or months nor something which has a determinable duration. Thus, for example, if there was no rain during the expected 'wet season' people would call it 'dry season'. They would wonder why it is not as 'wet' as their knowledge would lead them to expect. Thus it is the rising or lowering of water and the flowering of plants or their condition of dryness which determine the 'season', and not vice versa. These periods are not institutionalized, that is, they are not defined and measured by a calendar; they are seen, touched, and lived. By growing up into them they become part of people's *kay*. The symbolic possibilities of time are embodied in the material world including themselves, just as the entire counting system of the Iqwaye is materialized in their bodies and body parts (Mimica 1988). Such a conceptualization of 'period' is easier to comprehend when the term is used in other contexts, such as: period of fighting (*isakrarkɨr pɨsɨnɨm*), period of hunger (*kambian pɨsɨnɨm*), period of eating, i.e. plenitude of food (*amɨndɨ pɨsɨnɨm*) or at the same time as now (*pɨnda pɨsɨnɨmbɨna*, lit. 'at this [kind of] period'). *Pɨsɨnɨm*, then, denotes a temporal span between the beginning and the end of a process which is essentially characterized by the continuation of sameness of a definite kind. This distinctiveness of the process defines the span. *Pɨsɨnɨm*, whatever its contents, does not represent simply an external dimension of people's lives which then influences a person or is internalized, but is already merged with a person and his or her practices. The term *pɨsɨnɨm* is never used alone in discourse but always in combination with some other term denoting an event or practice which then 'brackets' *pɨsɨnɨm* into a limited period, e.g. hunger-times, flowering-times, sorrow-times. One has to be aware that it is not 'times' which is stressed but hunger, flowering, and sorrow. These events then influence people's practices and actions. I have never heard the term *pɨsɨnɨm* being used to refer to a period of ancestors or to the lifetime of a person, though there are no grammatical obstacles to doing so.

Identity in Ambonwari

Pɨsɨnɨm, along with the preceding discussion about the Karawari language, show us the variety of expressions which order temporality.[9] Linguistic expressions, however, represent only one dimension of the context in which we need to view *kay*, which, as I said in the first chapter, articulates with a number of other concepts to define the space of Ambonwari personhood and identity.

When referring to ceremonies, habitual practices, people's thought and speech, myths and stories, people's movements between wards, their marriages

[9] As Gell argues: 'One cannot attempt to construct a model of the psychic foundations of time cognition (psychological time universals) on the basis of natural-language grammars. In order to do that, time has to be considered more abstractly, as a feature of experience as such, not as a just [*sic*] feature of discourse' (1992: 145).

and exchanges, in short, anything they see as part of 'their way' (*imɨnggan kay*)—that which gives them individual and collective identity—Ambonwari use three key expressions: *kay* itself, *konggong* (path; marriage) and *mariawk* (speech, discourse, thought; story, myth). Because of the complexity and multiple meanings of each of these concepts I will discuss them one by one.

While I was observing rituals, talking about marriages, or recording myths people referred to them as, respectively, *kupambɨn kay* (the way of the ancestors), *kupambɨn konggong* (the path of the ancestors), and *kupambɨn mariawk* (the story of the ancestors). Closer scrunity of these terms will reveal their significance.

Let me examine more closely the word *kupambɨn*. *Kupan* (Pl. *kupam*) is the term for an elder brother (*kupanma* for an elder sister), but also for a big or great man, and for all those important men who have died, including ancestors. As one of only three adjectives in the Karawari vernacular (the other two are 'good' and 'other') it can be used for anything big or large (cf. Foley 1991: 93)—that is, of course, with the proper suffix according to the noun class (e.g. *yip kupo* 'large coconut palm', *kay kupay* 'large canoe', *sɨsɨng kupang* 'big tooth'). Taboo, that is, a prohibition on food, or the prohibition on women and children entering a men's house, is called *pan kupambi* ('very ancestral'). The suffix *-bɨn* (i.e. *kupam-bɨn*) means 'just after' (see above). Thus *kupam-bɨn* can be translated 'after the ancestors' or 'as the ancestors did', or simply as 'ancestral'.

Kay, *konggong* (path), and *mariawk* (speech) are continually used in everyday life without the preceding *kupambɨn*. Hence, I would like to explore the individual meanings of these terms.

The verb *kay-*, which is probably the most frequently used verb in the language, captures several meanings: to be, to exist, to live, to remain, to stay. But *kay* is also used as a noun.[10] It has the following meanings: habit, way, fashion, manner, as well as ritual, custom, law. People speak of the habits of animals, the manner in which plants blossom, sleeping postures, and eating manners, or the ritual of initiation and the ways of the village (traditions, customs) as *kay*. Every individual in the village has his or her own *kay* which may be either accepted (*kay yapakupan*, 'good manners') or rejected (*kay maman*, 'bad manners'). Thus, for example, if people see women sitting with their legs apart they would say 'their way of sitting is bad' (*sar kay mbunan maman*). People talk about 'the way of dying' (*mar kay*), 'the way of the moon' (*sunggwin kay*, menstruation), or 'the way of the men's house(s)' (*iman(bas) kay*, initiation). 'The way of the ancestors' (*kupambɨn kay*) refers to all those diacritical practices associated with ancestors, for example, the *kurang* ceremony (known as *naven* to most anthropologists since Bateson) or the ways people use dance decoration or beat the slit drums, including such mundane

[10] *Kay* might be translated as 'standing presence' in a similar way to Classical Greek *parousia*, 'being' (Heidegger 1987 [1959]: 61).

activities as processing sago or trapping pigs. But when used in an unqualified sense *kupambɨn kay* refers to the most important ancestral ritual, male initiation, or sometimes to the rites of first menstruation. It is as though only these two rituals, but especially male initiation, concern Ambonwari existence or being in its totality.

In comparison to the importance of *kay* in discourse about identity there seems to be less to say about *konggong* (path, road). However, when one considers that the term relates to such institutions as marriage and exchange, it is clearly very important. *Konggong* reflects the relational dimension of identity, that is, the particularities of individuals and groups in terms of their marriage paths and consequent relationships. The concept—like *kay*—is used of different levels of inclusiveness: individual households, lineages, clans. Ancestral marriages created paths which continue to define relationships between clans and their members in ritual context and the distribution of land. *Konggong* might be thought to be used metaphorically in designating affinal links between groups but in relation to the paths taken by their ancestors it is not. Ambonwari ancestors, in common with those of other Sepik groups, were peripatetic. These journeys are not simply of mythic significance, for they created the many relationships with different groups encountered on their paths which today are still important for trade, exchange, and visiting. These journeys terminated in Ambonwari, but *kupambɨn konggong* continued through the marriage paths created between groups as they arrived. These ancestral paths underlie the complex distribution of land between and within clans, and are embodied in the exact knowledge of every piece of land and its ownership. When one looks at the land of particular groups, one can grasp the connection between the stories which recall the arrival of the particular ancestors and the places owned by their successors. Evidence of past relationships between different clans and lineages can be obtained by looking at the distribution of land. The complexity of this relational dimension of identity is produced by the fact that each lineage also has its specific 'marriage paths' in addition to the *kupambɨn konggong* which it has by virtue of its clan membership. Because of the implications that marriage has for land use, the 'marriage paths' of lineages increase enormously the complexity of land distribution defined in relation to *kupambɨn konggong*. The distribution and ownership of land holdings reflect the ongoing formation of new relationships between groups; each piece of land implicates a complex marriage history conceptualized as *konggong*, and each marriage treads once more an established path or extends it, and at the same time has repercussions for the distribution of resources. These matters will be discussed later in more detail.

The final term of significance for the constitution of Ambonwari identity is *mariawk*: speech, talk, story, discourse, narrative. Ambonwari people say that to have understanding means that you have both the ability to listen to others and the ability to talk. People say: '*Arɨm sambɨs ngandɨkɨm kwandɨkas ngandɨkɨm* (lit. 'skin with/has eyes, with/has ears'). The 'body' (skin) has eyes to see and ears to listen. *Mariawk* is the means of sharing apprehensions of the world and also the means of affecting others. When someone says *ama mariawk ngandɨkɨn* (lit. 'I am with talk'), it means that a person 'has' a thought which he or she wants to share with an audience.[11] *Mariawk* in Ambonwari is both speech and thought. The neighbouring Yimas, have a special verb to express 'to think'. This verb in literal translation means 'to feel words'. For Yimas, thinking is internal speech (Foley, pers. comm.). Ambonwari concepts which we call belief and faith are expressed through the meaning 'listen-to-talk'. Thus *mi andɨnbɨn yarar* (*yarma*) can be translated as 'you are man (woman) who listens to talk', which at the same time means that you believe (in God, for example). In an even simpler way one can express falsehood or lying by *kambra mariawk* ('nothing-talk'), and truth by *pan mariawk* ('very-talk') or *panbi* ('very much like that'). This truth represents a kind of constant negotiation between people, not necessarily a commonsense truth but more of a temporary consensus or agreement.[12] The 'real' world is constructed and reconstructed through *mariawk*, through communication, through gossip, rumour, stories, and the powerful speech of orators. People may talk in parables, figuratively: *wapaysur mariawk*, 'talk (goes) above', known in Tok Pisin as *tok bokis* (lit. 'box talk') or *tok piksa* (lit. 'picture talk'). A short statement uttered with an air of secrecy, and a magical spell, is called *wasa mariawk* (lit. 'little talk'). Joking and teasing are referred to as *wurumɨndɨ mariawk* (lit. 'laughing talk'), clever speech as *mariawk yapakupan* (lit. 'good talk'), and swearing and cursing as *mariawk maman* (lit. 'bad talk'). Stories and interpersonal communication influence most powerfully the lives of Ambonwari people.[13] When *mariawk* is used together with *kupambɨn*, i.e. *kupambɨn mariawk*, 'talk of the ancestors', it means either a myth (story of the ancestors) or short recitations, spells, and stories which are considered as having their origin in ancestral times. *Kupambɨn mariawk* narrativizes historicity and identity.[14]

[11] The Ambonwari do not have a verb 'to have'. *Ngandɨk-* means 'being with', 'staying with', or simply 'with'. The agreement suffix changes according to the class and number of the noun, in the same way as it does in adjectives. Needless to say, not being a verb, *ngandɨk-* has no tense distinction. This is not unusual. Speakers of Hebrew, for example, express possession by the indirect form *jesh li*, 'it is to me' (Fromm 1995 [1976]: 32).

[12] There have to be other people around for truth to exist; they have to want, choose, and decide the truth (Sartre 1992: 27).

[13] In recent years such concepts as narrative truth, narrative time, narrative in psychoanalysis, narrative and historical interpretation, mainly generated by Hayden White and Paul Ricoeur, have brought some useful insights into the importance of stories in human lives. Nevertheless, independently of these works, myths and tales, oral histories, interpretation and reconstruction of events, the power of gossip and rumours, and the oratorical skills of leaders have long been recognized in Papua New Guinea ethnography (for recent accounts in the Sepik area, see Brison 1992; Gewertz and Errington 1991).

[14] The relationship between *mariawk* and temporality is in many ways similar to that discussed in several of Ricoeur's works on time and narrative. For Ricoeur, temporality is 'that structure of existence that reaches language in narrativity', while narrativity is 'the language structure that has temporality as its ultimate referent. Their relationship is therefore reciprocal' (1980: 169). With his stress on plot, Ricoeur also emphasized the intelligible whole of a story, that is its 'thought' (1980: 179; 1984: 65).

A child is born into the village, into a ward, into a clan and lineage, into a household (though it might be given away in adoption). A child obtains a name and is cared for. She or he is born into an intimate environment of flowering plants, the smell of sago pudding and smoked fish, the taste of soft sago grubs and refreshing green coconuts, into the sight of foggy mornings, rising water in creeks, into the painful touch of sago palm thorns. She or he is introduced into the ways of living within this intimate environment and the ways of dealing with it (processing sago, using fish traps, paddling canoes . . .). She or he acquires these practices not only mimetically but through a simultaneous development of understanding (*wambung*) of the world in which they live. While learning how to do things, children also learn how to speak about these practices. Practices and ideas about them are never finished but always open to further elaboration (see Toren 1988; 1990; 1993).

By being born, a person is defined by all those spatial, bodily, and narrative identifiers (descent, kinship, myths, names, language, land to use, land to live on, and so on) which provide the basis for a person's life in a particular community. These identifiers place an individual in the historical horizon of the community. An individual assimilates stories, familiar ones with which he or she can identify, or foreign ones which because of their 'difference' continuously build his or her identity.[15] In this way one's temporality begins, is, and goes on. 'To a large extent, in fact, the identity of a person or a community is made up of these identifications with values, norms, ideals, models, and heroes *in* which the person or the community recognizes itself' (Ricoeur 1992: 121; see also Young 1983). With such an identity a person enters village life, a familiar environment, and interpersonal and inter-group relationships. In the case of a family, relationships, reciprocity, intersubjectivity, and identity exist in a relatively closed system. To become aware of and to recognize identity in its wider meanings, the household must be transcended. Only after a child perceives the world beyond its own household can the latter be apprehended as providing the child with its essential identity. This echoes Lévi-Strauss's notion (1968: 51) that the elementary form of human kinship lies not in the isolated family but in the relations between several.

Identifying symbols, signs, names, or narratives may be forgotten, lost, or contested (see Harrison 1990a); people may become unsure about them, other groups may appropriate them and use them as their own. A group can thus become 'homeless'. Changes in power and hierarchy occur as other groups claim the land, primacy in ritual, or leadership.[16] Opposing groups tend to

[15] George Steiner writes: 'To experience difference, to feel the characteristic resistance and "materiality" of that which differs, is to re-experience identity. One's own space is mapped by what lies outside; it derives coherence, tactile configuration, from the pressure of the external. "Otherness", particularly when it has the wealth and penetration of language, compels "presentness" to stand clear' (1992 [1975]: 381).

[16] During disputes, especially about the land, people's identities rely on both 'a historical right to specific territory and a territorial right to a particular history' (Alonso 1988: 41).

generate different myths; they narrate variations to support their own political interests. In such situations, myth and ritual are not a 'chorus of harmony' but a language of argument (Leach 1965; see also Appadurai 1981: 202).

Personal Spirit and Heart

In this section I want to examine two further concepts which are crucial to Ambonwari ideas about what constitutes a person: *anggɨndarkwan-ar* (*-ma* for females), which denotes a person's spirit ('watchman'), and *wambung,* which literally means 'insideness'. Knowing, understanding, and feeling are the capacities of the *wambung,* which is disseminated through the organs of the upper abdomen, a seat which I call Heart.

Every person has his or her own spirit, i.e. protector, a guardian, *anggɨndarkwanar*, which means 'watchman of the light' (*anggɨndarkwanma* for a woman). *Kwanar* or *kwanma* literally mean 'a carer', someone who looks after somebody; *da-r* (*sa-r*) means 'to hold'; and *anggɨn*, means both 'to give' and a 'beam of light' which comes into the house through a slot, from the sun, the moon, or a torch. The 'give/light-hold-watch-man/woman' may be seen as a shadow, and nowadays as an image in a mirror or a photograph. It is the spirit of a person.[17]

The person's spirit travels at night by itself, and a person apprehends its travels as dreams (cf. Bowden 1984: 451). During such travel a spirit may be beaten by bush-spirits, in which case the person may become sick after the spirit returns. If a spirit travels a great deal, a person is tired on getting up in the morning. The 'watchman' can leave the body forever, which causes a person to die. So the old and sick always have to stay at the back of the house, and are protected by the young and healthy who sleep near the entrance. The 'watchman' of those at the door may stop the 'watchman' of the old or sick if it suddenly decides to leave the person. For the same reason, people do not abruptly awaken (and thereby frighten) small children, as their spirits might leave them.

When this spirit leaves a person at death, it is not called 'watchman' any more but *wundumbunar* if male or *wunduma* if female (a gendered spirit of the dead; see Chapter 6).When someone is seriously sick people might ask: 'Has he/she died?' If a person is still alive the answer might be: 'No, [there is still] a man/woman. He/she sleeps' (*Kaya, yarmasɨnar/yarmasɨnma. Mɨn kɨndipaykan*). By using the term 'man/woman' (*yarmasɨnar/yarmasɨnma*) the Ambonwari emphasize that her or his spirit is not yet 'a spirit of a dead person' (*wundumbunar/wunduma*). This means that the personal spirit is still with a person. If a person has died the answer would be: 'He/she died. Just now it (the spirit) has come out.'

[17] For similar concepts as well as for concepts of spirits in Melanesia generally, see Bowden (1984: 451), Harrison (1990a: 354; 1993: 106–11), and Lewis (1975: 155–80) for the Sepik area, and individual articles in Lawrence and Meggitt (1965).

A dead person's spirit keeps the *wambung*, 'insideness/Heart'. It retains the character and qualities of the living man or woman. Consequently, the spirits of hot-tempered 'bad' men are feared and blamed for sickness and misfortune, mostly in the very houses where they had lived. The 'watchman' who left the person cannot enter and possess some other being, not even a newborn child. What it leaves behind is skin only (*kambra arɨm*, lit. 'nothing skin'), skin removed from the rest of the body (*kambra arɨng*), or a half-man (*pinjingɨnar*).

A grammatical distinction in Karawari places objects such as trees and stones in a different nominal class from humans and spirits. Trees and stones also have shadows (spirits), *anggɨndar-kwa* (*kwa* means the same as *kwanar* and *kwanma*, 'carer', but is in a different nominal class). When a man catches and kills a pig in a trap, he will tie knots in a length of vine in his hand, thereby tying up its spirit. He will throw it far away, ensuring in this way that the pig's spirit will not inform other pigs where the trap lies. Spirits of animals are not dangerous to humans (people laughed when I posed this question, though they found it within the bounds of possibility).

As people realize that one can express his or her thoughts mainly through speech, everyone in the village in discourse with others repeats several times every day that he or she is a good person, that he or she is generous, that he or she is bashful, that he or she is not a person who can easily get angry. Such a person tries to convince others that he or she really is such and such a person, that he or she has all those virtues which are in their interpersonal relationship regarded as 'good thoughts' and 'good feelings'. This is expressed by good Heart (*wambung yapakupan*) as opposed to bad Heart (*wambung maman*). Thus to be good mainly means that one is generous (*warɨmbarar*), while to be bad mainly shows that one is stingy (*karisikɨn*, firm). A man who keeps everything for himself can be referred to as a man who eats alone all the time (*sandɨkambayn amɨndar*), as 'he alone is important' (*mɨn panatna* or *mɨn pasa*). If he happened to be very successful for a long period and did not change his stingy habits people would gossip about how bad he was; soon he would get sick, or some young boys or men would steal things from his garden, his betel nut, his tools, or whatever. People usually do not have enough food to fulfil conscientiously all their kin obligations, so they have to manage carefully the distribution and sharing of resources. If someone is forgotten several times in succession, he or she will begin to complain about the wrongs done to him or her. Such situations can result in open conflict.

To be generous, showing that one cares, is the most important characteristic of good people. Generosity and 'caring about' prevent resentment. But to be generous one has to be busy all the time (searching for food and, nowadays, money), which means that lazy people are derided and gossiped about. This applies to both men and women. Ambonwari have other labels for people, such as 'hot-tempered' (*arkɨsan*); such a person is feared, but still considered good if he or she is generous. On the other hand a person who is quiet (*yaprisipasɨkɨn*) has also to be generous if he wants to be thought good. People who talk a lot, *mariawk kusɨrar*, or laugh a lot, *wurumɨndar*, are admired, but only if they are generous.

Both personal labels 'stingy' (*karisikɨn*) and 'bad' (*maman*) can also carry positive moral connotations in particular contexts. For example, men or women who do not seek lovers and who are known to be faithful to their spouses are also referred to as *karisikɨn*, but this time in the sense of 'firm, restrained', spoken with moral approval and admiration.

There is no doubt that the Ambonwari notion of 'caring about' was given particular importance, a particular flavour, and a particular virtue during the colonial period through the influence of Christianity. It seems that in the past Heart was less concerned with the well-being of everyone and was reserved principally for one's self and closest kin, while outsiders especially were regarded with permanent distrust. I think this is the reason why 'bad Heart' (*wambung maman*) can also imply admiration (cf. Harrison 1993: 120–1). An aggressive, ambitious, energetic, and dominant man, who may be referred to as 'bad', is feared but admired and followed by others. He represents security and protection from the threat of potential enemies, sorcerers, and dangerous spirits. Such a person is often a successful, fearless hunter. Potency and self-assertiveness in such men, though in many ways antisocial, are none the less admired (see Harrison 1993). The Ambonwari call an aggressive young man who puts himself at the front of the fighting line 'a dog' and say that he is 'bad'. Whenever conflicts occur in the village—and these are far from rare—the 'bad men' are those who are most vigorous, most aggressive, and most feared. When a woman of such character is also called bad (*mamanma*) it is her self-confidence and productivity which are being admired. These are most often displayed through her production of mats, baskets, and grass skirts and her success in catching fish. If their husbands lack a similar character they usually avoid the arguments with their wives; it is said that such 'bad' women are those who 'hold the posts' in the household. If both husband and wife are excesively temperamental their arguments may be so clamourous that others fear that 'the house will collapse'.

In short, those clans which have 'bad men' feel safer and more confident than others, despite their own anxiety in the company of such men. Though conflict, greed, envy, sorcery, and betrayal will not receive much airing in this book it should be borne in mind that lies, frauds, arguments, fights and double-dealings occur as frequently inside the village as between villages.

Let me return to *wambung*. Heart represents the mindful and affective 'insideness' of a person.[18] While the anatomical heart is called a seed (*sɨsɨnɨng*), *wambung* relates to understanding, affectivity, perception, and imagination. In

[18] I have already discussed Heart and how Ambonwari refer to their own feelings, especially caring, and resentment in Telban (1993).

the Karawari language *wam-* is a verb stem meaning 'to go inside'.[19] At the same time the word *wambung* itself onomatopoeically denotes life in terms of the beating heart: *wam-bung*, *wam-bung* (Foley pers. comm.). Thus, Ambonwari can use *wambung* to refer to life itself, e.g. 'the pig is still alive' (*imbian wambung ngandɨkɨn*, lit. 'pig is with insideness/Heart').

The same term, *wambung*, is also used when referring to the insideness of plants, animals, the moon, and the sun. Forest-spirits have their own Hearts. The middle, the heart, the soft pith of a plant, the core is called *wambung*. When someone makes a new canoe he removes *wambung* from a tree trunk, as he does from a palm when processing sago. I was told a story about an old man who asked the children to go and get some betel nut for him. As they did not listen to him he went by himself. It was already dark and the betel palm was asleep. The old man was near the top when the palm woke up, shaking. The man fell down and died. People say that if they want to climb a palm at night they kick it first and tell it that they are going to climb it to get some nuts. People say: 'A tree has Heart. When you cut a tree, it shakes, because it worries, it wants to stand up. *Wambung mɨn sɨkan maray amɨndarin*, 'it worries about water it wants to drink'.

The Ambonwari can say of an ignorant or stupid person that he or she 'does not have Heart' (cf. Harrison 1990b: 354; 1993: 98). It means that he or she is not conscious of something in particular, has no skills, does not feel and think as he or she is expected to under the circumstances, or is senile. Such a person does not have worries as others have—does not, for example, mourn at a death. Such people are spoken of literally as 'not having Heart'. But paradoxically, perhaps, this expression is also applied to the highly esteemed person who gives away everything. Such a person does not care about things for himself or herself. Contrasting with the generous person is the stingy one, who worries all the time about his own well-being by keeping things for himself or herself, 'he/she feels Heart very much' (*mɨn wambung pan sɨrar/sɨrma*).

A 'truly' ignorant or stupid person is said to have no ears (*kwandɨkas kanar/kanma*, 'ears-no-man/woman'); he or she is unable to join in reasonable discourse, does not listen when asked to do something, and does everything in his or her own way. *Mariawk* (talk, speech, discourse) represents the chief criterion guiding people to conclusions about who 'has ears' and who does not. Thus a deaf or a mute person may have good Heart (*wambung yapakupan*), and yet will

[19] The Ambonwari expression *wam-*, 'going inside', in the conceptualization of understanding and its relation to speaking and narrativity, calls to mind Ricoeur's use of the term 'deepening'. In his investigation of narrative time and the problem of the 'illusion of sequence' in particular, Ricoeur (1980) tries to escape the usual dichotomy between the chronology of sequence and the achronology of models. Following Heidegger 's notion of *Wiederholen* ('repetition', 'recollection') as the origin of fate, Ricoeur argues that the highest form of narrative repetition is the equivalent of individual fate or communal destiny (1980: 186). Such a 'repetition' is characterized by 'existential deepening' of time and by a 'deeper experience of time' (1980: 169, 185).

be placed in the same category as someone who is mad—though, of course, treated as being good.[20]

The notion that someone has no Heart is mostly used in relation to small children, who do not mourn at funerals, who eat food by themselves and do not think about its distribution and sharing. This does not mean, the Ambonwari acknowledge, that a child does not really have Heart. Indeed it has. Ambonwari say that when a baby is born its Heart does not know anything, does not understand: if its mother leaves it, it cries; it exists only by depending on its parents. Its 'worry' is a worry about its own existence, even though it has no consciousness of this worry. It cares only for itself, but pre-consciously. Later the child acquires the capacity to smile in response to others, a stage the Ambonwari regard as highly significant. If a baby sees one of its parents or its siblings and smiles, others will say 'child knows' (*wambung ngandɨkɨn* ('it has/ is with Heart'). However, people say, a child's Heart does not really exist for anyone except itself until the age of 4 or 5, when it begins to look after its younger siblings. As it grows the child begins to realize that its own cares and those of others are in a reciprocal relationship. In this way the child becomes conscious and begins to acquire understanding. Existence can be recognized as such only because people have Hearts, only because they understand. The necessary condition of being a person is the capacity to understand being (*kay*).

What I call 'Heart' in Ambonwari, Harrison (1990b: 352; 1993: 97) calls 'Understanding' (*mawul*) in Avatip, a term which, I agree, most properly highlights Heart's main characteristics. Its aptness is increased if we accept that the emotional part attributed to Heart is 'a mode of existence of consciousness, one of the ways in which it *understands* (in the Heideggerian sense of "Verstehen") its "being-in-the-world"' (Sartre 1948: 91).[21] To have understanding means to be able to learn—to 'bring inside'—the way of the village and to follow this way. To express knowledge and skills, the Ambonwari say 'I know how' (*ama wambung ngandɨkɨn*, 'I have understanding', lit. 'I am with Heart'). As Harrison writes: 'To have an Understanding is to be rational, to possess all appropriate adult skills and knowledge, and to be aware of one's obligations and the rights of others. Above all it implies an empathic disposition and the capacity to identify sympathetically with other people' (1990a: 90). This 'empathic disposition' towards others in Ambonwari as well as Avatip suggests what Levy (1973: 324; 1984: 227) calls 'hypercognized social sentiment'. Heart in collective terms appears to be characterized more by the social dynamics of care, which includes both affectivity and understanding. This care represents morality. Thus, to have Heart means to have understanding of the needs of others and sentiments about them.[22] Heart in collective terms refers to obligations and, like

[20] Somewhat similar concepts as these are found also among Pintupi Aborigines (Myers 1979: 350).

[21] Sartre (1948) argued that emotion is a certain way of apprehending the world; it is transformation of the world, a phenomenon of belief.

[22] Read (1955: 255–7) says of Gahuku-Gama that moral values have to be internalized and accepted by the majority of the members of a particular group, and that they represent a generalized expression of the good (which has basically collective connotation). For Read, not only is this an emotional

Avatip *mawul* (Understanding), it is the source of sociability, compassion, and respect. 'Having Heart' means that a person is socially attuned.

For Harrison, selfhood in Avatip has two basic dimensions: Understanding and Spirit. These two aspects of personhood stand in counterpoint to each other (1990a: 88–90; 1990b; 1993: 96–111), as 'two distinct orientations of the self toward the world' (1990b: 354). Harrison links this dichotomy to another, that between 'ritual' and 'secular' aspects of Manambu life.[23] As the translations of Avatip terms *mawul* (Understanding) and *kaiyik* (Spirit) correspond to Ambonwari terms *wambung* (Heart) and *anggɨndarkwanar/kwanma* (spirit) I would make the following comparative remarks. When I earlier discussed the spirit of a person, I mentioned that when a spirit leaves the body at the moment of death it 'carries' with him or her the person's Heart. In my view of Ambonwari, understanding and spirit are not opposed dimensions of self—spirit has understanding. When a person dies, the spirit (with its understanding) leaves the body. Harrison associates 'hearing' with the Understanding and 'speaking' with the Spirit. Making this kind of connection, of course, one who 'does not hear' has no Understanding and can be qualified as insane, a view found among the Ambonwari too. Harrison attributes the notion of thinking (i.e. speaking) to the Spirit. From an Ambonwari perspective, separating the 'speaking Spirit' from the 'hearing Heart' has little to recommend it. Ambonwari people explicitly say that to have understanding (Heart) means that a person has both the ability to listen and the ability to talk. It is this *ability* (understanding) which characterizes Heart.

Harrison states that Spirit, an aspect of driving selfhood, is 'more or less tied to the politico-ritual organizational structures', while Understanding, drawing individuals together in empathy, is linked 'to the structures of the domestic-secular domain' (1990b: 362). For Harrison this became the central disjunction within Avatip society and not the dichotomy of the sexes. Again, from an Ambonwari perspective it is hard to accept this separation. Though I agree that the capacity to speak is important in men's public lives, and that both men and women also have their distinctive habits, manners, and practices, this differentiation would in Ambonwari be conceptualized as a function of *kay*. 'The way of the village' *(imɨnggan kay)* depends upon appropriate understanding, speaking, and hearing in both sexes. People are 'watched' by their embodied spirits. Rather than opposing spirit to Heart, I would say that people recognize and then emphasize the gendered *kay* of spirits. That is why the spirits of dead women (*wunduma*), as I will show in Chapter 6, anchor the paths between village groups and their relational identities. In contrast, male spirits associated

statement but it also has intellectual, ideological components. This is precisely what the Ambonwari mean by 'having Heart'. For a comparison with a New Guinea Highlands society, see Biersack (1990: 67, 69; 1991: 235).

[23] Thus he argues that in Avatip the 'hierarchy of ritual statuses is a relation between people's spirits and not between entire persons. Spirit is only one of two conceived components of social identity, and relates principally to the sphere of ritual' (Harrison 1990a: 90).

with the men's house (*saki*) protect the village, help men in fights, and 'participate' in their politico-ritual life.

Heart's Motions

For the situations when Heart 'directs' itself, the two most important verbs are: *sɨ-* for desires and feelings and *aykap-* for knowing, remembering, and learning. *Aykap-* can be used alone or together with Heart. The Ambonwari thus say: he/she does not remember (*mɨn wara yan aykapɨr*, i.e. he/she forgot) or, together with Heart, he/she does not understand (*wambung mɨnana wara mɨn aykapikaykan*). Yimas use the same verb but never together with *wambung*; the verb which is a compound of 'breath-give' means 'to know' (Foley, pers. comm.).

Following the Ambonwari verb *aykap-*, we could say that Heart as understanding is construed and maintained by learning, memorizing, and remembering. Thus, Heart can be said to refer also to the memory: 'your Heart still remembers' (*wambung minyana nggok yan kɨndi aykapikaykana*) or 'I hold it in my Heart' (*ama wambung amanana yaka saykan*, i.e. I keep the memory). Ambonwari emphasize Heart's collective foundation as being its main characteristic. It can be said that Heart's understanding and affectivity are in many ways constructed by collective memory.[24]

Like everyone else, the Ambonwari person would not exist without her or his collective identity, embodied as 'habit-memory'. 'The individual's concrete sense of self, which will presumably be different for each individual, is nevertheless essentially linked to the social past' (Carr 1986: 115). Identity as a process relates individuals and groups to their predecessors and successors.

Turning now to the second of the two main verbs used in relation to Heart, *sɨ-* combines several interrelated meanings: to do, to become, and to feel.[25] 'I feel Heart' (*ama wambung ama sɨkan*) is the expression most often used to characterize empathic relationships. This dictum has a great number of meanings, which are comprehended by fellow villagers who are familiar with the circumstances of a particular situation. Thus 'I feel Heart' is used to mean 'I care', 'I am anxious', 'I am worried', 'I am sorry', 'I am sad', 'I am compassionate', 'I am longing for', 'I am homesick', 'I love'. Though not explicitly expressed, these feelings are directed at someone. Whenever the feeling is strong, the Ambonwari express this by using the intensifier 'very' (*pan*). Direc-

[24] Following Ricoeur, Alonso argues that 'social memory is integral to the creation of social meaning: representations of the past are central to the symbolic construction of social groups and social identities' (1988: 33, 40).

[25] The verb *sɨ-* is used in the following contexts: what are you doing (*mi waria mi sɨkana*), it is becoming dark/night (*wi mi sɨkan*), the high water is coming up (*maray kupay mi sɨkan*). Besides Heart (*wambung*) a person can feel shame (*mɨrɨnggi*), hunger (*kambia*), sleepy (*piamɨng*), cold (*sarɨk*), deep pain (*irɨng*), a smart following contact between an injured part and water (*arambayn*), heavy (*yipisikɨn*, 'tired after hard work'), and light (*waringan*); all of these use the same verb, *sɨ-*.

tion can be more clearly expressed by a pronoun when the verb becomes transitive. What the above expressions have in common is care for (or anxiety about) somebody or something. This represents the proper response in the relationship between people. Anyone who does not care (who is indifferent, unconcerned, or irresponsible) is considered to be bad.

People respond with resentment and anger to the experience of 'not being cared for', which is primarily experienced when others do not give or share work, things, and food. By contrast, if someone cares for you, it may maintain your equanimity or even cause you to experience happiness, depending on who that person is. If you are happy because you have something and do not share it (which means that you do not care for others), others will experience envy and resentment. In such a state your happiness will be seen by others only as evidence that you are proud and/or stingy. For the Ambonwari, real contentment is 'caring for someone' and 'being cared for'.[26] Caring for people is a profound virtue.

'Caring about' can show itself in all kinds of situations: giving presents to a boyfriend or a girlfriend; being present at mourning; visiting those who are sick; sharing of goods and presentation of gifts; and through assistance in building houses, hunting, or making large canoes. But the relationship where care shows itself most clearly is that of parenthood. In a place where the rate of adoption is very high (see Chapter 4) this is not established by the facts of birth, but by giving a 'name' and by providing care.

'To feel Heart' genuinely represents 'the way of the village', its social dynamics, public sentiment, and ideology, and shows itself in all kinds of social arrangements. Caring is thus a consequence of and a condition *sine qua non* of being with one another. By taking care of each other, people diminish the anxiety of homelessness. Heart in Ambonwari, like the concept of family (*walytja*) among the Pintupi, characterizes the ideal world of support, generosity, familiarity, and warmth (Myers 1979: 352). It also seems to correspond to the condition of *fago* among the Ifaluk of the Caroline Islands, which Lutz translates as 'compassion', 'love', and 'sadness' and sees as the essence of positive human relationships (1988: 121).

When something discomfits or excites a person, Heart becomes sore, it jumps, it is set in motion until it returns to its proper state of quietness. Expressions such as 'my Heart sleeps on the side' (*wambun amanan kaynggian inga paykan*, i.e. I am angry), 'my Heart has run away' (*wambung amanan inga mɨnggaykan*, i.e. I am scared/afraid), 'my Heart is coming out' (*wambung amanan mɨn sikian*, i.e. I am excited or I feel generous), are typical examples of 'symptomatic metaphors' (Solomon 1980: 251).[27] As long as Heart is hidden

[26] Myers argues that emotional concepts are major constructs of the Pintupi view of what it means to be a person as well as of the political order of Pintupi life (1979: 345).

[27] See also Harrison (1990b: 353) for Avatip expressions; Stephen (1989: 164) for Mekeo; Heelas (1986: 244–5) for various cultures.

and does not reveal itself through the person's deeds or words, her or his feelings are also hidden. For this reason people often say, 'I do not know how she or he feels, or what he or she thinks'—a common refrain in the ethnography of Papua New Guinea (Fajans 1985: 383; Read 1955: 281–2; Young 1974: 66). But an appropriate social response of Heart is expected and recognized as a part of village sociality. Thus when such feelings are situationally proper—for instance, if a child cries because his father did not take him to the garden—people will not speculate but will quite confidently state their opinion about the child's feelings.

If a person does not care about someone for whom he or she should care, that person is considered bad. As a reaction to not being cared for by someone, a person experiences a sense of insult or affront, and will react to it with all those external expressions which correspond to *sukunang* or *mambara*, both meaning a kind of resentment, the latter a more transient form. Neither expression is used with the verb 'to feel' nor are they referred to as Heart's motion. They are used with a verb *kay-* (to be, to stay, to remain, to exist), thus representing a state in which a person is. I translate these states as 'resentment' or 'being affronted', but one should bear in mind that 'he/she is in a state of resentment' (*sukunang/mambara mɨn karar*) combines all kinds of feelings such as envy, jealousy, spite, self-pity, offence, frustration, depression, for which the Ambonwari do not have special words. This state of feeling bad arises when someone feels or thinks that he or she has been wronged (left out, forgotten, reproached for something that he or she did not do, gossiped about . . .). It is an affective and cognitive response to being left out of something others share, to being excluded from joint activity or excommunicated. It is an affective and cognitive response to 'not being cared for'.[28]

In many situations people experience *mambara*. In the village one can often hear someone saying 'me too' when he or she wants to join others who are going somewhere, doing something, or getting something. *Mambara* occurs among children all the time, and their parents and other relatives constantly have to counteract it by showing that they care. It is considered that younger children, 'not having Heart', are allowed to some extent to demand things at the expense of their elders. When a boy, for example, does not get enough meat on his sago pudding while others, in his opinion, do, a boy can try to extort it. Quite often a child threatens others (whom, he reasonably supposes care about him) by saying: 'All right, I will go straight to the doorway and fall down from the house'. When three-year-old Sanggɨrmari did not get from his mother something he wanted, he started to beat his 7-month old sister, Maya, who was breastfeeding in her mother's lap. Siblings and parents try to soothe this kind

[28] The parallel with *popokl* among the Melpa (Strathern 1968) and to *unuwewe* in Kalauna (Young 1983) is obvious. It seems to me that at a very general level this relationship between 'care' and its lack powerfully influences relationships and behaviour all over Papua New Guinea.

of resentment quickly, either by giving the child more or by making promises. A different case arises when a sibling eats all the food without sharing it with others. The others will talk loudly about how bad she or he is, how she or he does not care, how she or he does this all the time, and so on, and the child will shortly feel ashamed of her or his doings. But if only one child is left out, he or she will be on the verge of crying or will cry. When others call him or her to join them and get the same thing later, he or she will not even look at them and will not approach. He or she may pathetically say: 'It does not matter, you can take it all, I do not care about it', and will walk away.

Those who experience *sukunang*, a long-lasting condition of resentment, may come to desire revenge. Accordingly, someone else will always try to comfort such a person. Ambonwari, like people everywhere, find themselves caught many times between two people who are angry with one another, when their kin relationships are such that they cannot help either of them. Brothers can, for example, be divided, one on each side of the quarrel or fight.

The behaviour of people seems to follow the awareness of possible resentful reactions, and their actions and reactions are accordingly appropriated. Thus people afraid of visitors' demands hide most of their possessions such as money, clothing, fuel, tobacco, i.e. mostly introduced goods. They do not want to give them away and they are afraid of the resentment of those who might see them. When visitors come near the house (never entering without invitation), the house-owner would usually reapeat several times that there is no food in the house, 'very-nothing' (*pan kaya*), 'we stay nothing' (*apia kambra apia san*), that they have eaten plain sago pudding, 'nothing pudding' (*kambra karis*). And the visitor would reply that it does not matter, 'no talk' (*wara mariawk*), that it is all right. They will share betel nut and depart with good will.

Generosity is not something which is taken for granted. There is always a conflict, a tension, a disjunction between one's desire to keep for oneself and share with one's closest kin and the need to share with one's more remote kin. People have to make their *kay* and *wambung* manifest, materialized in their deeds.

Young boys and girls learn generosity from an early stage. Their compliance with social expectations or norms is noticeable in situations when children organize themselves. For example, when they go fishing together or looking for eggs of wild fowl they return with such stories as the following: 'I found an egg first and gave it to Sanggɨrmari. Kapun found two, gave me one and one to Imbianmari. Sanggɨrmari found four and each of us got one. I got another two and gave them to Kapun and Imbianmari. Thus each of us had two eggs. When we found two more we decided to cook them in leaves and eat them.' In everyday life such stories are heard all the time among children, and among adults too. In this way one's sense of self is enhanced through the eyes of the receiver; one is perceived as an individual capable of finding, making, having things that others want.

Identity as a Process

Having presented the different concepts which constitute Ambonwari persons and their identity, I would like now to bring them together.

The first chapter has already established the central role of *kay* in Ambonwari conceptualization of both personal and collective identity. In this chapter I have presented those concepts which articulate with *kay* in the discursive construction of Ambonwari being. *Kay*, when used with 'ancestral' (*kupambɨn*), refers to the identity of the whole village, being thus the most important aspect of 'the way of the village' (*imɨnggan kay*). In contrast, 'ancestral stories' (*kupambɨn mariawk*) and 'ancestral path' (*kupambɨn konggong*, 'ancestral marriage') are specific for each clan, differentiating them from one another. Stories and paths always remain the two dimensions of identity which differentiate clans, lineages, and households. Neither path (*konggong*) nor talk (*mariawk*, speech, stories) are used together with 'of the village' (*imɨnggan*).

When we turn to the concepts specifically related to individual existence, it is significant that *kay* appears here too. *Kay* (way, manner, habit) becomes the main characteristic of an individual, whose existence is secured by spirit and whose openness to the world and historicity is a consequence of Heart (*wambung*). Through Heart *kay* is transformed; but since Heart itself is an aspect of *kay*, Heart is the means of *kay*'s self-transformation. Heart itself is not the personality or identity. It is 'spirit' (*anggɨndarkwanar/kwanma*) together with Heart which enables a person to act and in such a way construe her or his body, thought, and relationships—thus constructing her or his *kay* (being) and thereby identity. By having Heart, one's own as well as others' life processes are comprehended and accommodated. Heart is what makes people conscious, enabling them to perceive, memorize, and anticipate, to reflect on and understand their own existence.

I wish to stress the processual nature of all these concepts. *Mariawk* (speech, story, thought), *konggong* (path, marriage), and *kay* (ways of doing things, manners, habits) are not timeless structures but are processual and thus temporal in themselves. Together they form historicity, temporality, and identity. While *kay* as bodily dimension of identity could be said to represent habit-memory, *wambung* represents psychic memory. *Kay* and *wambung* are thus two interrelated dimensions of spirit. When spirit leaves a person it takes away both *kay* (being) and *wambung* (understanding) and leaves behind only a 'skin'.

Through socialization—which has to be seen as an active process, a process, moreover, that is an aspect of *kay* and *wambung*—every person is subjected to the collective village practices, habits, and manners, in short, 'the way of the village'. For Bourdieu, practical belief is not adherence to a set of instituted doctrines or beliefs nor even a 'state of mind', but rather a state of the body (1990: 68). A recent study of 'bodily social memory' by Connerton also argues that 'habit cannot be thought without a notion of bodily automatisms' (1989: 5;

see also Foucault 1979: 135–70).[29] However, the Ambonwari evidence suggests to me that these 'bodily automatisms' should not be construed as atemporal: they are themselves the product of a definite history implicating both a human body and a specific social history. Like *kay*, of which it is an aspect, human understanding (*wambung*), or 'making sense' as Toren puts it, 'is a historical process because it constitutes—and in constituting inevitably transforms—the ideas and practices of which it appears to be the product' (1993: 461–2).

For a deeper understanding of the village we must now survey the different kinds of practices which constitute the collective history and identity of the Ambonwari.

[29] Connerton goes on to say that 'embodied experience, of which habitual practices form a significant part, has recently been subjected to a cognitive imperialism and interpreted in terms of a linguistic model of meaning' (1989: 94).

3

Social Institutions

Certain features of 'the way of the village' (*imɨnggan kay*) are represented as quintessentially primordial—as if they were unchanged and unchanging. In constructing their present, however, people also construct their past. '[W]hat constitutes a living tradition may reveal an extra dimension to the past—one whose validity is not a matter of "what happened" but of how it may be understood' (Toren 1988: 696). By collapsing 'the way of the village' and 'the way of the ancestors' into one another, both can be made to manifest the present or to reveal the past.

In this chapter I deal with one set of these features, specifically those pertaining to organization of persons in social groups (which is at the same time the production of persons *by* social groups), the principles of their continuity, and the practices that constitute the relations between them. So this chapter includes clanship and lineage affiliation, the naming system, 'fathers' and 'mothers' of the village, exchange partnerships, and the system of 'ritual inaugurator' (*sɨmɨrinyar*) relationships. I shall discuss kinship, marriage, and adoption in the next chapter.

Totemic Clans

I call the totemic patrilineal descent group the 'clan'. Its members claim descent in the male line from the same founder (see also Errington and Gewertz 1987: 18 for Chambri, and Stanek 1990: 268 for Iatmul). Ambonwari village is visibly divided into clusters of houses separated by paths, ditches, coconut palms, or patches of open land. These village sections I shall call wards (see Map 3). Of the twelve totemic clans only five are the owners of their own wards. Each of these five clans has given land in its ward to some other, usually smaller, clan to live on. Such non-owning, co-resident clans preserve their own identity, however, and retain their totems, names, spirits, ritual paraphernalia, men's houses, and so forth. Thus, we have the following situation:

Owner clans	*Co-resident clans*
Bird of Paradise	
Crocodile-1	Eagle-1
	Hornbill
	Cockatoo
Pig-1	
Wallaby	

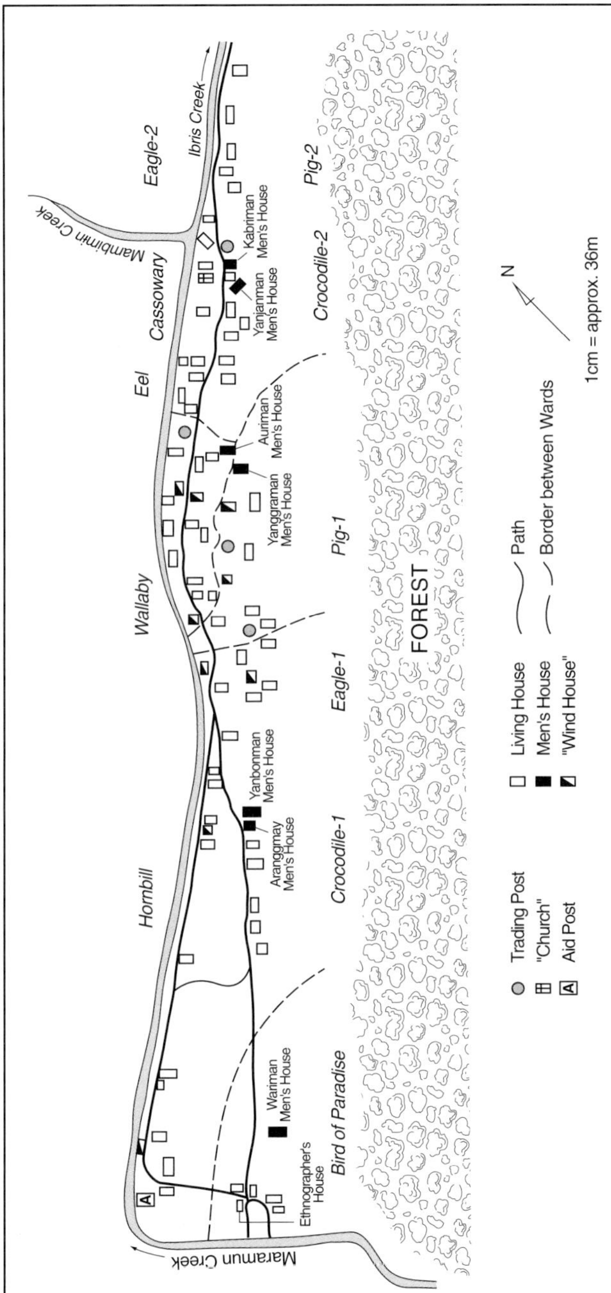

Map 3. Ambonwari village in January 1992

Crocodile-2	Cassowary
	Eel
	Pig-2
	Eagle-2

As among the Manambu, these 'ritually specialized descent groups' (Harrison 1990a: 66) are associated with an ideology of rank based on precedence and an institutionalized status based on totemic cosmology. Land ownership and distribution reflect the order of arrival of their mythical ancestors, and the relations set up between them. Thus Eagle-1, Hornbill (including only one Cockatoo member), Cassowary, Eel, Pig-2, and Eagle-2 are not landless, even though they do not live in their own wards. They are the owners of lands and waters along the Konmei Creek. At the clan level, marriage is likewise believed to reflect ancestral patterns of affinity.

Clan members are the common owners of totems, myths, stories, and sets of names, and have relationships with the same spirits. People from other clans refer to their living area by the name of their totem, their main men's house, their most significant spirit-object, or by other descriptive and spatial terms. Within the clan all men refer to each other either as 'brothers' or as 'fathers' and 'sons', inheriting these terms from their 'fathers'. Residence is patri-virilocal, though whenever trouble occurs within a particular household a man with his family can go to live temporarily with his wife's brother or father. The social structure of the village reflects the importance of affinal relationships, and those with maternal groups. Harrison argued that:

> the key to Avatip social structure—including agnatic descent—lies in the structure (less explicitly conceptualized by the villagers) of alliance relations set up by marriage. From this perspective, myth can be shown to portray a perfectly accurate replica of the social system, but—in keeping with the atemporal quality of myth—distorted along its time dimension. (1984: 389)

Harrison goes on to say that the mythical social system of the totemic ancestors in Avatip resembles social systems in the New Guinea Highlands to the extent that 'the key to village social structure is its marriage system' (1984: 389). He concludes that both 'myth and social structure can be explained as an effect of the difficulties of projecting a *temporally oriented* social structure into a medium lacking a time dimension [myth]' (1984: 399, emphasis in original). Harrison's argument can also be applied to the Ambonwari, allowing the notion that the eternal character of myth represents the eternal nature of the totemic clan with its men's houses, spirits, and hierarchical relationships with other clans which reproduce past village formations. One can also look at the marriage avoidances between particular clans, and the alliance relationships between others.

To summarize: *imɨngga*, the village, is a characteristic creation of humans and spirits, and the centre of Ambonwari's social cosmos. The wards, occupied by totemic clans, are organized according to the order of arrival of the patriclan founders. Several smaller clans, which came later, were incorporated into these wards under the 'supervision' of the main ones. To understand the allocation of land to totemic clans inside and beyond the village, it is necessary to look at the myths of origin, paying special attention to the period of arrival of the ancestors of individual clans.[1] The main myth of the arrival of the founders of the village, the ancestors of Crocodile-1 and Bird of Paradise clans, will be presented in full in Chapter 5. What follows here is a shortened version of the arrival of every clan's ancestor and the formation of the first ties between all the clans of Ambonwari. Maps 3, 4, and 5 and Table 1 may help to clarify the village structure at the level of patrilineal totemic clans. The numbers indicating clans on Map 4 indicate the order in which they arrived in Ambonwari. This order is also used in the following account of their arrival.

Crocodile-1, Bird of Paradise, and Crocodile-2 clans. Of the four sons of Angguru, Kapi was the third. The sons left Kumbranggawi Mountain and Aibom near Chambri Lakes and wandered around in search of sago and a place of their own. Kapi was followed by the oldest brother, Akɨmbrɨkupan, and later by the second, Mamanggamay. The former two brothers are the main figures in the formation of Ambonwari. Kapi was the first to cut the virgin jungle at the entrance of Konmei Creek. He marked the land and took over the main men's house, Yanmbonman, from his elder brother. He founded his own Crocodile-1 clan. Akɨmbrɨkupan started the Bird of Paradise clan, at that time still without his own men's house. Mamanggamay became the founder of the Crocodile-2 clan.

Wallaby and Hornbill clans. During the same period other ancestors followed their own paths (see Maps 4 and 5). Ambrɨpi, who travelled as a man, a wallaby, and some say also as a great hornbill, left the Marukamari Mountain near the Karawari River and headed towards Konmei Creek. There he met Kapi. They spoke the same language; in some versions Ambrɨpi was even the brother of Akɨmbrɨkupan's wives. He had his own men's house, Auriman, and his own spirit, Warimɨndan. He divided his descendants into two groups: Wallaby and Hornbill. While moving around what they believed was their land, they came across a tree which had already been marked with red clay.

Cassowary clan. Ambɨn, the first ancestor of the Cassowary clan, arrived from Manam Island. With a *sɨmɨndanɨng* vine he made ten knots on a *kambuka* (*Cordyline* sp.) leaf (to mark a period of ten days), put it into his basket, and hung it above the mark on the tree. He kept one leaf with knots for himself. When he returned to the tree he noticed that someone had painted the tree trunk with white clay, removed his *kambuka* leaf, and

[1] The long versions of myths of origin were told to me by Bob Kanjik Anjapi (Sapɨndam, Crocodile-1), Tony Simiwariya Andiyapi (Kumbranggawi, Bird of Paradise), Luk Kinggapan Ambɨnawi (Kunbaynjam, Crocodile-2), Elias Wapun Kandanggway (Yimasɨndam, Wallaby), Tino Apasi Kapari (Imanmari, Wallaby), Yoachim Opisay Karuop (Wanbanimbuk, Hornbill), Yambos Kapayn (Wayanggara, Cassowary), Godfried Yanggus Saramambi (Mɨndapmarpinjing, Eagle-1), Christian Yapɨsa Warɨng (Singgaimarijam, Pig-1), Willy Awa Kwanggai (Warɨpɨndam, Eel), Aunjak Wakariwan and Raymond Anggurapi Yakopia (both Murimari, Eagle-2), and Murimari Kwanggiambi Maraul (Yawyamari, Pig-2).

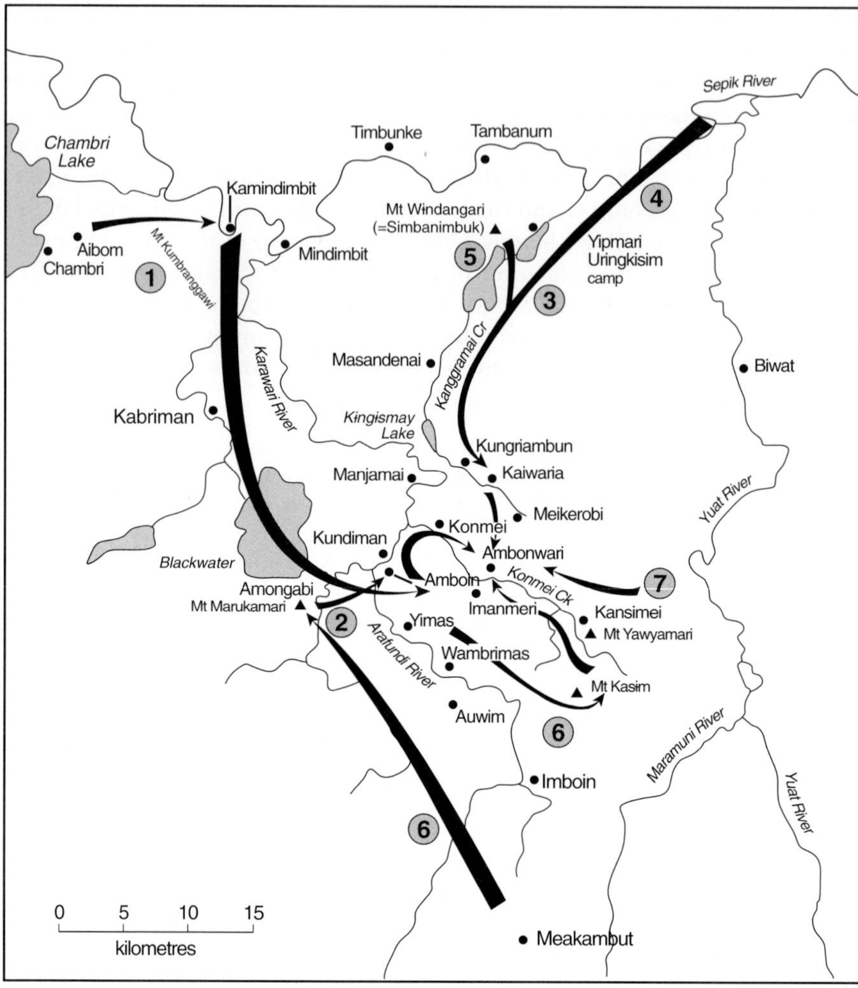

MAP 4. Paths of the ancestors

1 Crocodile and Bird of Paradise Clans
2 Wallaby Clan
3 Cassowary Clan
4 Eagle-1 Clan
5 Pig-1 Clan
6 Eel Clan
7 Pig-2 and Eagle-2 Clans

replaced it with one with only five knots. He painted red clay over the white mark and decided that he would check the place every second day. After they had missed each other several times, one day Ambrɨpi and Kapi finally saw Ambɨn coming. His shoulders were covered with black feathers and he made the sound of a cassowary. Kapi and Ambrɨpi jumped on him, overpowered him, and brought him to their place. He spoke

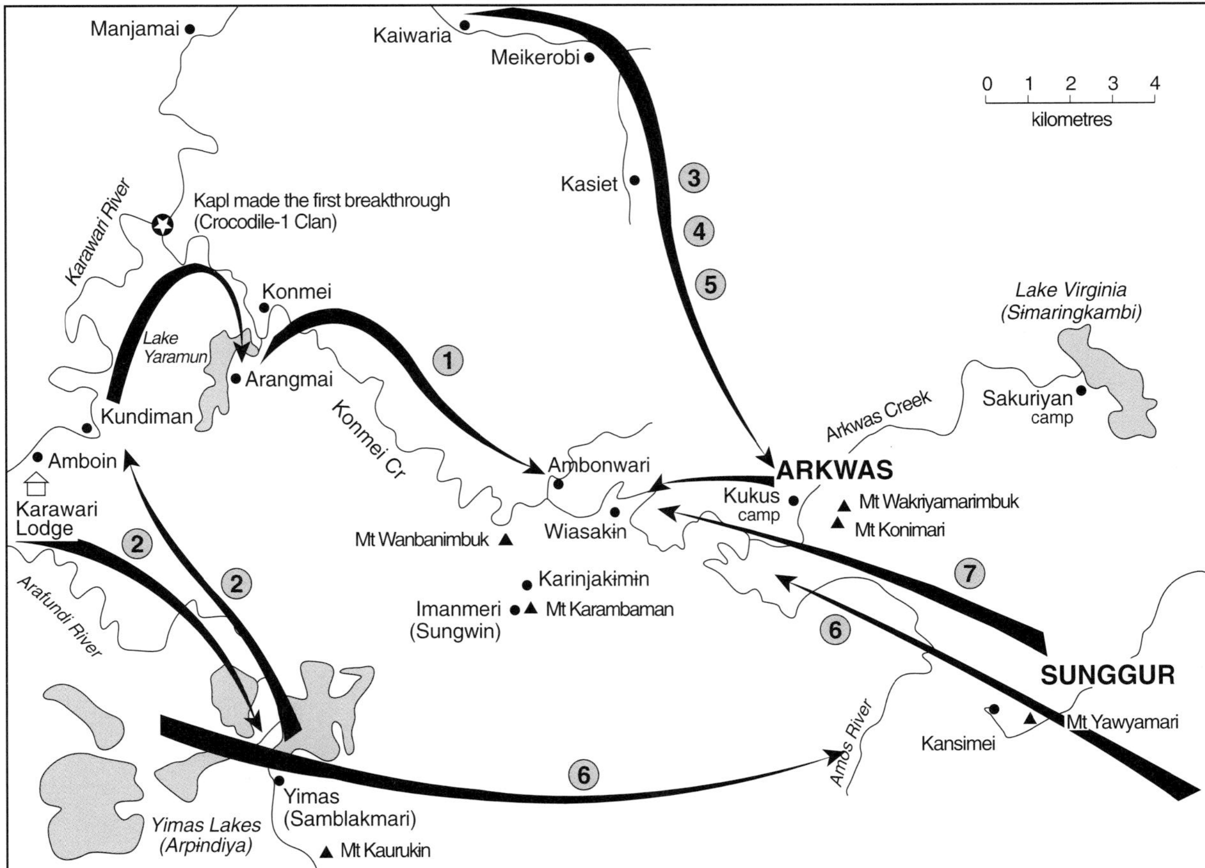

MAP 5. Place-names accompanying paths of the ancestors

1	Crocodile and Bird of Paradise Clans	3	Cassowary Clan	5	Pig-1 Clan	7	Pig-2 and Eagle-2 Clans
2	Wallaby Clan	4	Eagle-1 Clan	6	Eel Clan		

a different language. He did not tell them that he had left his two daughters upstream in a camp; they later married the two ancestors of Eagle-2 and Pig-2 clans. Ambɨn brought with him a spirit-house, Yanjanman, and a slit drum, Primari.

Eagle-1 clan. The ancestors of the Eagle-1 clan were already on their way when Kapi, Akɨmbrɨkupan, and Mamanggamay were still in Aranggmay near the Yaramun Lake. Kwandɨkan and his younger brother Ambanyarɨm came from the Sepik, and whenever someone asked them about their origin they replied: '*Apia Suwas* (we are Sepik or Sawos people)'. Some people from their group stayed at a small place called Ambramai, near the current village of Manjamai on the Karawari River. The two Eagle brothers met Kapi in Aranggmai and later followed him up Konmei Creek. After arriving in the village they settled in the ward of the Crocodile-1 clan and were allocated the back part of Kapi's men's house.

Pig-1 clan. The ancestors of Pig-1 clan, Warɨng and his older brother, Kwangɨn, shared the camps with the ancestors of the Eagle-1 clan after they entered Kanggramai Creek from the Sepik River. After being accepted into Ambonwari village, the two brothers presented Akɨmbrɨkupan, who was still without his own men's house, with the leaves from their own men's house, Wariman. They also gave him the carved crocodile-spirit Kangris, whom he renamed Konggunyapan. In this way the men from the Bird of Paradise and Pig-1 clans began to relate to each other as brothers.

Eel clan. Pamɨngapi, who travelled as an eel and as a crowned goura pigeon, arrived from the Meakambut area to the south. He met Mamanggamai (Crocodile-2), who took him to the village. Kapi exchanged a clay bowl for a *kina* (TP) shell with Pamɨngapi (today the descendants of Eel clan act as *sɨmɨrinyar* to Crocodile-1 clan, see below). As there were only a few people in the Eel clan in recent times, they joined the Cassowary clan, live in their ward, and use the back of their men's house.

Pig-2 and Eagle-2 clans. The older Anggurapi and the younger Ambunapi, the ancestors of the last two clans, arrived from Manam Island and settled down not far upriver from Ambonwari at the place called Arkwas. They divided the land all the way to the large lake Sɨmaringkambi ('sunny lake'; today known as Lake Virginia) between them. They still lived there only a few generations ago when a disaster destroyed the village. The few who were saved were brought to Ambonwari, where they re-established their Pig-2 and Eagle-2 clans. They were allocated a place in the part of the village furthermost up-river.

It is impossible from the myths of origin to reconstruct the beginning of the village in terms of years or even of epoch. Except for the last two, all other clan founders are said to have arrived in Ambonwari during the lifetimes of Kapi and Akɨmbrɨkupan. What seems important is precedence: who came before and who after, how the first relationships were formed, and which places ancestors came from. One can speculate that Ambonwari initially had a moiety system (see Chapter 5), Akɨmbrɨkupan representing the Moon moiety and Kapi the Sun moiety. Following the arrival of their brother Mamanggamai they broke the village into three parts, making him the 'father' of Sun. When the ancestor of the Wallaby clan arrived, Kapi made him the 'father' of Moon. From then on the people of the Crocodile-2 clan are called 'sun people/men' (*sɨmarin yanmas*) and those of the Wallaby clan 'moon people/men' (*sunggwin yanmas*), using Sun (*Sɨmari*) and Moon (*Sunggwi*) for their personal names. The Ambonwari say that sun people 'die' in the daytime and moon people in the night-time. The organization of ceremonies is still a dual one which 'encourages the villagers to think in terms of a moiety system' (Harrison 1990a: 44). As I shall explain in more detail later, in every dance, either in a men's house or dwelling-house, the most prominent front places of the two rows of dancers are reserved for the descendants of Kapi and Akɨmbrɨkupan, that is, men from the Crocodile-1 and Bird of Paradise clans.

The most usual way of addressing each other in clanship terms is by using the names of the men's houses. But there are several other ways of referring to clans in the village. These ways refer to the places from which these clans came or the sequence in which they arrived (in the sense of 'earlier' and 'later'). Thus, the term for the members of the Bird of Paradise and two Crocodile clans whose ancestors started the village itself is *yanmas*, 'the ones'. The term may be extended from the members of these three main clans to the members of the Wallaby and Eagle-1 clans. While Wallabies can be called 'behind people' (*sambrambimbɨn yanmas*), those from the Eagle-1 clan can be referred to as those who 'walked in the sun' (*sɨmarinya kuraykɨn*). Only the people from the Crocodile-1 and Bird of Paradise clans can be called 'night people/men' (*wurin yanmas*), which brings us back to the original name of the village, *Amban-wurin*, discussed in Chapter 1. It often happens that a girl calls her boyfriend from one of these two clans simply *wurian*, 'night man'. Ambonwari call those from the Pig-2 and Eagle-2 clans 'latecomers', *yambias*, or *Wakriamari yanmas*, 'the ones from Wakriamari mountain'. The Ambonwari call all those who trace their origins on the Sepik River *Suwas*, 'people from the Sepik', the term derived from Sawos. Cassowaries can be called *yarɨngi*, 'people from the sago swamp', while neighbouring Imanmeri are lumped with other Highlanders and referred to as the 'hill people' (*kamban*, 'the hill man' or *kambanma*, 'the hill woman').[2]

Sometimes Ambonwari use the term *yayn* for a clan. It was translated for me as 'tree trunk' but it is also cognate with *yay*, 'blood'. Though seldom used in daily life, the 'tree-trunk' metaphor is significant for an understanding of the position of those lineages (and smaller clans) which do not build their houses in the wards of their own clans. In the past, larger clans sent some of their own lineages to live in the wards of other totemic clans when, for instance, the membership of the latter was too small to perpetuate itself. Saying that these lineages were 'sent', as the Ambonwari say, is not to imply a dominant lineage which could make decisions on behalf of other 'subordinate' lineages of the same clan. It means rather that a young man or even two brothers were given,

[2] The terms used for the 'hill people' have their origin in the hill found up-river called *Kambanimbuk*. *Kamban* is used as a term for all 'highlanders', while *(y)imbuk* means 'hill'. It can be seen from Table 1 that several lineages have this lexeme as the second part of their name.

together with the lineage name, by the elders of their clan to the depleted clan. They would already have had some relationship with that clan and would have been willing to stay in its ward. After being 'adopted', the 'dislocated' lineage then used the land of its foster clan and assumed the roles and institutions which pertained to this clan. Such a group is called *yapan*, 'a branch'. For example, the Warinjam lineage of the Pig-1 clan has over many generations (some say since the period when the mythical ancestors founded the village) lived in the ward of the Crocodile-1 clan. Thus everyone from this lineage cooperates in day-to-day living with those from the Crocodile-1 clan. A man from that lineage joins discussions in and presents food to their men's house, helps in house-building, assists in fetching tree-trunks for canoes and in carving them, and supports them in fights. But the lineage still retains its own identity. People say that when Kapi took the main men's house for himself he 'gave' to his elder brother his lineage Yanbanjam to live in the ward of the Bird of Paradise clan.[3] Another lineage displaced in this way from its own Pig-1 clan is Sanggutsɨmɨnɨng, which was 'given' to support the Bird of Paradise clan, on whose land it has lived since 'ancestral times'. When the other '*yapan* group' (i.e. the whole Hornbill clan) built their own men's house, Kɨkrayman, in the ward of the Crocodile-1 clan, they gave the back of it to the Warinjam lineage (Pig-1 clan), whose members used their own men's house name, Angrɨman, for their back part. There are a few more cases. Cockatoo clan, for example, which has only one member, is obviously too small to be independent. Adopted by the Hornbills, Cockatoo clan is at the same time a 'branch' of the Wallabies. The sole living member also lives on the land of the Crocodile-1 clan.

Clans seem to be open, and able to incorporate other groups and individuals from elsewhere. These adopted groups or persons can keep their own lineage names. If they do not have a name, the adopting clan can give them the name of their own extinct lineage. The newcomers adjust to the kinship terminology of the adopting clan and to its particular marital arrangements with other groups. In the same way as people do not talk about the adoption of children, they are expected not to talk about the adoption of groups. Such adopted groups and individuals are often asked to take over, temporarily, important institutionalized roles in the clan or lineage, while those whose descent gives them the right to assume them step aside and behave as though they were nobodies. This is done only for one generation, and the roles have to be returned in the following generation to their original owners. Thus, by giving and then retrieving the social roles people consolidate the relational dimension of identity of the clan or lineage as a whole. An example of such an adoption is presented below.

[3] The time aspect, in the sense of the actual 'when', seems meaningless for Ambonwari. The name 'Kapi' can thus represent the Crocodile-1 clan itself, and whenever (in terms of years) they actually gave their lineage to the Bird of Paradise clan, it was still 'Kapi', i.e. the Crocodile-1 clan, who gave it.

The adoption occurred between lineages inside Eagle-1 clan. The father of Godfried Yangus from the Mɨndapmay lineage adopted Rudi from the Ambaynggmari lineage. As Godfried's father died soon after the adoption, Godfried, as the elder brother, took care of Rudi. When Rudi married, Godfried transferred his own lineage roles to Rudi so that he became the 'father of the lineage'. Later, when Godfried fell sick, the village healer discovered that he had not transferred all his roles. Godfried still acted as *sɨmɨrinyar* ('ritual inaugurator'). He told me that he could not transmit the lineage name to Rudi just by saying so, but that he also had to give him one of his children. So Rudi bestowed a name on Godfried's youngest daughter because Godfried had only one son but several daughters and was not prepared to give him away. While this daughter still lives with her consanguineal parents, Rudi will arrange her marriage and will share the bride-price with Godfried.

This 'adoption' of social roles is formally similar to the adoption of children, a very important practice in Ambonwari, as I shall make clear in the next chapter.

In daily life, relationships between people only appear to be egalitarian. There is, however, a hierarchical structure of people and social institutions, including the cosmology, the knowledge of which reflects the views of the dominant groups. The power and domination of the leading clans and individuals is based on age, gender, knowledge, proficiency in communication, group size, ability to claim a particular past, land, name, and so on. Clan membership, for instance, becomes important in ritual contexts and on the occasions of disputes over land, names, or women, and whenever there is sickness, death, or other crises. Clan membership becomes important in all those situations where the identity of an individual or a group has to be reconfirmed for whatever reason. This is when hereditary inequalities in status are most visible. We will see in the following chapters how in all important ceremonies the foremost positions are reserved for descendants of the Crocodile-1 and Bird of Paradise clans. Whenever other clans want to perform a ceremony or ritual they have to borrow the decorations from these two main clans. The preparation of an initiation ritual can take several years, and requires carefully prepared strategies to be put into action. During my stay, for example, seven boys in the village were supposed to be initiated. They belonged to different clans. In every large meeting and on many occasions, when someone had died people talked about the necessity of staging an initiation ritual. Those from the main village clans were, of course, the most supportive, and tried to persuade others to take the initiative in its organization. They tried in particular to influence the members of those two clans which arrived last in Ambonwari. The latter had a suitable boy, and they supported the initiation ritual itself. But for years they avoided taking the first step, which would require them not only to present the main village spirits with a large amount of pork but also to acknowledge the ritual hierarchy of the village. Staging a ritual would have obliged them to negotiate with the main clans of the village and present them with gifts. Through this activity they would acknowledge them as the village

TABLE 1. *Ambonwari Village in 1990*

Clan	Men's house	Lineage	No. of households	No. of hearths	Men	Women	Children (under 16)		Total
							Male	Female	
Yamara (Bird of Paradise)	Wariman	Yimaria	2	2	4	3	1	3	11
		Kumbranggawi	1	1	1	1	2	4	8
		Ambarmarɨng	–	–	–	–	–	–	–
		Singgayjam	1	2	2	–	5	1	8
TOTAL			4	5	7	4	8	8	27
Manbo (Crocodile-1)	Yanbonman	Mambinjam	2	2	2	2	1	5	10
	Aranggmay	Sapɨndam	8	9	13	9	19	9	50
		Ikwari	–	–	–	–	–	–	–
		Yanbanjam (live in Bird of Paradise ward)	–	–	2	–	–	–	2
TOTAL			10	11	17	11	20	14	62
Manbo (Crocodile-2)	Sanggwayman (not built)	Kunbaynjam	3	3	5	4	6	2	17
Warakapwi (Wallaby)	Auriman	Imanmari	3	4	5	1	6	7	19
		Sambrakmari	3	4	7	4	12	3	26
	Wapaman (not built)	Yimasɨndam	2	2	2	3	5	4	14
		Karpanjam	–	–	–	–	–	–	–
		Marikapinjam	1	1	2	1	–	–	3
TOTAL			9	11	16	9	23	14	62
Sɨpɨrɨ (Hornbill)	Kɨkrayman (not built)	Wanbanimbuk	2	2	4	3	5	5	17
		Kanjimarpinjing	2	2	4	2	6	5	17
Andɨma (Cockatoo)		Adopted by Wanbanimbuk	–	1	1	–	–	–	1
TOTAL			4	5	9	5	11	10	35

Awa (Cassowary)	Yanjanman	Wayanggara	3	4	9	11	4	2	26
		Kasɨmanimbuk	1	2	3	3	2	–	8
	Wayman (not built)	Aunjam	3	5	4	8	4	5	21
		Mayprukong	1	1	2	–	2	3	7
		Paymbɨndam	–	–	–	–	–	–	–
TOTAL			8	12	18	22	12	10	62
Warɨpa (Eel)		Warɨpɨndam	2	2	2	4	2	1	9
Imbian (Pig-1)	Yanggraman	Simbanimbuk	1	3	4	3	5	4	16
		Singgaimarijam	1	1	1	1	4	–	6
	Kukumbruman (not built)	Konggɨndam	1	3	4	9	5	5	23
	Anggrɨman (not built)	Warinjam (live in Crocodile-1 ward)	1	1	1	2	1	3	7
		Sanggutsɨmɨnɨng (live in Bird of (Paradise ward)	–	–	1	–	–	–	1
TOTAL			4	8	11	15	15	12	53
Yambari (Eagle-1)	Ambarsoa	Ambayngmari	3	3	6	–	3	5	14
		Mɨndapmarpinjing	1	2	2	–	1	4	7
	Mɨndapmay (not built)	Yuwanjam	–	1	1	–	3	1	5
		Kɨkrinjam	–	–	–	–	–	–	–
		Yambarinjam	1	1	1	–	2	–	3
TOTAL			5	7	10	–	9	10	29
Yambari (Eagle-2)	Kabriman	Murimari	3	3	6	2	3	7	18
		Konimari	1	2	2	1	6	4	13
TOTAL			4	5	8	3	9	11	31
Imbian (Pig-2)	Yanggraman	Yawyamari	4	6	8	1	12	14	35
TOTAL			57	75	111	78	127	106	422

Note: The first lineage listed for each clan is the leading one, that of the clan founder.

'leaders'. Knowledge of songs, names, stories, as well as the staging itself, in which flute-players are ordered according to the ranking of clans, endorse the hierarchical structure. This shows that ritual hierarchy cannot be taken for granted as something automatically inherited, as existing independently of initiation rituals which make it manifest. This is in accordance with the Ambonwari notion of *kay* in which people are seen as who they are. By participating in the ritual in such a way, the two clans would demonstrate that they recognized and accepted the hierarchical structure of the village. The members of these two clans, however, refused to acknowledge the ritual leadership of the main clans, and thereby placed themselves outside the community of Ambonwari. An elder of one of these clans went back to their place of origin and built a new men's house there. In this way he 'materialized' the two clans' seclusion from Ambonwari. After several deaths had occurred in their clans (the deaths were, of course, attributed to their aberrant *kay*, that is to their devious 'way of doing things'), the men began to reconsider the possibility of rejoining Ambonwari by initiating a large-scale initiation ritual. Despite the talk, however, it did not take place until December 1993—long after I had left the village.

'House-Insideness': Lineages in Ambonwari

I call the smallest named group in Ambonwari a 'lineage'. Ambonwari people call it 'house-insideness'. The male members of such a patrilineal group build separate houses which are nevertheless subsumed under the name of their lineage. All the houses of the lineage occupy a specific area of the clan's ward. Whenever I use the phrase 'house with a name' or place a 'house' in quotation marks, I refer not to the physical structure of the house itself but to such a named patrilineage. The number of named ascendants in these lineages is usually four, while the maximum is seven. But it is the common name which defines the lineage rather than agnatic descent. Table 1 summarizes the group composition of Ambonwari by listing all the clans, men's houses, lineages, and number of households and hearths. The figures for inhabitants include in-married woman but exclude out-married ones. In-married women retain their own natal or adopted affiliation.[4] The table shows that three of the twelve clans

[4] The number of people in Table 1 is calculated from Oct. 1990. During my stay in the village 32 women gave birth, 18 to girls and 14 to boys. Four women were visibly pregnant at the time of my departure in 1992. Of seven deaths which occurred during my stay in the village 3 were adult men, 3 were young boys, and 1 was a young girl. A census from 1954 shows that there were only 187 people in Ambonwari at that time (compared to 422 in 1990). Of these 187 people 76 were adults (only 25 of them men) and 96 were children (52 boys and 44 girls), while 15 people were absent (Copley 1954). This was the highest figure among the Karawari villages at that period. The National Population Census of 1980 already shows a large increase in population: of 304 people in Ambonwari 175 were male and 129 female. The second largest Karawari village was Kundiman with 218 people, and the third Kaiwaria with 159 people.

(Crocodile-2, Eel, Pig-2) have only one lineage. The Cockatoo clan has only one member (who does not remember any of his clan's lineage names). Other clans have between two and five lineages; the largest lineage is Sapɨndam of the Crocodile-1 clan, with fifty people including in-married women. The smallest lineages are Sanggutsɨmɨnɨng of Pig-1, with only one member, and Yanbanjam of Crocodile-1, with two members, both of which live in the ward of the Bird of Paradise clan and 'help' its three small lineages which in total comprise only twenty-seven people. In all I collected the names of thirty-five lineages, five of them with no living members.

The members of the same lineage refer to themselves as *yam-ɨn-ɨng* (house-inside + noun suffix) which can be translated as 'house-insideness'. People say *apia yamɨnɨng mbang*, 'we are from one and the same house', or *gra yamɨnɨng*, 'all the people of one house', though one should not imagine them actually dwelling inside one house. The term *yamɨnɨng* translates better as 'patrilineal descent group', for the same term, *yamɨnɨng*, is used to refer also to all those who are members of the same men's house or the same patrilineal totemic clan, as in *warakapwi apia mɨna yamɨnɨng* (we are of Wallaby 'house-insideness'). Women can use the term *yamɨnɨng* for the house, lineage, and clan they are married into and the house, lineage, and clan into which they were born or adopted.[5]

Every totemic clan contains one lineage considered to be the one founded by their first ancestor. The 'father' of this lineage acts as the 'father' of the ward, of the clan's main men's house and of the clan's spirit-things. This does not mean that some other lineage cannot temporarily take over leadership within the clan. This happened during my stay when elders of the leading lineage of the Bird of Paradise clan died and their sons were still not married and too young to be able to take up their responsibilities.

People identify with a particular lineage name even when they live elsewhere or when they build separate dwelling-houses. This happens when two or more brothers build their own houses, usually in a cluster around the house with a name in which the oldest (and possibly also the youngest) brother lives. Brothers can also be members of different lineages of the same clan; this occurs after a lineage has died out and the name has remained 'vacant' for some time. Such a name can then be 'adopted' and revived. If a lineage becomes too large and its members persistently argue, they might decide to split. Thus, as will become clear below, one of the middle brothers inherits the name of the 'vacant' lineage along with the property (land, spirits, and so forth) which it owned in the past.

[5] Bateson (1932: 257) translates the Iatmul word *nggai(va)* as having a double meaning: a dwelling-house and a totemic group. Most Ambonwari lineages are very small, while a few large lineages, such as Sapɨndam of the Crocodile-1 clan, resemble those groups in Avatip which Harrison called 'subclans'. This means that the Sapɨndam lineage consists of two segments, each claiming descent from the same apical ancestor (see Fig. 2). One group acts as younger and the other as elder. Because there are not many such lineages in Ambonwari, I do not know if they claim descent from the same ancestor while being of different mothers, as the Manambu do (see Harrison 1990a: 69).

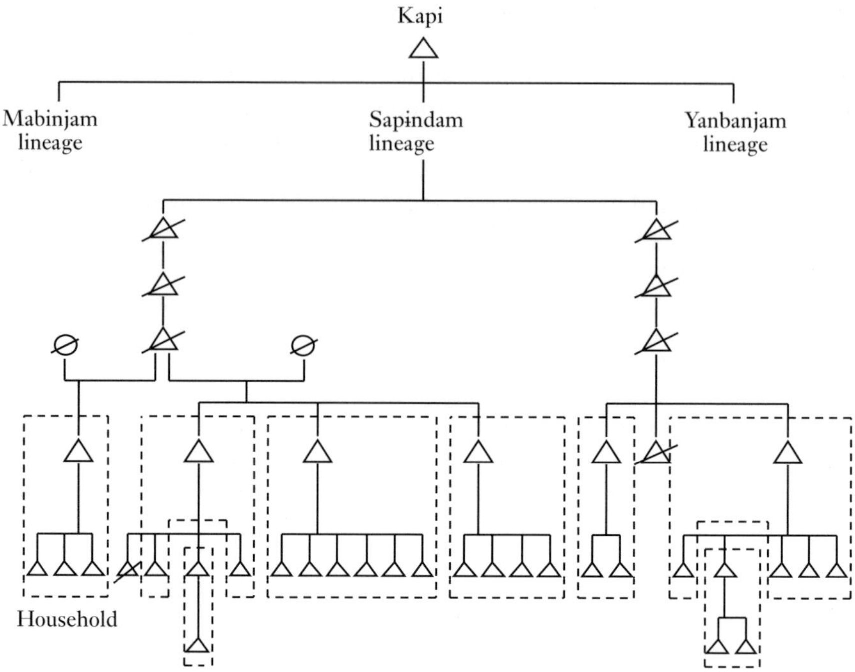

FIG. 2. Households of Sapɨndam lineage of the Crocodile-1 clan

A small ceremony (*kay*) secures ancestral approval for the revival of an old lineage and a new identity for the middle brother. The oldest and youngest brothers stay with their original lineage.

A house with a name represents all its component dwelling-houses. As may be seen from Table 1, I refer to these dwelling-houses as households. People remember how in the past the whole lineage used to live in one large house. As many lineages are small today and as their genealogical span is not very wide, it still happens that the male members of a lineage live together in a single household. These lineages rely on other lineages of the same clan for help in work, marriages, and quarrels. Larger lineages are far more self-sufficient. In all important events the members of different households of the same large lineage will come together to discuss matters and to help each other. However, simply because a lineage is large, arguments will tend to occur between its members.

The lineage is a unit which organizes the marriages of its members, but individual households are the groups which maintain closer and more intimate ties with their particular affines. Daily activities are matters of the household and not of the whole lineage. It is the affinal relationship which mainly directs people to the houses of others. This is particularly true for brothers-in-law and sons-in-law. For instance, Godfried is married to Tony's sister. He and his children often come to Tony's house; they even lived there once for almost a year. But Godfried never enters the house of Tony's classificatory brothers from other lineages of the same clan. Again, Samson, who is married to Bob's daughter, often comes to his house and eats and sleeps there. But he never visits the house of Bob's consanguineal brother, Andrew, nor has he any other relationship with him, though both Bob and Andrew belong to the same lineage.

While the name of the lineage is important inside the village and for the relationships with neighbouring villages, it is their totemic clanship which represents people's identity beyond the Karawari language area. When Ambonwari people travel on the Sepik River and stay overnight in a foreign village, they always seek out those people whose totemic group is related to their own or to one of their affines. Throughout the middle Sepik area, important social relationships across language and geographical boundaries are sustained through the sharing of totems, ancestral origins, and inheritance of trading partnerships (see Harrison 1993: 41). For example, when Iatmul people had a dispute with some villages up the Sepik they told my 'brother' Tony to mark his canoe so that he would not be mistakenly attacked by Iatmul men on his way to Angoram. Wisely, Tony wrote in white paint the name of his lineage, Kumbranggawi, on both sides of his canoe. The Iatmul not only acknowledged the name of his canoe but also recognized the traditional friendship between their groups from the times of their ancestors (see Bateson 1932: 253–4).[6] The name Kumbranggawi in this case did not simply refer to the name of the lineage but to the hills in the middle Sepik where Tony's Bird of Paradise totemic group originated.

Personal Names

By taking a name an individual receives a personal identity which establishes her or his membership in a pre-existing social group (Lévi-Strauss 1966: 181). The socially recognized personal names in Ambonwari are crucial for people's identity in terms of their lineage and clan membership. Moreover, just as totemic groups and the names of lineages link together people throughout the wider Sepik area, so do personal names.[7] Names relate living people to the

[6] This traditional friendship, as it turned out, became very important for my own personal safety both in the village and whenever I went to Angoram. During my fieldwork the menace of *raskal* gangs was at its height, and they were confronting anyone who was likely to have money. As some of the members of these gangs were Iatmul men, they respected the past friendship with Karawari people. As one of the *raskals* said while we were travelling on the same truck while waving his gun: 'We people down-river always drank your shit and urine.'

[7] Sepik studies in which the complexity and importance of personal names have been discussed are: Bateson (1932; 1958); Errington and Gewertz (1987); Gewertz (1977b); Harrison (1985; 1990a); Tuzin (1976); Wassmann (1991); and others (see Harrison 1990a: 59 n.). Ethnographers, at least in the Sepik

people, places, events, and relationships of the past while at the same time, being part of a collective identity with its source in the past, names open up possibilities for and anticipations of the future.

> If we treat names as merely names, points of reference, then symbolism becomes a matter of reference; a microcosm of names is counterposed to a macrocosm of referents. But if we treat 'name' as relationship, the microcosm of names is no longer a microcosm; it becomes immersed in a macrocosm of analogic construction. Not only do we have an analogy that encompasses name and named, but that analogy suggests, and tends to enter us into, analogic relations among macrocosmic constructs. (Wagner 1986: 15)

Names in Ambonwari are owned by the patrilineal clans and lineages. Every alternate generation is named according to the set of names used, or preserved from two generations before. Thus we might say that there are two sets of names: those of 'fathers' (and their 'sisters') and those of 'sons' (and their 'sisters'). While most of the names are not used, their generational level is always remembered so that a child of the second, fourth, sixth (or first, third, and fifth) generation will get her or his generationally appropriate name. Every person notionally carries not only the name or names but also the relationships, events, roles, characteristics, and deeds (including wrongdoings) of her or his past namesake. These features are embodied by a new person on assuming the same name. This 'sameness' means that a person belongs to the same descent group, uses the same land, has the same marriage paths, has the same knowledge and stories, and the same ways of doing things. It means that the main aspects of identity are those which characterized his or her namesake. Thus, when a man receives the names of his grandfather who was the leader of a clan, everyone knows that he is the oldest son of the leading lineage and will also become the clan leader. Of course, as I will show below, this is only the ideal situation and rarely occurs in reality.

By being given a set of names, then, a person receives her or his identity in terms of lineage, men's house, and totemic clan. It is not so much the place of birth, or the household and lineage into which you were born, or even the mother who gave birth to you or the father who contributed towards your conception which determine your identity as the names you were given and the life-world of those who care for you during your childhood. Only then do you find the relationship with the past upon which you can build expectations about the future.

region, are frequently incorporated into the society by receiving the name of the lineage or clan which 'adopted' them. After I had been in Ambonwari 2 months and after consulting Bob Kanjik Anjapi from the Crocodile-1 clan, an old woman, Marta Yambakɨnda (whom I called 'mother'), a widow of Abram Kambanggay from the Bird of Paradise clan, gave me a name from the set of names used for a generation below or above her husband. In this way the Ambonwari recognized my relationship with Tony Simiwariya Andiyapi, who was already my 'brother'. From then on I was called Kapraymari, from the Kumbranggawi lineage of the Bird of Paradise clan, and maintained the closest relationships (in terms of sharing, reciprocity, and communication) with my 'relatives'.

These names are totemic, for they refer both to the ancestors who once held them and to the . . . objects, territories, species and processes—owned and controlled by these ancestors. Correct utterance of these totemic names, particularly those which are esoteric, enables a man to establish an identification with his ancestors in which he incorporates their totemic power . . . These names are the basis and indicators for both men and women of the social networks which afford them and their children basic identity and protection from coercion and assault. (Errington and Gewertz 1987: 19, 47)

Adoptions of children are usually arranged immediately after birth so that new parents can bestow a name on the child from their own set of names. If a child is already grown up when adopted, she or he can bear names from two different clans. For example, a woman who married into the Cassowary clan gave her daughter Acustaria to her brother from the Crocodile-1 clan. Acustaria, who was already a marriagable woman, brought with her the name Apɨsiki from the sets of names belonging to the Cassowaries. In the Crocodile-1 clan they gave her the name Ipungmay from their own sets of names. Her husband, who is from the Bird of Paradise clan, calls her Apɨsiki, probably in recognition of ancestral marriage between the Cassowaries and the Bird of Paradise clan, while those from the Crocodile-1 clan call her Ipungmay.

As the roles of women differ from those of men, women's names carry different significance to men's. Men's names are important for gaining power over others and provide the basis for political competition.

Since, for men, the social relationship between—and within—clans are manifestations of relative power, and since social relationships constitute personal identity, men define themselves primarily in terms of relative control over names—over that which gives them power. (Errington and Gewertz 1987: 47)

This power of men is extended outside the village in their relationships with powerful spirits and with the people from neighbouring villages—potential enemies. Though women's names cannot shape social relationships in the way that men's names do, they not only 'ensure reproduction' as Errington and Gewertz (1987: 47) argue, but reaffirm *konggong* (paths) between groups and thus secure the relational dimension of identity. The female spirits represented in carved objects and all women (in particular the 'mothers of the village') 'anchor' these paths between groups within the village. In the same way as totemic spirit-objects (masks, for example) which bear female names are exposed to the eyes of everyone, so the names of women are not hidden. Since they and their names can be 'seen', they represent transparent familiarity and stability. Having one name, a woman is 'fixed'. Her *kay* is attached to her 'bodily' presence and to her activity at every particular moment. She visibly ensures reproduction, continuity, and the future.

The word for 'name' is *wi*, with homonymous meanings of 'night' and 'dream'. The semantic connection between these different meanings is the consequence, I suspect, of their common characteristic: they make people invisible and unrecognizable if they do not identify themselves. No one can see or hear my dream if I do not tell it; no one can recognize me at night if I do not identify myself; and no one can see or hear my name if it is not spoken by me or by someone else. In everyday life, and even more so on special occasions, Ambonwari men keep hidden from everybody at least one of their several names and in this way protect themselves. If a man were attacked by spirits or by a sorcerer who in his spell repeated a person's name, that man would be at least partly protected by hiding himself under his secret name. A similar situation occured whenever a man killed an enemy: his name was not called. The killing was usually attributed to spirit-crocodiles, and during the celebration he was substituted by his 'dancing partner' (*warɨmbar*) or his classificatory father's father (*wasamari*), who had different names. In this way he could not be recognized by the spirit of the man whom he had killed.

The name by which people daily refer to a person is called the 'up-river name' (*pambɨn wi*). These names probably refer to those used in daily life at the old site of Ambonwari, not far up-river in the sago forest. It is the name available for public use, and identifies namesakes and paternal grandfather, and the individual's brothers and sisters with their grandchildren.

Because all the men from the same men's house and totemic clan can be grouped in two generations of 'brothers', they also share two sets of personal names. Some totemic clans, especially those who do not marry each other and are in a notional brotherhood relationship (Pig-1 and Bird of Paradise, for example), can also share a large number of names. Many clan names become characteristic of particular lineages as they are continually transmitted from paternal grandfather and his sisters to his grandchildren. In contrast to women, who usually have only one (but see below), each man has several names, on average between two and five, which he receives all at once. One name connects a man to his predecessors, the others give him special relationships and powers. When a child is born its possible generational names already exist, waiting to be bestowed. The names are always given as a set; this means that a boy gets all of the names of one particular person of his grandfather's generation. It is important who gives the names. For a child of either sex this fact determines who is going to be her or his parent. Within the same clan a father might ask his 'brother' from a different lineage to bestow one of the clan's names on his child. Such an act of quasi-adoption brings the classificatory father closer to this child and results in a warmer relationship between the two lineages.

A boy cannot bear the names of his father or his father's brothers, and a girl cannot bear a name of her father's sister, least of all of her mother. Ideally, the firstborn son of an oldest brother will be named after his father's father. The second son can be given the names of his father's father's brother if the latter has no son's son of his own; otherwise he can be given any set of names which belongs to his generation. If the youngest brother stays in his father's house as the future 'father of the lineage' (see next chapter) he can use his father's names for his oldest son. The names of grandparents are bestowed on children while they are still alive. The 'middle' brothers are not expected to use their father's names for their sons. If the oldest brother, who is supposed to inherit all his father's roles, has only daughters, he can adopt a boy and name him after his own (and not the boy's) father. Daughters receive the names of their father's father's sisters. When a person dies, her sister or his brother takes over the names of the deceased. Those who are familiar with the names can easily recognize who is older and who younger in the family. They can call a person by these several names, but in reality one or two names prevail, while the others are remembered but not used. One of them can be kept as the secret name of the person. In addressing each other, the Ambonwari love to shorten the names of those with whom they have close relationships and whose names end with the usual female (*-may*) or male (*-mari*) suffixes: thus Kuri(may), Sanggr(mari), Kapray(mari). The suffixes *-wi* (Akawi), *-wariya* (Simiwariya), and *-pi* (Apungapi) for men and *-da* (Yambakɨnda) and *-ja* (Wapanja) for women are also common. There are exceptions—for instance, when the names are taken from everyday life: Kambonma (Hill-woman), Sɨmari (Sun—male name), Imbian (Pig). After the villagers were baptized in the 1950s the Ambonwari began using Christian names as an additional way of addressing each other. Christian names are without any power, however, and do not carry with them any sense of mythical and predecessors' past. Beyond their pragmatic daily use, they are unimportant for people's identity. Ambonwari do not preserve or remember such names in their generational sets; and, as people say, they are useless for spirits and sorcerers, because a Christian name cannot be identified with an Ambonwari person.

A father cannot bestow his clan's names upon his daughter's children. She is married into some other clan with its own sets of names: her husband or some other relative from his clan will name his children. Women in general rarely name their children. It has happened only occasionally that a woman who was familiar with her husband's clan's names gave a child a proper name. The following example shows how names form an important part of the conceptualization of identity.

Sabina (Pig-1) was adopted by the Bird of Paradise clan. She married Warwari (Crocodile-1) and they had four children. After her husband died she left two of her sons with her adopted mother in the ward of the Bird of Paradise clan. After they grow up they will return to the ward of their father's Crocodile-1 clan. Sabina married again, this time to Johannes from a different lineage of the same Crocodile-1 clan. They had two sons. After frequent quarrels, Sabina left Johannes and took their children with her. She complained around the village about her hardship with all these children. She gave the second son to Nelson (Wallaby clan) who had already named the child when he was born. [Nelson's wife could not bear children and Nelson had enquired about possible adoptions.] He became the boy's 'real' father. Johannes had already bestowed the name Bob Kanjik on their firstborn son, using the name of his own father. After separation Johannes wanted Sabina to return the boy's name. She did so. She gave the boy a new name, Yanggambi, from the lineage of her first husband from the same Crocodile-1 clan. The boy retains his Christian name, Bob, as it has no importance. In any case, as the boy's eye was damaged in an accident, everyone calls him 'One Eye', in this way avoiding possible embarrassment and shame by the use of any of his recently changed names. Johannes says that he will not tell his children about his fatherhood. He even feels that he is not their father now, because they do not have the names of his lineage and because he is not the one who looks after them.

This example illustrates how the name of a child is considered as the chief factor determining who its parents are. I mentioned earlier the identifying 'sameness' between a person and his or her namesake. Here, I present an example of a changed identity. With his change of names and the people who care for him, 'One Eye' receives a new identity. The paths that he walks on, the land on which he plays, the relationships with other children from different wards, his future marriage possibilities, all constitute his life-space. He is acquiring habitual practices distinctive of his new lineage, and his future knowledge will be transmitted to him by the elders of his new lineage.

A maternal clan, including the mother herself, rarely names children. When a single woman becomes pregnant she may not want to reveal the father, or she may be unsure herself because she had intercourse with several men in a short interval of time. In such cases the identity of the child's father remains unknown. The woman's real or classificatory brothers will give a name to the child and will care for it as one of their own. Such children stay with and belong to their maternal lineage.

A mother, mother's brother, or maternal lineage can lend a name to a child for a period of time. This happens only in cases where there is persistent sickness, attributed to the spirits of the dead of the lineage or when a child's namesake had killed someone in the past. People fear that the spirit of the victim might take revenge on the child who, having the same name, would be identified with the killer. It is only in such cases that the maternal lineage bestows their name on a child.[8] The spirits of the dead are unable to recognize

[8] The naming system in the Sepik region is not uniform, though it seems that some kind of relationship between paternal and maternal names exists everywhere. Thus, among the Ilahita Arapesh, whose kinship is a modified Hawaiian type ('Omaha' in Ambonwari), a male infant receives a name from each of his mother's brothers, though the name is later recalled (Tuzin 1976: 137–8). It is also unacceptable in Ambonwari for a sister in naming her brother's daughters to prefer her own names to come first, which is the case among the Ilahita Arapesh (Tuzin 1976: 141). The Ambonwari naming system also differs from those used among the Iatmul and the Chambri, where each child gets names from both maternal and paternal clans (Bateson 1958: 127; Errington and Gewertz 1987: 31). On the other hand, Wassmann (1991: 35, 242–7, 277–83) shows that the Iatmul from Kandɨngei get matri-names only under exceptional circumstances, and in this way resemble the Manambu (Harrison 1985, 1990a) and the Ambonwari.

the child bearing this 'foreign' name and leave him (rarely her) in peace.[9] When a person grows up and gets married, he returns the name to his maternal lineage and pays for it. This practice is best understood through an example.

Gerard is the first-born son of Karuk, the 'father' of the Ambayngmari lineage (Eagle-1 clan). Because Karuk's father's elder brother had killed someone in the past, Karuk was afraid of giving the same name, which was the proper one, to his son. When Gerard was born, his maternal clan took his umbilical cord and tied it at both ends so that the spirits of the dead were unable to smell the blood. They put the cord into a piece of bamboo and plugged it with a leaf. After a month, when they saw he was going to survive, the maternal clan gave him the name Kambunggawsi. Throughout his childhood everyone called him by this name. When Gerard got married, his maternal clan came to his house with clay pots and plates. They returned his umbilical cord and took their name back. Gerard buried his own cord and paid his maternal clan for lending him their name and for the gifts he received. At that time, people say, his paternal name, Arukum, 'came out from hiding'. Sometimes, the maternal name stays with a person until he dies. However, it cannot be used in the succeeding generations of his patrilineal clan.

The following example represents a different context in which a person gets a name.

In the past, when Ambonwari still lived in the nearby sago forest, a man from the Cassowary clan wanted to marry a woman from the Bird of Paradise clan. The latter tried to postpone the marriage, protesting that they did not have the clay hearth and other necessary items to 'decorate the girl' upon her leaving. The man, thinking that they did not want her to marry him at all, got angry and burnt their house of the Ambarmarɨng lineage. The Bird of Paradise clansmen attacked him with spears but did not kill him. While he was running through the village the men of the Hornbill clan, who were the maternal clan of Ambarmarɨng lineage, killed him. The Bird of Paradise clan indicated to the Cassowaries that if they had wanted to fight they should have thought first where were they going to live and eat, as this was the land of the Bird of Paradise clan and its men were the 'fathers of the village'. The Cassowaries did not take revenge and the men of the Bird of Paradise clan did not compensate for the killing. As they said, everything, including the spirit of the house, skulls, shells, and personal belongings, were burnt in the fire and that was compensation enough.

Tony Simiwariya Andiyapi from the Kumbranggawi lineage of the Bird of Paradise clan was born forty or so years ago. His father, who feared the revenge of the spirit of the man killed in the event described above, asked Kapɨkay from the Pig-1 clan to name his son. What is interesting here is that Kapɨkay was not from Tony's maternal clan. Furthermore, the Pig-1 and Bird of Paradise clans relate to each other as brothers and share a certain number of names which could not be used in this case. But because the

[9] Harrison (1985) compares the naming systems of Iatmul and Manambu and finds that Iatmul matri-names make their bearers vulnerable to attack by spirits, while the Manambu matri-names protect their holders. It is when someone is bereaved that his maternal clan bestows upon him their name and in this way protect him from the spirits of the dead, which are familiar only with his patri-names. These matri-names are replaced each time someone dies in the clan.

Pigs also have their own names, Kapɨkay was able to bestow a name which was 'foreign' for the Bird of Paradise clan. He called him Kambadɨma and the spirits of the dead were deceived. When Tony was married and already had five children, he and Kapɨkay's son Kaspar cooked food and exchanged it with one another, together with betel nut. Tony paid ten kina to Kaspar, who gently beat him with a small broom 'to clean him' of the old name.

It is interesting to look at the multiplicity of inherited relationships between Kaspar's and Tony's lineages. First, Tony's mother was from the Cassowary clan (the dead man's clan), which means that Tony's father could not ask them to name his son. Second, the relationship between the Pig-1 and Bird of Paradise clans is one of 'brotherhood' according to which people do not use money but exchange goods. But, third, Kaspar's lineage, Warinjam, was actually adopted by the Crocodile-1 clan, who represented the possible maternal clan of the lineages in the Bird of Paradise clan. Because of this, Kaspar's lineage, Warinjam, took over from the Crocodile-1 clan the practice of acting in all important events as *sɨmɨrinyar*, the 'ritual inaugurator' (see below) for Tony's Kumbranggawi lineage. In this kind of relationship one would not expect someone to pay his *sɨmɨrinyar* (the latter is the one who has to pay whenever he opens a ceremony, while he receives food). Following this, Warinjam and Kumbranggawi even practised marriage exchange (though marriage between other lineages of these two clans is avoided). After all these explanations one can conclude that Tony's father, being unable to give Tony the name of his wife's clan (Tony's maternal clan), asked Warinjam to name his son, for they are his 'brothers' (in view of the Pig-1 clan of origin) but also the lineage with whom they can marry (in view of the adoption by Crocodile-1). When Tony returned the name he followed the same practice as he would have done if his name had been bestowed on him by his maternal clan.

As Ambonwari do not keep pigs or any other large domestic animals, only dogs are named. Ambonwari like to use the names of spirits for dogs to increase their powers (see Table 2). Each dog gets only one name, either from its owner's own or maternal lineage; in the latter case the name must be returned to the patrilineage which owns it. As is the case for people, a change of suffix makes a name applicable to either female or male dogs. Dogs do not have generational sets of names, so a son can use the same names for his dogs as his father did. On the other hand, it is understandable that he will not use the names of his father's mother's lineage, but will instead ask his own maternal clan to give a name to a dog. The reasons for this practice lie in the desire for successful hunting. If someone in the hunter's lineage or clan had done something wrong in the recent past, those dogs with names of the same clan may not find any game. On the other hand, problems in a hunter's clan will not affect those dogs which carry the names of his maternal clan. These matri-names are not paid for.

TABLE 2. *Names of dogs in Ambonwari village*

Clan	Men's house	Lineage	Owner's name (Christian)	Dog's name	Sex	Meaning
Wallaby	Auriman	Sambrakmari	Augustin	Kanggrima	F	Name only
				Kamɨnggɨmay[a]	F	Name only
				Wanjirɨm	M	Name of a hill
				Krisisikamɨng	M	Crocodile-spirit (also humans)
				Anggɨmara[b]	M	Name only
				Sawnandimari	M	Crocodile-spirit
				Wuringgi[b]	M	Name only
Crocodile-2	Sangwayman	Kunbaynjam	Rius	Makiyapan[b]	M	Name only
				Makiwapanma[b]	F	Name only
				Kamasapan[c]	M	Crocodile-spirit
				Kamasmarinma[c]	F	Name only
				Simbrikɨn	M	Crocodile-spirit
				Wambaymari	M	Crocodile-spirit
Crocodile-1	Yanbonman	Sapɨndam	Daniel	Kamasmarinma[c]	F	Name only (also humans)
				Kamasapan[c]	M	Crocodile-spirit
				Mariakari	M	Crocodile-spirit
				Mariakarima	F	Name only
Crocodile-1	Yanbonman	Sapɨndam	Andrew	Kwarkɨmbi	M	Name only
				Yanggunmari[b]	M	Bush-spirit
				Wandas	F	Flute name (female ancestor) (also humans)
				Sapɨn	M	Mythical ancestor's name (also humans)
				Mambi	M	Mythical ancestor's name (also humans)
				Sapɨnda	F	Mythical ancestor's name (also humans)
				Mambinja	F	Mythical ancestor's name (also humans)
				Singarmari	M	Bush-spirit

TABLE 2. (*contd.*)

Clan	Men's house	Lineage	Owner's name (Christian)	Dog's name	Sex	Meaning
Pig-1	Yanggraman (Sukurismay)	Konggɨndam	Rowi	Marinjawun	M	Crocodile-spirit
				Sunggwandɨkariamari	M	Crocodile-spirit
				Mariwapanma	F	Name only
				Mɨndɨmbit	F	Name of the village where the dog came from
				Sanggimari	M	Crocodile-spirit
				Kanggɨmarɨma	F	Bush-spirit
				Kiata	M	Foreign name
				Kambandɨmanɨng	M	Crocodile-spirit
				Kamɨndiarmari	M	Crocodile-spirit
				Kambɨsmambaw	M	Crocodile-spirit
Eagle-2	Kabriman	Murimari	Raymond	Kanjimarwapanma	F	Bush-spirit
				Sunggurmari	M	Bush-spirit (also humans)
				Yarɨkɨrakmay	F	Bush-spirit
				Sanggimari	M	Bush-spirit
				Yarkaymbari	M	Bush-spirit
				Kanjimarisaun	M	Bush-spirit
				Wakɨnamungway	M	Bush-spirit
				Aypakmari	M	Bush-spirit
				Aypakmay	F	Bush-spirit (also humans)
				Wanggɨrmari	M	Bush-spirit
Eagle-2	Kabriman	Konimari	Benjamin	Rokisangi[b]	M	Bush-spirit (also humans)
				Wapɨsaun[b]	M	Crocodile-spirit
				Kambraman	M	Place-name, given from Kundiman village
				Wanggɨrmari	M	Bush-spirit
				Kanjimai	F	Bush-spirit, also name of the village

[a] The name bestowed by the other clan.
[b] The name bestowed by the owner's maternal clan.
[c] Shared names with the other group.

People often borrow dogs from their owners.[10] If a hunting party is successful, the hunter ties some money (between two and ten kina) to a dog's legs as a sign of gratitude. If the hunter did not catch anything he does not pay. When someone is in urgent need of pork (for example, to feed those who help him with house building, or to give as a mortuary payment after the death of a close relative) the owner of the dogs may feel sorry for a man whose hunt was unsuccessful. He might go into the forest himself. If he catches a pig he may give it to the unsuccessful hunter, who repays him with money. As people say, this payment is not only for the meat, but for 'all the thorns that pierced the hunter's feet'.

'Fathers' and 'Mothers' of the Village

Many ethnographers have emphasized the egalitarian aspects of social relationships in the Sepik River area. Harrison presents a more complex view when he argues that ritual and the division of ceremonial powers create hereditary inequalities in status and represent not only the model but also the lived experience of hierarchy (1990a: 4, 6, 85). Contrasting ritual hierarchy with secular equality, which for Harrison constitute two coexisting dimensions of Avatip social organization, he argues:

> The 'same' cosmology can remain, the same ritual symbols enduring timeless and unchanged over history, but their political *meaning* can be progressively transformed by a reordering of the key social roles and relationships among the actors in ritual. Depending on the degree to which ritual powers are controlled by a minority or dispersed equally among all descent groups, they can 'signify' either rank at one extreme or complete political equality of groups at the other. (1990a: 7)

Harrison maintains that whilst marriage, nurture, and reciprocity between kin form a more or less egalitarian domestic sphere, the control of ritual, totems, spirits, names, myths, spells, and songs enables a group to institutionalize power in the political sphere while having little influence in mundane social activities (1985: 417–18; 1990a: 7, 8).

Looking at the same aspect from a different perspective, hierarchy in ritual and hierarchy in secular life are more related than one might at first think. From my observations in Ambonwari, it seems that those at the top of a hierarchical scale secure their power by temporarily giving up their ritual positions, by being continuously generous, by helping newcomers and minor groups. This of course has its influence also in daily life. This is possible as long

[10] Because the Ambonwari do not have domestic pigs, the hunting of wild ones is the only way of securing the large amounts of pork necessary for maintaining relationships between people and, of course, for the commencement of any large work, ceremony, or ritual. It is not surprising, then, that Ambonwari men are very concerned with hunting magic. They pay as much as 15 kina for a small root of powerful ginger traded from the Highlands. Before a man goes hunting he chews up pieces of ginger and sprays them, mixed with powerful *ingasan* bark, into his dogs' snouts.

as people recognize and agree with those who claim the institutionalized power. Thus, the role of *sɨmɨrinyar*, 'ritual inaugurator', is temporarily given to those who are, for example, adopted into the lineage; some names may be shared between two clans; the dancers from different clans can borrow the decoration from the main clans and take the lead in ritual. This is possible as long as those having authority are confident about their secret, sacred knowledge and as long as the order of key social roles and relationships between the actors in ritual are not questioned or transformed.

Hierarchy in Ambonwari is based on the order of arrival of the first ancestors of every clan and the knowledge, spirits, relationships, and practices (especially ritual) they brought with them. This can be seen in the organization of wards in the village, the central site (the middle of the 'snake-spirit': see Chapter 1) and importance of the main men's house, Yanbonman, of the Crocodile-1 clan, the ownership of land and creeks, the ownership of powerful spirits, the distribution of supervising roles in disputes within their village or between villages, and, of course, in ceremonies and rituals. All these aspects of Ambonwari life have to be retold and reconfirmed through rituals and other collective practices. Only through lived experience are all these notions brought into being as if they were eternal. Ambonwari do not recall any disputes about the primacy of arrival, men's houses, or personal names (but see Harrison 1990a and Wassmann 1991). Leadership, which in Ambonwari is expressed in such terms as fatherhood and motherhood, is the privilege of the three clans whose members are descendants of the three brothers who were the first to settle in the place today known as Ambonwari village. While the importance of these clans, their spirits, their myths, and their roles in rituals are discussed throughout the book, I will concentrate here mainly on those roles known by the terms 'the father of the village' (*imɨnggan anay*) and 'the mother of the village' (*imɨnggan yandang*).

Earlier in this chapter I presented a short version of myths of origin concerning the arrival of the first three brothers: the youngest, Kapi, who was the first to arrive, the eldest, Akɨmbrɨkupan, who joined him shortly after, and the middle one, Mamanggamay, who arrived last. In this way three totemic clans were formed: Crocodile-1 (Kapi), Bird of Paradise (Akɨmbrɨkupan), and Crocodile-2 (Mamanggamay) clans. Each of them became a 'father of the village'. Their descendants remained *imɨnggan yanmas*, the 'ones of the village', or *imɨnggan anunggum*, the 'fathers of the village'.

To become *imɨnggan anay*, a 'father of the village', a man has to be first *yamɨn anay*, a 'father of the lineage', and his lineage has to be the main one of his clan's men's house. In this way he is also *iman anay*, the 'father of the men's house'. If his men's house is the main one in his totemic clan, and his totemic clan is one of the three leading clans previously mentioned, he is a 'father of the village'. These statuses are inherited by the first-born son—providing he is married. His wife is then a 'mother of the village'. The role of a 'father of the village' can, for many reasons, be temporarily transferred to a 'brother' from the same totemic clan. This happens, for example, when a person is not old enough or is not yet married; then the oldest brother gives the status to a younger brother or to a 'brother' who was adopted into the clan. It seems almost like an obligation for those who are in possession of any title of ownership to give these titles to those who otherwise have minor status. This reduces bad feelings, such as envy, resentment, and jealousy, which could otherwise seriously affect the stability and well-being of a particular clan. Such a temporary transfer of a social status usually lasts through the lifetimes of these men. Later, the status is returned to the lineage which originally held it.

During my stay in Ambonwari the situation was as follows. The village 'father' in the Bird of Paradise totemic clan should be the eldest brother from the Yimaria lineage. As the man who held this status had died and because his two sons (twins) were not yet married, the title of 'father of the village' was transferred first to their eldest 'brother', Tony Simiwariya from the Kumbranggawi lineage. However, because Tony and his wife did not follow the proper behaviour (see below) required from those who hold this role and because this was believed to be the cause of the frequent sickness suffered by his family, the role was transferred to the younger 'brother', Jacob, from the second house of the Yimaria lineage. Jacob keeps the old painted paddle, or rather its replica, of Akɨmbrɨkupan. Again, Jacob's wife Veronica neglected to perform the role of the 'mother of the village' correctly: she did not 'sit well'. Jacob's house was struck by persistent sickness. Veronica died of tuberculosis in 1992.

Another 'father of the village' is Joe, a young man of the Mambinjam lineage of the Crocodile-1 clan. He and his wife observe the correct performance of their roles. But because he is a very quiet man and not an orator, the lineage is led by his energetic, generous, and cheerful wife, Cecilia, who, because of her speaking abilities and her joking attitude, is admired around the village. Though Joe keeps all the ritual sacra, such as a piece of crocodile jaw, a stone breast, and shells, in his house, the knowledgeable old man Bob Kanjik from the Sapɨndam lineage (the second lineage of the same Crocodile-1 clan) takes care of rituals and does all the important talking on Joe's behalf.

The third 'father of Ambonwari' at the time of my arrival was a man from the Kunbanjam lineage of the Crocodile-2 totemic clan. Though he was the younger of two brothers, the elder one temporarily appointed him 'father of the village'. At the beginning of 1991 he passed the title on to the newly married oldest son of his elder brother.

If a dispute about land arises in the village, for example, when someone intends to cut previously untouched bush, the villagers are soon reminded who are *imɨnggan anay*, 'fathers of the village'. The three fathers of the village are the custodians of the most important sacred objects: the shells, clay hearths, bowls and plates, paddles and stools of their first ancestors. They keep the body decoration of their ancestors and have a prominent status in rituals. They are the owners of the most important spirits, names, songs, and spells. In this way they secure the power of the village and its protection. People respect their status. They have the main say in all important decisions, as illustrated by the following two examples.

A 'father of the village' can end a fight. In the past, he would crack his *as kɨmbi*, 'lime gourd container', and say something like: 'I broke this gourd for fighting to stop. Now!' People regard such an act by a 'village father' as something which has to be obeyed, both because of shame and because of the fear of what might happen (punishment by spirits) if they disregard the 'father's' wish. Later, when the two fighting groups make peace by having a feast, they would make a new gourd container, decorate it with shells, and fill it up with lime. The 'village father' would take it and rattle it with *kupisɨrkɨnɨng*, 'a lime stick from the gourd', as a sign of approval and contentment. He would reiterate the wise decision made by both groups.

While I was in the village a young man decided that he would leave the village permanently because his mother did not approve of his new marriage. One real and two classificatory 'fathers of the village' jumped fully dressed into the creek and stopped the departing canoe. The man felt ashamed and returned to his house. Later the man's parents invited the real 'village fathers' to their house where the latter made speeches on the man's behalf and were given food. If he had ignored the action of the 'fathers' and left the village, people said the snake-spirit of the village would have followed him wherever he went. He could then be punished by accident, sickness, or even death. As the old men said, the snake-spirit would push its tail deep into the ground, stretch out its head, and bite.

A wife of the 'village father' is called *imɨnggan yandang*, 'mother of the village'. Ambonwari do not use the kin term *asay* (mother) for this role. *Yandang*, I think, derives from *yan* (the one, 'who') and *san* (pronounced by many as *dan*) which means 'to sit' and 'to hold'. Thus, the literal meaning of *yandang* is 'the one who sits' or 'the one who holds'. Her position can be compared to that of a queen bee. She is not supposed to work: she should not go fishing or processing sago and she should not work in the garden or fetch firewood. Whenever she and her husband move from one house to another, he is the one who should carry everything: baskets, mosquito nets, clay pots, personal belongings, and so on. The main obligation of the 'mother of the village' is to sit well (*yapakupan kandamba*) inside the house. In this way she anchors the whole village. It would be quite improper for her to sit near the entrance and gossip or just look around, or, even worse, to have an affair with some other man.

The circumscribed spatial role of women in the sense of place, household, and home finds its extreme realization in the behaviour of a 'village mother'. With her *kay* she holds together all Ambonwari groups. While the 'fathers of the village' are involved in *mariawk* (discourse), the 'mothers of the village' assure that the village relationships 'sit well'. These three women can do some work inside the house, such as cooking and sewing mats and baskets. To disregard their prescribed conduct would bring sickness and misfortune to their lineage and clan members. Worse, a disaster (storm, earthquake, epidemic) could befall the whole village.

'Mothers' of Ambonwari, in cosmological terms, hold a huge fishing-basket inside which they hide and protect their 'children', the villagers. According to other explanations given to me, these women 'hold' a *kanda* (TP) vine which connects and protects all the villagers by tying them together. They are the only women in the village who are allowed to use the same lime gourds as the men, and they can even rattle them with the lime sticks. All other women can only use bamboo holders, *as kaman*, or bottles and cans. Each of the 'mothers' has her own secret tube of bamboo in which she hides the old, 'ancestral' grass used for mats and baskets. Every new 'mother of the village' inherits this bamboo tube and its contents from the previous one. During the initiation ceremony of young boys the elders advise those who are going to become 'fathers of the village' not to search through their wives' baskets and bundles, even if they should be hiding something given to them by other men. As 'mothers' of the village are not supposed to fish or process sago 'fathers of the village' have to work much harder than others to provide their families with a sufficient amount of food.

People say that in the past when the 'fathers' formed the first line in fights their wives released the cord but still kept it tight. They sat in their houses pressing the head, the middle, and the tail of the snake-spirit firmly down, thus keeping the village firm. During the 1950s, because of a serious dispute in the village, two lineages (Simbanimbuk from the Pig-1 clan and Yimasɨndam from the Wallaby clan who were related to each other through marriage exchange) left the main village ground and built their houses just over the creek.[11] In the following years sickness struck the village, and the cause was easily identified as the separation of lineages from the area protected by the snake-spirit. The snake-spirit was either in an unnatural position—that is, crooked in the middle—or it was unable to protect the relocated lineages. All protests by the village fathers were ignored. After several deaths the two lineages left their new place across the Maramun Creek and returned to the village area.

Though 'fathers of the village' are respected—they are consulted before initiation rituals, for example—everyday mundane power tends to be in the hands of those who are skilled orators (*mariawk kusɨrar*). As in other Sepik societies, village moots are an important part of village politics. These men, who are not necessarily *kupan* ('big' or older men), though they may be called that, are recognized because of their ability to interpret the rhetorical speeches of others, the meanings 'between the lines', and to elucidate them. The same

[11] The late Anthony Forge, who visited Ambonwari for a couple of days in May 1959, provided me with a map of the village as it was at that time: there were 16 houses, including 3 men's houses, built just over the Maramun Creek. Five of these houses belonged to Yimasɨndam (Wallaby clan) lineage, 7 to Simbanimbuk (Pig-1 clan), and one to an unidentified lineage from Cassowary clan.

talk can be repeated in different ways by several speakers over and over again. The most powerful men are those who can manipulate the speeches of others to turn doubtful situations to their advantage. These are not men whose knowledge is fixed, static, or timeless, but those whose knowledge is adept and quick to adjust to new situations. They are able to seek possibilities 'beyond the reach of habitual or reflex thinking' (Gell 1992: 284). But because tradition is oriented towards the stability of the village, even these powerful men have to be advised by knowledgeable old men on such matters as names, myths, rituals, drum signals, land ownership, and so on. While ancestral tradition as presented by dominant individuals and groups structures many aspects of village life, the politics of interpersonal relationships are a continuous process which shape the ethos of the community. All leaders are consequently restrained by the obligations of giving and exchange and by gossip and rumour.

Brison (1992) cogently argues that it is not only initiation, exchange, and secret knowledge which are important elements of leadership, but also rumours and speeches that manipulate people's impressions of events. Everyday talk and 'impression management'—constant negotiation and speculation, spreading of rumours, and telling of stories in private conversations or in public meetings about sorcery, causes of misfortune, wrongdoings, or everyday events—are, according to Brison (p. 93), central to the constitution of power and leadership in Melanesian political systems generally. In people's everyday life rumour and gossip are not simply escape from or distortions of reality, nor just the narrativization of events (such that they are distilled from 'real life' into 'story'), but are themselves extensions of events (Carr 1986: 16).

Most of the orators are known as 'hot-tempered' (*arkɨsan*), their ability to talk matched by their ability to fight. Ambonwari people resent being told what to do. The villagers need to have their own self-interest addressed in the proposals of their orators, for they are not interested in the ambitions of others. People quickly resent the success of others, and regard them as showing off. A very important quality of a leader is generosity. Admiration for generous men (*warɨmbarar*), those who continually give food and goods to others, and condemnation and disapproval of those who are stingy (*karisikɨn*, 'strong, hard'), ensure that one cannot accumulate power through wealth. One who has more must give away more.

In addition to 'fathers' and 'mothers' of the village, knowledgeable men and skilful orators, there are other men who manipulate the relationships of daily life and influence decisions concerning the staging of any large event. During my stay I was able to observe first a decline and then an increase in the importance of the two traditional healers in the village. Their performance was first gossiped about as being merely trickery which brought them money. After a few months had passed, persistent sickness induced people to forget their previous allegations and seek the healers' help. The influence of those who are skilled in divination following death and those who are experts in sending slit-drum signals was also discernible. Since Australian colonial administration introduced village councils, the Ambonwari have also elected their own councillors and their assistants. They act as arbiters in village courts and represent the village outside it. After colonial times Ambonwari village has seen other 'important' men emerge: school representatives; *dokta bois* (TP medical orderlies); those who own large canoes necessary for long journeys and those who own outboard motors; those who try to succeed in trading by selling petrol and kerosene or by setting up a trading store; those who produce and sell rubber; those who work at Karawari Lodge or on a large tourist boat, *Sepik Spirit*, and have a permanent source of income; and, not least, the young man who returned after twenty months' sojourn on Norfolk Island, where he was invited by a couple of tourists who had met him at the Lodge. One could look even further and find all kinds of engagements which enable Ambonwari people to adjust their own positions in relation to social change in Papua New Guinea. All these postcolonial 'positions' are filtered through the conceptual framework of Ambonwari 'way of the village', through the traditional knowledge and ideology of the time, through the common understanding and sentiments of the village, coloured by the envy and jealousy of some and the enthusiasm and support of others.

Warɨmbar: *Exchange Partner*

During Bateson's stay among the Iatmul, Sepik men already worked in distant places and were familiar with basic Tok Pisin, which spread rapidly throughout the East Sepik. Terms such as *poroman* and *pren* began to replace local lexemes while at the same time concealing the vernacular concepts, meanings, and contexts in which these terms were used. Their use was not restricted to intra-village relationships but crossed village borders and linguistic boundaries. As Mihalic (1971: 159) correctly states, *poroman* is 'comrade, companion, partner or mate', and the word takes its origin from the English 'foreman'. *Pren* is Tok Pisin for 'friend'. As the whole context of friendship, at least from the Ambonwari perspective, is based on either kin or trading and exchange relationships, it is most useful to look at these aspects if we want to understand indigenous meanings. Bateson (1932: 264–5) was doubtful about the Iatmul term *tambinien*, which people translated as *poroman*, for he was unable to find any genealogical or clanship connotations in its meaning. *Tambinien,* who were socially identified as siblings, danced together as partners and helped each other in everyday life. Their children were expected to marry by brother–sister exchange. Bateson was also perplexed about another type of relationship between those who called each other *tshambela*, the term being used in a classificatory sense. People said that they assisted each other in initiation ceremonies and, as shown below, duplicated one another's misfortune (1932: 270–1; see also Forge 1972: 535). Mead, when writing about Mundugumor (today called Biwat), also noted a kind of hereditary ceremonial friendship between individuals who stood in a real or classificatory brother-in-law relationship and performed ceremonial exchanges similar to those between affinal relatives or between a man and his maternal group (1970: 26–7). As in Bateson's study, Mead's distinction between two different types of relationships, *kamain* and *wareun,* was also ambiguous.

It was not until McDowell had done research among the Bun (close neighbours of the Mundugumor) that the role of *kamain* became clearer. Indicating that an adequate translation of *kamain* is not possible, McDowell defined it as referring to an 'exchange partner', based on part-kin and part-exchange relationships (1976: 37; 1991: 198). The other type of relationship, *wareun*, is reserved among the Bun for 'the familiar Melanesian trading partner found in remote villages, with whom transactions resemble barter' (McDowell 1980: 64). She excludes *wareun* from the village context by using this term only for non-kin trading partners (1991: 198). Thus, only *kamain* is applicable in the village and the kin-based relationship.

In several publications McDowell has presented a lucid description of *kamain* behavior which allows comparison with a similar kind of relationship among the Ambonwari. There are significant differences between the Ambonwari and Bun systems of kinship and marriage, however. Bun have 'ropes', which are not descent groups but represent ties between intermarrying classificatory brother–sister pairs and their children over time, reuniting in every fourth generation (McDowell 1991: 266, 269), though McDowell has stated that in Bun named matrilineal clans are unimportant. On the other hand, as we have seen, the Ambonwari have named patrilineages and totemic clans. Bun kinship terminology is 'Iroquois', but Ambonwari is 'Omaha': this means that the features of social structures and kin systems are harder to compare. I will first present *kamain* among the Bun as described and analysed by McDowell and then follow this presentation with a description of *warɨmbar*, which is, I believe, the Ambonwari version of *kamain* on the Yuat River. Finally I will point out similarities and differences between the two.

McDowell found that those who are cross-cousins, or whose ancestors were once cross-cousins, can be transformed through ritual into *kamain*.[12] 'Any classificatory cross cousin is a potential marriage partner and simultaneously a potential *kamain*' (McDowell 1976: 46). All transactions between two *kamain* are expected to be balanced. Whenever food is distributed, one's *kamain* gets the largest and choicest part. There is no competition between them. The most remarkable aspect of their relationship, recorded also by Bateson, Forge, and Mead, is one of reciprocal duplication of misfortune as a sign of the essential equality between two *kamain*. If one falls out of a canoe the other will jump out too; if one destroys something in anger the other will do the same. Exchanges

[12] The following discussion about the *kamain* in Bun, as well as the information in Table 3, is based on several writings by McDowell (1976; 1978; 1980; 1990; 1991).

between *kamain* are usually prescribed, typically on such occasions as birth, death, initiation, and marriage. A good *kamain* gives as much as possible and reciprocates quickly and meticulously. There is great respect and a certain shyness between the two. This also means that they would feel ashamed to quarrel with someone in the presence of the other. A person is forbidden to speak the name of his *kamain*, and the behaviour between the two is characterized by avoidance. When people are in a *kamain* relationship, the taboo on names is extended to persons such as mother's brother or father's sister.

The *warɨmbar* relationship in Ambonwari exists mainly as the recognition of an ancestral relationship between two lineages. People are unsure about its origin, though some suggest that *warɨmbar* reveals the ancient connection between two lineages from different totemic clans whose ancestors married women, either real or classificatory sisters, from a common lineage or clan. They further suggest that the sons of such marriages, related as matrilateral siblings, became *warɨmbar*. This relationship was then transmitted from fathers to sons. Two *warɨmbar* call each other *kupan* (elder brother) and *wasapuk* (younger brother). As their mothers came from the same group, the two of them shared the same maternal clan and therefore blood. Such 'one-blood' relationship is in daily life referred to as 'two of us are "brothers" by blood' (*kapa warɨ-kara*). These terms are nowadays extended to classificatory kin and even to the people from other villages in the wider Sepik area whose men's house or totemic group is related to either of them. Nevertheless, as one of Bateson's informants stated, only a particular man was his 'own' *poroman* (1932: 264). The informant was obviously referring to what Ambonwari call 'the ancestral *warɨmbar*' (*kupambɨn warɨmbar*), the institutionalized ancestral relationship between two lineages.

The wives of two partners follow their husbands' relationships. The wives of two *warɨmbar*s are also 'partners' (*warɨmbɨnma*), and they can refer to one another as sisters. There is no doubt that the terms *warɨmbar* and *warɨmbɨnma* connote generosity, as the related terms *warɨmbarar* and *warɨmbarma*, which are used frequently in the village, literally mean 'the one who sends (food)', that is 'a generous man' and 'a generous woman'. People who are called by these latter terms are known as those who always share whatever they catch. One can also look at the way the Ambonwari refer to bride-wealth. There is no single word for it. People simply say *krambɨn warɨmbararin,* 'while cutting one wants to send (be generous)', meaning that while a man cuts the tie between his bride and her parents he also sends shells to them, or, when sister exchange occurs, food.

Every lineage knows the other lineage from which their ancestral *warɨmbar* came. Brothers share their *warɨmbar* partner(s), with priority given to age. If the same marriage path were always followed there would be no problem: both *warɨmbar* would, every second generation, marry women from the same lineage. As this is not the case, and because the ancestral marriages are seldom repeated, *warɨmbar* remains a fixed relationship from the past and is transferred from fathers to sons (and not from fathers to grandsons as names are, for example). New *warɨmbar* relationships are not everlasting and are secondary in importance to the 'ancestral' ones.[13] New *warɨmbar* can be good 'friends'; they can give each other as much as possible and they can help each other. However, the events and ceremonies in which *warɨmbar* are expected to act require the presence of *kupambɨn warɨmbar*. I do not know if new *warɨmbar* relationships can be institutionalized.

It often happens that children, after being beaten by their parents, run off to their father's *warɨmbar*'s house for protection. Their father is too ashamed to vent his anger in front of his *warɨmbar*, and even gives some food to him for the burden.

One of the most visible examples of companionship between two *warɨmbar* is that they are dancing partners. In overnight dances they are decorated identically, with the same body and face patterns in the same colours. If one is tired and has to sit down, the other will follow him. In a few cases several men share *warɨmbar* partnership: the man whose 'mother' was the oldest will dance with the one with the youngest 'mother'. *Warɨmbar* and *warɨmbɨnma* are present at all the important events of their partners. Thus, for example, on the last day after the several months' seclusion of a young girl during the main part of her first menstruation ritual, her *warɨmbɨnma*, who is also her dancing partner, brings a large plate of cooked food to the house. Her partner's mother, who is busy cooking food for the visitors, gives her in return the best food that she can find. If a man is prohibited from eating on the occasion of a big feast when, for example, he organizes the *kurang* ceremony with an all-night dance, his *warɨmbar* will also refuse to eat. In the past, when a man killed an enemy, his *warɨmbar* decorated himself as a succesful killer. He became the centre of attention in the all-night celebration, while the man who actually killed an enemy stayed secluded in the men's house for five days. We can see from all these examples that *warɨmbar* identify with each other by performing the same acts. In this way they achieve a common 'bodily' dimension of their identity, that is, the same *kay*. The *warɨmbar* relationship is another illustration of how Ambonwari see the 'body' of a person and associate it with existence and being. Not only the outside look of a person in terms of decoration but also the behaviour and ways of doing things have to be identical between those who are often referred to as being 'one', having one *kay*, or having one 'body'.

When an important man dies his *warɨmbar* sits on a stool next to the body, the first in a line of the dead person's 'father's fathers' (*wasamarinji*), 'son's sons' (*wunymas*), and namesakes. All of them represent special kinds of 'brothers': each of them 'is' a man who has died. They are all covered with white clay, the colour of mourning, and decorated with cassowary feathers,

[13] An example of two *warɨmbar* whose relationship in the far past went terribly wrong is given in Appendix B.

necklaces, and shells. In this way *warɨmbar* and 'father's fathers' actually represent the dead man by showing how he looked, by 'being' him. If the dead man was a councillor, his *warɨmbar* can also carry his badge (as happened during my stay in Ambonwari). If his *warɨmbar* is also dead, the latter's patrilateral grandson dresses up in the clothes of his grandfather's *warɨmbar* and is the first to mourn. He holds all those objects which had particular importance for the dead man. Real and classificatory brothers of the dead man, or his sons, roll cigarettes, put them between split sago stems, and give them to those sitting in a line, without close personal contact. In this way they avoid contact with a dead person who is identified as being each of those who 'are' whom he was. They avoid becoming an extension of his skin ('body'), which could be especially harmful for his sons. The women whom the dead man called 'mothers' dance around, 'catching' him, that is, his *warɨmbar*, in fishing-nets and baskets.[14] *Warɨmbar*, 'father's fathers', 'son's sons', and namesakes pay their 'mothers' with money and shells. On the floor next to *warɨmbar* is a basket with betel nuts, piper betel, lime, fried sago, tobacco, the dead man's plate, spoon, hat (or whatever he wore), spear, and paddle. At dawn *warɨmbar*, 'father's fathers', 'son's sons', and namesakes make a last circle around the coconut at the front of the men's house, so that 'the eye of the men's house and spirits see the decoration representing the victory over his enemies'. After the funeral they and the sisters of the dead man take all his things to their own houses. Throughout the mourning period, which can last for a year until the mourners 'wash their hands', *warɨmbar* and 'son's sons' will bring food to the mourning house. The last obligation is the hair-cutting ceremony, performed a year or more after the death. The ceremony ends with the burning of the dead man's clothes. That is when his *warɨmbar* will also put his own clothes on the fire.[15]

In Table 3 I present those features which are typical of the relationship between two persons who relate to each other by this term as found among the Bun (and to a large extent among the Biwat) and among the Ambonwari. Other differences between the two societies are also indicated. I want to stress again that the kin origin of *warɨmbar* relationship was suggested to me by Ambonwari themselves. Despite my scepticism, I decided to follow the Ambonwari's own explanation. Other institutionalized relationships, such as with a 'ritual inaugurator' (*sɨmɨrinyar*) and the one known as taboo relationship between *kasɨnma* (daughter's husband's mother, daughter's son's wife, and others) and *akimban* (son's wife's parents, husband's mother's parents, and others), will be discussed later in this book.

[14] This practice is very important in Ambonwari. Whenever people dance, e.g. in the *kurang* ceremony, the 'mothers' dance in front of their 'son' holding a fishing-net or a fish-trap in front of his face in a cosmological gesture of succour. In this way they show their 'sons' their continual care and protection. Dancers should pay their mothers on such occasions (see Ch. 7).

[15] Because of the complexity of mortuary ceremonies I reserve them for discussion elsewhere. We may note, however, that the burning of clothes (as extensions of one's body) implies burning the body itself.

TABLE 3. *Similarities and Differences between the Bun* kamain *and Ambonwari* warɨmbar

<table>
<tr><th>Bun
kamain</th><th>Ambonwari
warɨmbar</th></tr>
<tr><td colspan="2">inheritied relationship</td></tr>
<tr><td colspan="2">a special kind of sibling</td></tr>
<tr><td colspan="2">balanced exchange</td></tr>
<tr><td colspan="2">give goods and food to each other</td></tr>
<tr><td colspan="2">duplicate misfortune</td></tr>
<tr><td colspan="2">respect and shame</td></tr>
<tr><td colspan="2">do not quarrel</td></tr>
<tr><td colspan="2">help in fights</td></tr>
<tr><td colspan="2">take up each other's roles in rituals</td></tr>
<tr><td colspan="2">the role is transmitted to children</td></tr>
<tr><td colspan="2">the spouse is treated in a similar manner</td></tr>
<tr><td>'ropes' (matrilineal clans unimportant)</td><td>patrilineal clans</td></tr>
<tr><td>'Iroquois' kinship terminology</td><td>'Omaha' kinship terminology</td></tr>
<tr><td>intergenerational Z-exchange marriage</td><td>marries where FF married (with FMBSD)</td></tr>
<tr><td>descendants of cross-cousins</td><td>(?) descendants of matrilateral parallel cousins</td></tr>
<tr><td>new kamain relationship inaugurated through ritual</td><td>no ritual for new warɨmbar relationship</td></tr>
<tr><td>avoidance, even taboos on names</td><td>no taboos</td></tr>
<tr><td>not dancing partner (?)</td><td>dancing partner</td></tr>
</table>

One might note that in *kamain* and *warɨmbar* relationships there is no reference to payments in the form of money or, in the past, shells. The Bun, as do the Biwat, prefer exchange to payment, and marriage exchange to the payment of bride-price. It is impossible to exchange both goods and sisters at the same time. 'Any sort of exchange, goods or women, implies the kind of relation characterizing that between affines and *kamain*, both of which are derived from the context of cross-cousins' (McDowell 1976: 46). But this is only one, intra-generational mode of exchange with an exchange of identical items. The other mode is inter-generational, with an exchange of disparate items between mother's brother and sister's son and father's sister and brother's daughter (McDowell 1991: 252). The unique system of 'ropes', as McDowell (1991) has shown in her fascinating analysis, where a male ego calls his mother's father by the term for elder brother (in Ambonwari a male ego calls his father's father 'elder brother') and a female ego calls her father's mother by the term for elder sister (in Ambonwari she calls her father's father's sister 'elder sister'), makes the exchange of goods (and 'sisters') central to Bun and Biwat social organization.

The importance of exchange in general is much less prominent among the Ambonwari. The fact that money is not used in the *warɨmbar* relationship is understandable, as two 'partners' refer to each other as 'brothers' or 'sisters', between whom payments are not part of the relationship anyway. All the ceremonies and all relationships initiated through marriages involve two groups: those who give food, goods, or women and those who pay with money (shells in the past). Among the Ambonwari this can be understood through another hereditary relationship, between a man and his *sɨmɨrinyar*.

Sɨmɨrinyar: *Ritual Inaugurator*

The other hereditary role based on an ancestral relationship between different lineages is *sɨmɨrinyar*. It is hard to translate this word, and impossible to trace precisely the ancestral connections between those in this kind of relationship. However, people say that the members of a lineage are 'mothers' of their *sɨmɨrinyar*'s lineage. Thus, it seems that the ancestral sister's husband acts as *sɨmɨrinyar* for his wife's lineage. As is the case with *warɨmbar*, such a role is inherited from father to son and remains the same over generations, despite new marriages and new relationships created in every generation.

The *sɨmɨrinyar* is the inaugurator of a 'blessing' ceremony. This is performed whenever an important object such as a canoe or a men's house is completed, or when an outboard motor or gun is acquired. In this way the thing receives a name and becomes a being itself. A 'blessing' ceremony is also performed at the beginning of the initiation ritual. For this reason I refer to *sɨmɨrinyar* as a 'ritual inaugurator'. He is the one who, in Ambonwari Tok Pisin, 'opens the custom'.

Every lineage in the village has its own *sɨmɨrinyar*, and if a lineage is very big (Sapɨndam of Crocodile-1, for example) it can have two, in which case on the occasion of a large feast both may play their roles. His or her role (though only one woman acts as *sɨmɨrinyma*) is to 'open' the ceremony or the feast when the long-term work is (or at least is almost) completed: when the basic structure of a house or a large canoe are finished, the post of the men's house carved, the men's house built, or a new outboard motor brought to the village.[16] People say that in the past, when they hung a homicide victim at the front of the men's house, the *sɨmɨrinyar* walked around the body with an empty bag telling the dead person that he would put him into the bag the next day. On such an occasion he placed shells at the front of the men's house for its 'father'. Everything which has its 'father' also has its *sɨmɨrinyar*. The *sɨmɨrinyar* opens the ceremony by putting money, which he or she attaches to leaves of ginger, at the front of or on the new 'object'. At the same time she or he takes the betel nuts and food prepared by those who have built or carved something and distributes them between the members of her or his clan. The *sɨmɨrinyar* is also involved in the last mortuary obligation, the hair-cutting ceremony. In this way the mourning ends and the new beginning is properly acknowledged. As the role is inherited, it can be transferred also to a daughter when her father has no sons. Then his lineage—or, as was the case in Ambonwari during my stay, a second house of a lineage—dies out. While a woman can in daily life act as a *sɨmɨrinyma*, she is restricted in relation to the men's house and sacred objects.

[16] This should not be mistaken for the *naven*-like behaviour which is typical of a person's first significant achievement.

In such situations her husband or her 'brother' from the same clan represents her. Later the role will be conferred on her eldest son and the original name of her lineage will be remembered.

In Table 4, where I present all the *sɨmɨrinyar* relationships in the village, I

TABLE 4. *Lineages and their* sɨmɨrinyar

Clan/lineage	Lineage of *sɨmɨrinyar*	Clan of *sɨmɨrinyar*
Bird of Paradise		
Yimaria	Yanbanjam	Crocodile-1
Kumbranggawi	Warinjam	Pig-1
Ambarmarɨng	Yanbanjam	Crocodile-1
Singgayjam	Sapɨndam	Crocodile-1
Crocodile-1		
Mambinjam	Warɨpɨndam	Eel
Sapɨndam	Wanbanimbuk	Hornbill
Sapɨndam (No. 2)	Mɨndapmay	Eagle-1
Yanbanjam	Kumbranggawi	Bird of Paradise
Eagle-1		
Ambayngmari	Konggɨndam	Pig-1
Mɨndapmay	Warɨpɨndam	Eel
Yambarinjam	Warinjam	Pig-1
Hornbill		
Wanbanimbuk	Kasɨmanimbuk	Cassowary
Kanjimarpinjing	Sapɨndam	Crocodile-1
Wallaby		
Imanmari	Singgaimarijam	Pig-1
Sambrakmari	Singgaimarijam	Pig-1
Yimasɨndam	Simbanimbuk	Pig-1
Karpanjam	Konggɨndam	Pig-1
Marikapinjam	Singgaimarijam	Pig-1
Pig-1		
Simbanimbuk	Yimasɨndam	Wallaby
Singgaimarijam	Marikapinjam	Wallaby
Konggɨndam	Wayanggara	Cassowary
Warinjam	Kanjimarpinjing	Hornbill
Crocodile-2		
Kunbaynjam	Wayanggara	Cassowary
Eel		
Warɨpɨndam	Simbanimbuk	Pig-1
Cassowary		
Wayanggara	Kanjimarpinjing	Hornbill
Kasɨmanimbuk	Imanmari	Wallaby
Aunjam	Simbanimbuk	Pig-1
Mayprukong	Yimasɨndam	Wallaby
Paymbɨndam	Yimasɨndam	Wallaby
Eagle-2		
Murimari	Yawyamari	Pig-1
Konimari	Yawyamari	Pig-1
Pig-2		
Yawyamari	Kunbaynjam	Crocodile-2

follow the spatial organization of the village, from downriver upwards. As can be seen from the table, these sets of relationships closely resemble their myths of origin. The lineages of those clans which travelled together and arrived together, or those lineages which incorporated new clans into the village by marrying their 'sisters' or by giving them their own, are a majority of those in a *sɨmɨrinyar* relationship. Moreover, the arrival in the village, and the first marriage arrangements can be to some extent deduced from the spatial organization of the village (see Map 3). For example, in the myth of origin of the main two clans, people say that Akɨmbrɨkupan (Bird of Paradise) gave his daughter to Kapi (Crocodile-1). If we look at the *sɨmɨrinyar*s of the lineages from the Bird of Paradise clan we note that they are from the Crocodile-1 clan (recalling that Warinjam from the Pig-1 clan lives on the land of the Crocodile-1 clan). The *sɨmɨrinyar* of the lineages from the Eagle-2 clan are all from the Pig-2 clan, which is again understandable considering that they lived together in Arkwas before coming to Ambonwari. Also, as the two *sɨmɨrinyar* of the two Eagle-2 lineages are from the same lineage, the younger brother takes up his role for the main Murimari lineage while his elder brother acts as *sɨmɨrinyar* for the second Eagle-2 lineage, Konimari. Upon his arrival in Ambonwari, Warɨng, the younger of two brothers from the Pig-1 clan, married a woman from the Eel clan, Kambiak. Thus, the 'father' of the proper lineage from the Pig-1 clan acts as *sɨmɨrinyar* to the Warɨpɨndam lineage from the Eel clan. Another relationship can be discerned from Table 4: Wallaby and Pig-1 lineages are related through their past marriage exchanges, as the men from the two lineages from different clans act as *sɨmɨrinyar* to each other. The great majority of *sɨmɨrinyar* relationships can be traced in this way, while some, of course, remain a mystery both for the Ambonwari and for their puzzled ethnographer.

The entire *sɨmɨrin kay* ('the way of *sɨmɨrinyar*', the opening ritual) can last from anywhere between an hour to a whole day, depending on the kind of feast or ceremony. For example, Albert, from the Simbanimbuk lineage of the Pig-1 clan, worked for a month on his new large canoe. His 'brothers', sister's son, and his wife's and his mother's clansmen all helped him with carving. Several times Albert went hunting, and caught two large pigs. They arranged a day for the *sɨmɨrin kay*. Godfried, from the Mɨndapmay lineage of the Eagle-1 clan, attached three kina to the leaves of ginger and put the small bundle on the prow of a canoe on which a man from Albert's maternal clan had carved the head of a pig. People said: *marɨk yanapɨkan sɨmɨrinyar* ('*sɨmɨrinyar* put ginger on it'). Godfried took a cluster of betel nuts which Albert had previously put there. After Godfried began to chew, others could join him. Albert's brothers brought two large dishes of sago pudding and two pigs' heads. Godfried had to begin eating first and others followed him. Those from Albert's lineage and their associates were not allowed to eat (as they were the ones who had brought the food). The joking between the classificatory brothers of Albert's wife and Albert's classificatory brothers was then at its height (see Chapter 4). After the feast Godfried distributed the meat between his own relatives and took half a pig's head for himself. Albert spent another month finishing his canoe. When he brought a new outboard motor to the village the whole ceremony was performed again. In the meantime Albert went hunting several times to provide the meat necessary for the feast.

Similar inaugurations happened on occasions such as the opening of a new house and the last act in mourning, the hair-cutting ceremony. In all these events the *sɨmɨrinyar* is the one who attaches the money (shells in the past) either to ginger leaves or to the sago stem (before a dance or the cutting of hair begins). The main obligation is, of course, on those who organize the ceremony. They have to provide enough food for the participants as well as money to pay their maternal and wives' groups. These and other kin relations will be discussed in the next chapter.

4

Kinship and Marriage

To enter into a discussion on kinship in the 1990s involves re-engaging arguments which had already dominated anthropological discourse for 100 years when Lévi-Strauss (1965) presented his Huxley Memorial Lecture more than thirty years ago. The cultures of New Guinea have been crucial in many of these discussions. Although it would be far beyond the scope of this book to consider in any detail kinship systems elsewhere in New Guinea, as neither Ambonwari nor Karawari as a language group has been studied before, I find myself under an obligation to present a concise description of the Ambonwari system. This will, I hope, shed new light on other kinship systems of Sepik societies.

Many would consider the kinship system of Ambonwari to be a variety of the 'Omaha' system. Despite several different approaches (for a summary of these see McKinley 1971a; 1971b), there have been no entirely convincing answers to the question as to why Omaha (and Crow) kinship terminologies exist. Moreover, many have questioned the very existence of such systems generally (Needham 1971: 14–15; Barnes 1976: 387). When he called for a more complex approach to Crow–Omaha systems, Lévi-Strauss (1965: 18–19) proposed that 'a time dimension should be added to the spatial ones' if we want to understand them.

The division of the social world by kin terms is an aspect of *kay*, which is through 'the way of the village' (*imɨnggan kay*) embodied in every individual.[1] It is both acquired practically and learnt by understanding. Each of these terms has multiple associations which are specific for each individual while at the same time reflecting the collective habit-memory and the taken-for-granted *kay* of the village. Of course, terms in themselves are timeless, but from ego's perspective they both identify and differentiate groups and individuals. The model of kinship, inherited from generation to generation as 'the way of the ancestors' (*kupambɨn kay*), frames the possibilities of Ambonwari social life. Both through idealized and actual kin relationships, kin terms enable things, events, and relationships to reappear again and again at different times in apparently identical ways. But just as time represents the essential dimension for the reproduction of such apparently identical ways, it is itself produced by these processes of identification and differentiation. Social relationships

[1] Jadran Mimica argues: 'The configurations and categories constituting human social reality (e.g., language, kinship or clan organization), are the products of human bodily life: the multiple structures of the human psyche are inalienably the structures of embodied existence' (1991: 35).

between the holders of kin terms produce the identities of persons and groups and are temporalized through the process of actualization; and also vice versa, for they are actualized through time.

Patrilineal descent in Ambonwari, as discussed in the last chapter, takes a specific form in which father–son periodicity means that the characteristics (names and marriage paths) of the father's father's generation are reproduced in ego's own. Because of this periodicity all men in ego's descent group are fathers or brothers. Consociates who are sons can be alternatively called 'fathers' by their own fathers and vice versa. Thus, there are only two 'generations', which 'replace' each other alternately. Ambonwari, accordingly, recognize only two levels of marriage: the paths of 'fathers' and the paths of 'brothers'. Marriage (endogamous to the village, exogamous to a clan) represents paths—*konggong*—between places and groups within the village.

The Ambonwari 'Omaha' system (see Figs. 3 and 4), especially the generational skewing of kin terms for cross-cousins, closely resembles that of the Iatmul (Bateson 1932; 1958: 305; Wassmann 1991: 38).[2] However, while the Ambonwari call WBW, FFZ and FMBW a 'sister', the Iatmul do not. Thus the only distinction indicating asymmetry (Korn 1971: 111) among the Ambonwari is that father's sister is not called the same as mother's brother's wife, which is also the case among the Iatmul. Another, more important similarity is that all women of the father's mother's clan, which is one of possible and preferable marriage, are grouped under the single term *iai* or *yai* among the Iatmul, and under *arkay* among the Ambonwari. Thus a man can marry 'a woman of any generation provided she be of the same clan as his father's mother' (Bateson 1958: 265; see also Forge 1971: 134; Korn 1971). Though there were and are cases of marriages with FMBD in Ambonwari, it is not considered to be a 'correct' marriage. Other similarities between Iatmul and Ambonwari terminologies are as follows:

W (before marriage) = FFW = FMBSD (*arkay*)
M = SW = MBSD (*asay*)
ZH = ZSS (*sakuri*)
MB = MBSS = SWB (*away*)
FZS = DS (*lamɨn wasan*)
FZD = DD (*lamɨn kyawi*)

[2] In the following discussion I use common abbreviations: F = father, M = mother, D = daughter, S = son, Z = sister, B = brother, W = wife, H = husband, P = parents. Because MB = MBSS ≠ MBS, 'Omaha' systems of this kind were called 'alternating' (Ackerman 1976: 555, 570; Needham 1971: 19). Omaha-type kinship and marriages have received quite a lot of attention (e.g. Lévi-Strauss 1969: pp. xxxv–xlii; Lounsbury 1964; Muller 1980; Titiev 1956) and have often been discussed in the context of Sepik societies such as Abelam (Kaberry 1941; 1967; Forge 1971), Iatmul (Bateson 1932; 1958; Wassmann 1991), Kwoma (Whiting and Reed 1938–9: 200–2, 206; Bowden 1983; Williamson 1980), Chambri (Gewertz 1977a; 1983), Avatip (Harrison 1984). Other studies in Papua New Guinea, such as those by Gell (1975) and Schieffelin (1976), though dealing with variants of the 'Omaha' system, differ too much to be included in this discussion.

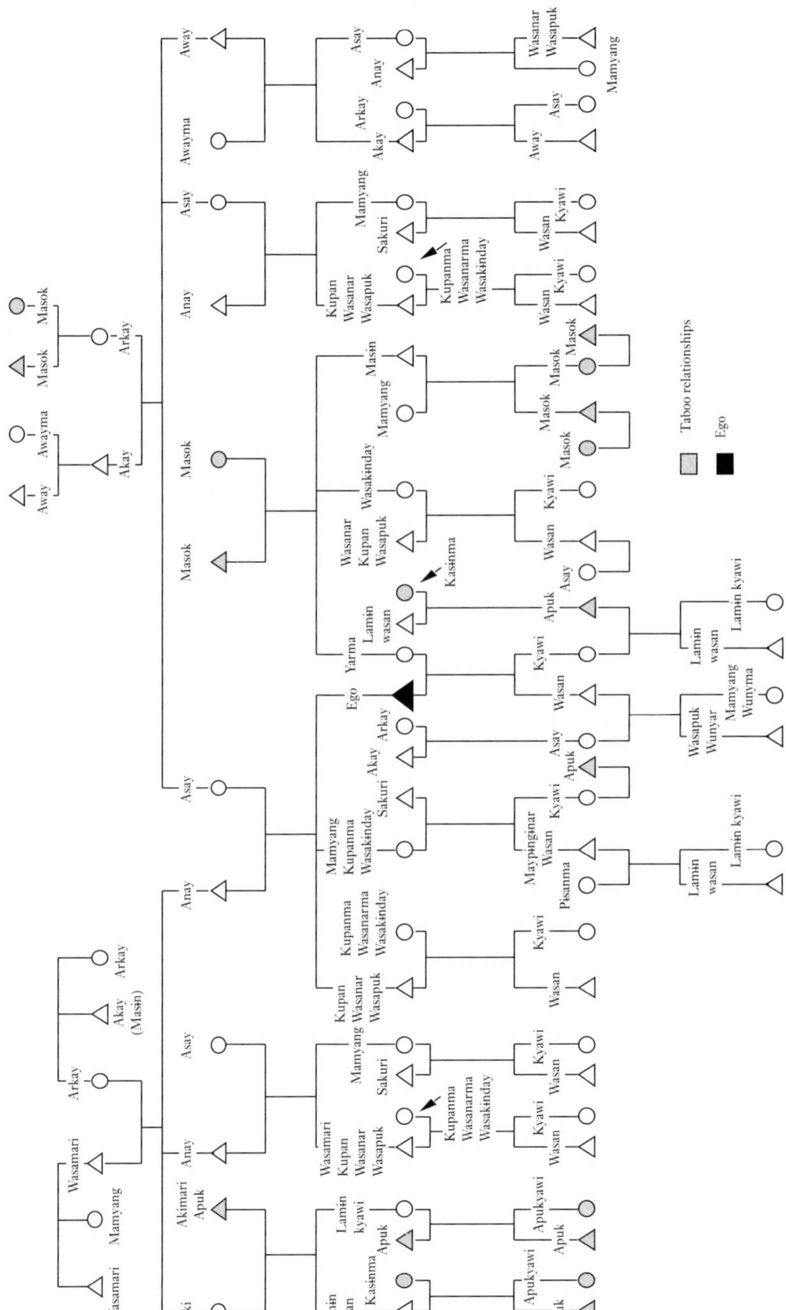

FIG. 3. Kin terms (male Ego)

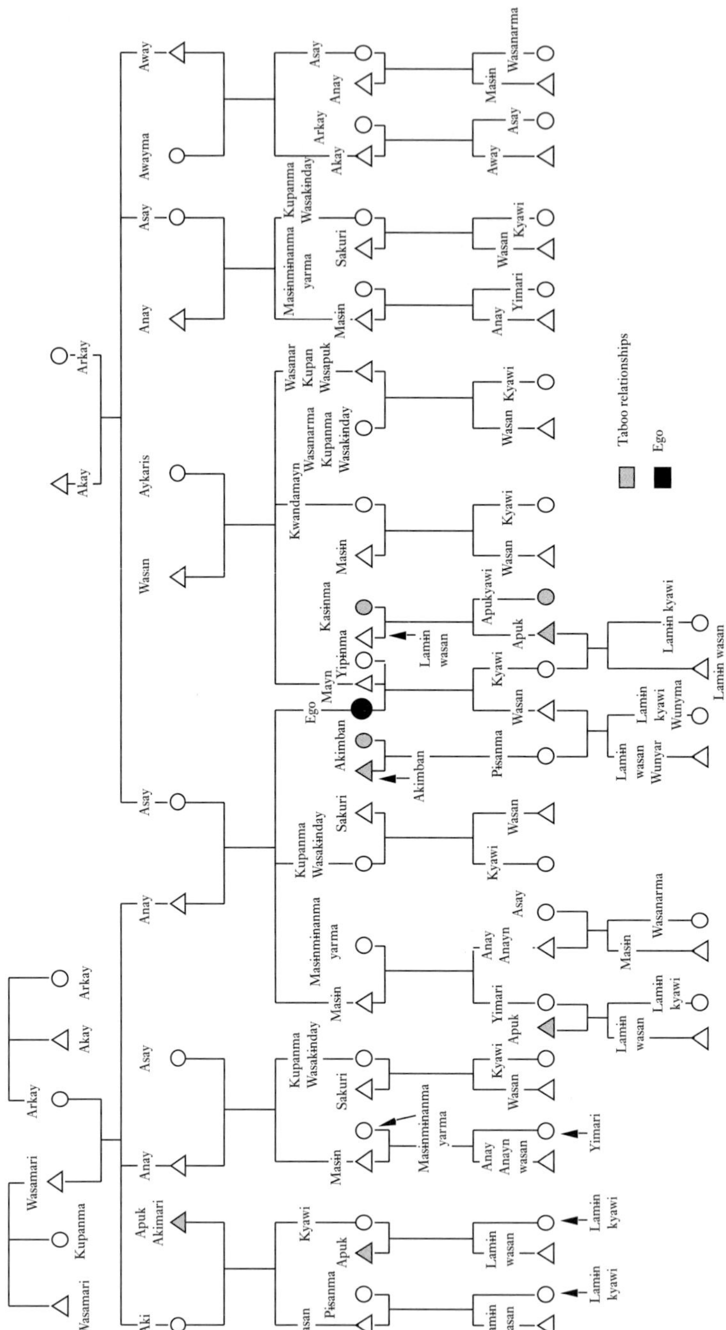

FIG. 4. Kin terms (female Ego)

Some observations concerning Ambonwari cross-cousins can be made at this point. The terminology for cross-cousins differs from that for siblings and parallel cousins; patrilateral cross-cousins are distinguished from matrilateral cross-cousins; patrilateral cross-cousins are 'lowered' two generations for a male ego and only one if ego is female; reciprocally matrilateral cross-cousins are 'raised' a generation (for an MBD) or two (for an MBS). As among the Iatmul (Bateson 1932: 263), all women in the mother's patriline are called 'mother' (*asay*), which is typical of the 'Omaha' system. The men in a maternal patriline are either mother's brothers (*away*) or maternal 'grandfathers' (*akay*).

In Ambonwari matrilateral cross-cousins are termed *asay* (mother) or *akay* (grandfather) because of 'the principle of the equivalence of alternate generations' (i.e. a man takes his paternal grandfather's identity in the form of his names and marriage paths). The way men refer to women in their own as well as in their marriageable lineage then conforms to this logic.

A person's patrilineal grandparents are called 'brothers' (i.e. *kupan*, 'big brother', along with the alternative term *wasamari*) and 'sisters'. A man calls his father *anay* (father) and his son *wasan* (son). As the latter two are 'brothers', according to the two-generational logic, a man can call his own son *anay* (father) which usually occurs during the period of marriage arrangements if not before (cf. Mimica 1988: 87–9). At the same time, a man's father and a man's son receive names from the same set of generational names, which means that a paternal grandfather and his grandson can have the same name. When the son marries, his father, while calling his son *anay* (father), calls his son's wife *asay* (mother) (see Fig. 3).[3]

We must now include the possible marriage arrangements in order to understand the use of ascending and descending terms for cross-cousins. The most probable marriage in Ambonwari is with FMBSD (as also among Iatmul and Abelam). A man already calls his MBSD *asay* (mother) as she could become a wife of his son whom he calls *anay* (father). Because of this, the father of *asay* (mother), that is, the MBS, has to be *akay* (mother's father). By the same logic, if a male ego calls a man *akay* he also calls that man's patrilineal grandson by the same term because the two are 'brothers'.

On the other side, the man's FZSS is a potential husband of the man's daughter. So he calls him *apuk* (FZSS, DH). The father and the mother of *apuk* thus have to be *lamɨn wasan* and *kasɨnma* respectively. Viewed from a different direction, the result remains the same. While the man's father becomes *wasan* (son), the man's father's sister, who is otherwise called *aki* (FZ), becomes *kyawi* (daughter). Because of this a man calls her children *lamɨn wasan* (daughter's

[3] A man calls his son's son either *wasapuk* ('little brother') or, in ritual contexts when he acts as *wasamari*, *wunyar* ('son's son'). In a similar way he calls his son's daughter *mamyang* ('sister') or *wunyma* ('son's daughter'). A woman, on the other hand, calls her son's children 'grandchildren', *lamɨn wasan* ('grandson') and *lamɨn kyawi* ('granddaughter'), or she may follow her husband's ritual address by calling them *wunyar* and *wunyma*. These terms are obviously related to *wun* ('sago grub').

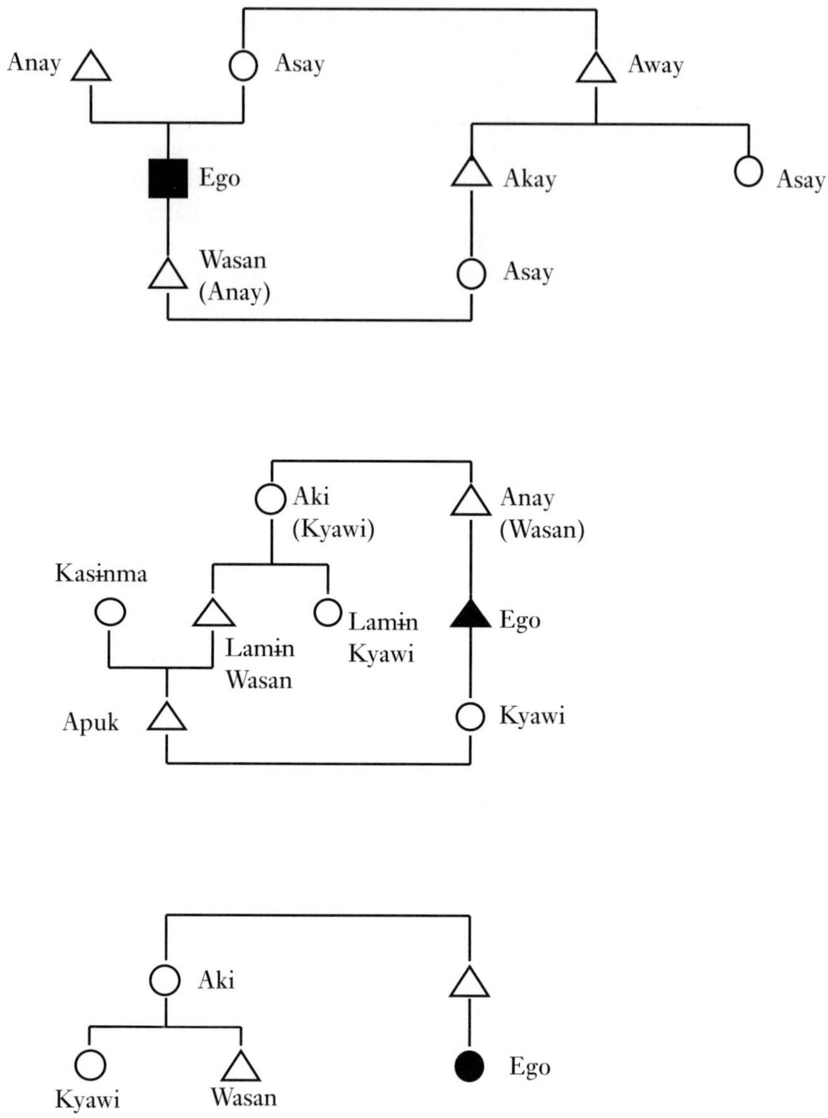

FIG. 5. Kin terms for cross-cousins

son) and *lamɨn kyawi* (daughter's daughter). They are not called *akay* (MF) or *asay* (M) because he did not trace them through the marriage of his son and his son's wife (whom a man calls 'mother') but through his 'daughter'.

In the case of a female ego, she may be called by her father *aki* (father's sister) as she is a sister of her brother (her father's son) whom he calls *anay* (father).

Thus while her father calls her *aki* (FZ), she and her brother call their FZ *aki*, and because both are women of the patriline, they can call each other *aki*. A woman simply follows the terminology of her FZ. Also, if we look at the woman's FZS we can see that he calls his MBD *asay* (mother) and she in turn calls him *wasan* (son).[4] A woman follows her brother's terminology when referring to her matrilateral cross-cousins (see Fig. 5). It remains to try to explicate why all the women of the mother's brother's lineage are terminologically equivalent to mothers.

It is as though Ambonwari push this logic as far as possible: if a man can call his son 'father' and can call his father 'son', generational differences can be made to disappear, even though, ironically, the whole emphasis of the system is on the duality of adjacent generations. Thus all the men of a patriline are equated and so too are the female members as sisters. If all our women are 'sisters' regardless of generation, by the same logic all the women of our marriageable lineage are 'mothers'. Tenuous though this explanation may seem, it receives some support from the fact that all lineages with which a group has affinal paths (*konggong*) are referred to as *yarɨmbawkum*—wife-women. It is perhaps significant, too, that there are vestiges of a moiety system (see Chapters 3 and 5).

Marriage

The marriages which follow the grandparental paths offer a set of possibilities for young unmarried men and women against any of which none can protest. In this sense they can be called preferred, at least from the point of view of the parents of the marriageable man or woman, who would like their son or daughter to maintain already established relationships.[5]

Ambonwari clans are exogamous and postmarital residence is patri-virilocal.

[4] The reader should note from Figs. 3 and 4 the differences between sexes in some terms of address: a woman does not call her brother *kupan* (big brother) or *wasapuk* (little brother) but *masɨn*. She calls her son's wife *pɨsanma*, while her husband calls her *asay* (mother). She uses the terms *aykaris* (HM) and *akimban* (SWP) which are not used by a man. Only a man uses the term *mamyang* for a sister.

[5] The spatial dimension of marriage has been recognized in other parts of New Guinea. Thus e.g. McDowell argues that because people talk about marriage arrangements in terms of 'roads', 'these relations are not conceptualized as structures or positions but as movements and process' (1977: 181). Biersack also notes that 'bush paths, if used, articulate affinally related areas. Marriage opens up a "road" between wife giver and wife taker' (1991: 238). In many African societies the metaphor of pathway applies not only to social relationships but to individual anatomy and physiology. Jackson (1983) presents African examples in which movements of 'blood', 'heat', and 'breath' are referred to in terms of pathways in the body and where the vital organs are seen as crossroads. He also views the physical movement of people within the community as being 'determined by social, economic and moral imperatives such as visiting and greeting, sharing food, cooperating in food production and in festivals, and giving gifts or services according to commitments of in-lawship, neighbourliness, or kinship . . . Such metaphors [like 'path'] are thus the verbal correlates of patterns of social interaction and bodily disposition within the *habitus*' (1983: 135–6). In a similar way, Ambonwari see *konggong* (path) as something which articulates with *kay* (being, existence, habit) and where change in one affects the other.

Every young person knows those clans into which a woman cannot marry, and from which a man cannot get a wife. A young man also knows that he can obtain his wife from the same lineage as his FF did, and a young woman knows that she can find a husband in the same lineage that her FFZ did. With these lineages his or her own lineage keeps good relations, maintaining a possible option for their offsprings' future marriages. In this way young people from these lineages get to know each other more intimately. In the past, people say, parents had more influence on their children's marital choices, and old men more often married young women. In recent times children themselves decide. This is called *laik marit* (love marriage) in Tok Pisin. Nevertheless, such marriages are not random, and paths which are known to have been avoided traditionally continue to be so. What differs from the past is that young people more often choose with whom they will have more intense relationships in the future. Parents' wishes to strengthen ties between clans which are already related, and thus to create even more powerful alliance relationships, are often ignored by the young.

In the past, Ambonwari men rarely married women from other villages, nor did Ambonwari women marry out of it. There were a few exceptions, however, when a Maikerobi woman (same language) married into Ambonwari and a few Ambonwari women married Imanmeri men (different language) and moved to their husband's village. In the recent past there were a few more cases of marriages with women from other villages: two men married women from Konmei, three men married women from Kaiwaria, and one man married a Biwat woman (different language). The average bride-price within the village is nowadays somewhere between 200 and 400 kina, but when a marriage is arranged with a woman from outside, the amount is between 400 and 1000 kina (which is still not a large sum compared to bride-price in the Highlands).

Polygyny in Ambonwari is practised on a very small scale. During my stay there was only one viable polygynous marriage; another man who married two women lives only with one; two other men were each married to two women but their wives had recently died. I will show below how co-wives act as sisters. One wife calls the brothers of the other wife 'brothers', fathers 'fathers', and so on, while their children have two maternal kin groups.

Concerning 'traditional' marriage the Ambonwari have two possibilities: one is 'one-way' marriage and the other is exchange marriage. A man cannot get a wife from the same clan as his father did, but he can follow the path of his father's father. Father and son thus have different 'marriage paths'. From the man's maternal group his son is again able to get a wife. It is obvious that a man cannot marry a woman from his own lineage nor from his own clan. Two or more brothers can take their wives from the same lineage or clan, though if there are not enough marriagable women the first-born son takes priority as does the first-born daughter. A girl has the possibility of marrying into the same lineage or clan as her father's father's sister did and thus of marrying her FFZSS.

When people talk about marriage they refer to both marriageable persons as *yan sarikaykan* ('she or he holds the other and stays') and to marriage itself as *kupa isa asararin* ('you two want to hold each other').

Exchange of women in marriage is important, in the sense of strengthening the ties between two lineages and two clans. Nevertheless, 'one-way' marriage is considered to be the older form, prior to exchange. As can be seen in myths, such marriages represent the beginning of relationships between lineages and clans. In Karawari vernacular people refer to such marriage as 'one way' (*mbanggor*), 'the path before' (*kanggrain konggong*), or 'the ancestral path' (*kupambɨn konggong*). Following this path, a man marries his FMBSD, whom he calls *arkay* and she marries her FFZSS, whom she calls *lamɨn wasan*. After a marriage a man calls his wife 'ancestral wife' (*kupambɨn yarɨma*), and she calls her husband 'ancestral husband' (*kupambɨn mayn*). When one asks about a particular lineage and clan from where men can get wives, people would say either: 'This is the ancestral path of the father' (*kɨnda anay mɨnak kupambɨn konggong*) or 'This is the ancestral path of the son' (*kɨnda wasan mɨnak kupambɨn konggong*).[6] Figure 6 represents the 'ancestral path' of marriage.

A man can in theory marry his FMZ, whom he calls *arkay*, but because of the likely age difference this, to my knowledge, does not happen.[7] The kin terminology in itself already points towards marriage between a man and his FMBSD. As I mentioned above, a man follows the terms used by his brothers including his *wasamari* (father's father). The same is true for a woman. After the marriage between a man and his *arkay*, his FMBS becomes his WF, terminological equivalents—so he still calls him *masok* (a taboo relationship: see below). A woman calls her FFZS *wasan* (son) before and after marriage when he becomes her husband's father. The only change in kin terminology occurs with the woman's reference to her husband's mother (her FFZSW). Before marriage she calls her *pɨsanma*, while *pɨsanma* calls her *aykaris*. After marriage these terms are reversed and a woman calls her husband's mother *aykaris*, while the husband's mother calls her son's wife *pɨsanma*.[8]

[6] Korn, analysing Bateson's statement that marriage with the *iai* (FMBSD) is consistent with the relationship terminology, says that this kind of marriage would lead to a closed system of alliances and to 'an asymmetric system defined by five patrilineal descent lines' (1971: 115–16). In reality there is no closed system: new alliances are formed all the time while old ones are kept as future possibilities.

[7] As I have mentioned earlier, young men call their possible wives *arkay*. In modern love-songs they call their girlfriends *akay*, which is otherwise the term used only for men (FMB, MBS, MF). This resembles Iatmul, where all the women from FM's patriline whom a man has a chance to marry are called *iai*.

[8] It is tempting to speculate on the suggestive similarity between two terms: *pɨsanma* (HM) and *pɨsɨnɨm* (period, 'season'). Before marriage a possible HM is called *pɨsanma*, which could mean 'periodical woman', the one who continues the periodicity of the lineage. After the marriage the HM calls her SW *pɨsanma* because she took the task of the continuity of the lineage from her HM.

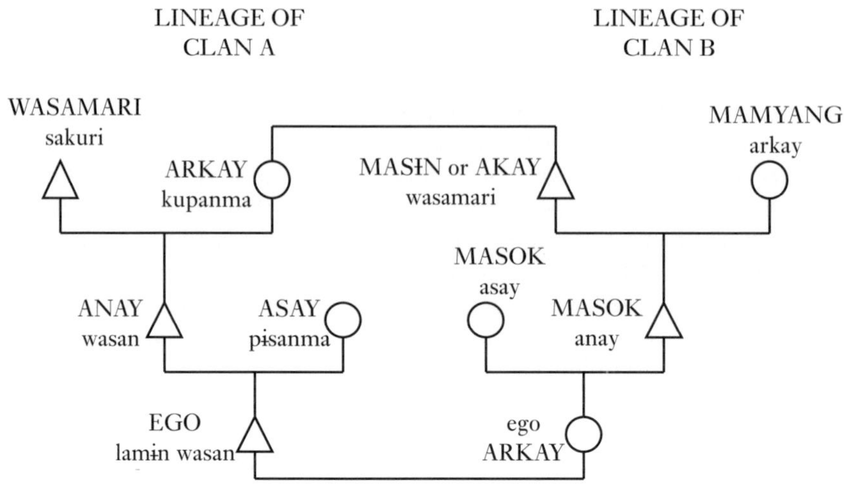

FIG. 6. 'One-way' marriage

In such a marriage it can actually happen (especially if both partners are first-born) that a couple have the same names as the husband's paternal grandparents, as he was named after his father's father and the wife after her father's father's sister, who was at the same time her husband's father's mother.

The second traditional way of marriage is by the exchange of women. The marriage exchange can be referred to as *yanggyamangginya* in Karawari, for which I cannot offer a translation. People explained by reference to particular people: 'I went and she came' or 'she went for him and she came for me'. On the other hand, people would say 'the two will exchange' (*nggri isa kumbukɨr*, 'the two will balance') or 'the two must lower their heads and come outside' (*kanggri isa yapakumbɨn apasɨmba*), meaning that one girl will go outside the house and the other will come inside and take her place. If two lineages of different clans have marriageable sons and daughters, and if no avoidances are involved, the classificatory brothers and sisters of a newly married couple might arrange a second marriage to 'balance' the first. If a lineage has no daughters for a direct exchange other lineages of the same clan can help. Thus, for example, a classificatory FBD from some other lineage can take the place of an actual daughter, but not FZD or MBD, of course.

If the girl's father rejects the wooing of the young man's lineage, either because he still wants his daughter to help in his house or because he has not yet prepared earth ovens and other things for his daughter's marriage, or because of a previous dispute between the two lineages, the young man can experience such shame that he will leave the village and reject further talks about the marriage. In such a case the girl's lineage will fear for her future marriage chances. A similar situation can arise when a hot-blooded young man continually presents a girl with small gifts but she does not 'repay' them and rejects his attempts to form a relationship.[9] She may gossip about his persistence and her dislike of him. He will then feel ashamed and angry. Such a girl could have problems finding other suitors, as young men would be wary of the rejected man's anger as well as fearing the possibility of being rejected too. Both situations occurred in the village during my stay.

After marriage a woman moves to her husband's house and her children become members of the lineage, men's house, and clan of their father. But if a couple separates, a woman may leave her ex-husband's lineage and, if the latter had not paid the bride-price, she may take her children with her and bestow on them the names of her natal lineage. This is often a reason for serious dispute between two lineages. If her husband dies and a widow is old she usually stays with her deceased husband's lineage even if she has no living children. If she is still young, and if she wishes to, she can marry again. But the expected choice for her is that she will marry a brother of her deceased husband: not a real but a classificatory one (who should not be the first-born 'head son' of his own lineage). A new husband's lineage will pay for her 'transfer' from the lineage of her previous husband. Children from her first marriage will retain membership of their father's lineage despite their new 'father', who from now on looks after them. When a daughter of such a family gets married, the payment goes to her deceased father's lineage, which then gives 'half' of the payment to her new 'father's' lineage. The following example presents some facts mentioned above.

Acustaria (her Christian name) was born in the Cassowary clan but her mother gave her to her 'brother' from the Crocodile-1 clan (see the discussion about her names in the previous chapter). She married Kamok, a man from the Bird of Paradise clan, thus following the preferred marriage path of both Cassowary and Crocodile-1 clans. Payment for her was made to Crocodiles. When Kamok died they had no children. The men from the Bird of Paradise clan talked to the men from the Pig-1 clan saying: 'You are our family, you will marry her.' One recalls from the myths of origin that the Pig-1 clan reciprocated the Bird of Paradise clan for being invited to live in Ambonwari by presenting them with a whole men's house. Thus they became 'brothers' and do not marry between themselves; otherwise, they say, they could die. So Yasɨpas from the Pig-1 clan married Acustaria and paid the bride-price to the lineage of her previous husband. She gave birth to two daughters: the first-born, Saunbi, married a man from the Wallaby clan, and the second daughter, Kusi, married a man from the Crocodile-1 clan (this despite Acustaria's adoption into the Crocodile-1 clan; they followed the traditional marriage path of particular lineages). When Yasɨpas died Acustaria was widowed a second time. Again people from the Pig-1 and Bird of Paradise clans proposed that she marry someone from the Bird of Paradise clan. So Noah (Christian name) from the Bird of Paradise clan married Acustaria, and the bride-price went to the lineage of her previous husband from the Pig-1 clan.

[9] Ambonwari people are very punctilious about the exchange of things between two young people. If a gift is not returned there is no relationship to speak of: the two cannot marry even if they want to.

When I tried to comprehend the actual pattern of marriages within the village, that is, how marriages were contracted over the years in every lineage in the village, I was faced with several problems. First, an individual would know the clan and lineage of his FM but rarely those of his FFM. If I persisted with my questioning, it happened several times that people thought of FFM and FFF as being from the same clan, or they would even confuse these ancestors with the mythological ones. Second, some men from the same clan but different lineages were of the same age but related to each other as 'fathers' and 'sons'. This often made an inquiry difficult, as the marriage path of one was not the path of the other. Moreover, as I explained in a previous chapter, real brothers may be members of different lineages; for example, in the Eagle-1 clan Gerard is the 'father' of the Ambayngmari lineage and his younger brother, Kenneth, is 'father' of Yambarinjam; in the Hornbill clan Camilus is the 'father' of the Kanjimarpinjing lineage and his younger brother, Acus, 'father' of Wanbanimbuk. A third problem began when I discovered that after I had written down all the possible marriages of one lineage, people of other lineages did not always agree that they could marry into the first lineage (though they might mention other lineages from the same clan).[10] I have to emphasize that I questioned people about marriages in both directions, that is, where they got wives and where they sent their sisters and daughters. At least there was more consistency at the level of clans. As lineages of the same clan help each other, as a man from a different lineage often gives a name to his BD and later arranges her marriage, such perplexity and confusion was to some extent understandable. Fourth, adoption of both individuals and groups, past and present, added to my difficulty.

After looking at all the existing marriages in the village and all those possible ones that I was told about, I concluded that I should look at those clans which do not marry each other (see Fig. 7). One reason for some of the avoidances of intermarriage lies in the ancestral brotherhood of two clans. This relationship exists, for example, between the Bird of Paradise and Pig-1 clans: a widow who was married to a member of one clan can remarry into the other. Such 'brother' relationships are also found between the Wallaby and Hornbill clans, and between the Cassowary and Eel clans. As in other situations, such as the transmission of institutionalized roles discussed earlier, priority is given to more remote classificatory brothers rather than to consanguineal brothers. Other reasons for clans avoiding intermarriage can be found in their past adoption practices, such that people have to consider very carefully their marriage options while somehow trying not to talk about adoption at all. Again, some men told me secretly that they did not want their son to marry the daughter of a woman with whom they had had sexual intercourse in the past.

[10] Such inconsistency shows that there is more contingent 'history' associated with the rigid social structure than one might think.

	Crocodile -1	Bird of Paradise	Crocodile -2	Wallaby	Pig-1	Eagle-1	Casso-wary	Pig-2	Eagle-2	Eel	Hornbill
Crocodile -1	■		•					+	•		
Bird of Paradise		■	•		•			•	•	•	•
Crocodile -2	•	•	■			•		•	•	•	
Wallaby				■					•		
Pig-1		•			■						
Eagle-1			•			■			•	•	
Casso-wary							■			•	
Pig-2	+	•	•					■			
Eagle-2	•	•	•	•		•			■		•
Eel		•	•			•	•			■	•
Hornbill		•							•	•	■

+ Recent marriages only, before no marriages
• No marriages

FIG. 7. Marriage avoidances in Ambonwari

Even if they were not the real fathers of the daughter, she still carried a small part of them within her.

The individual lineages in large clans such as Cassowary and Wallaby have different marriage options and also different avoidances. Thus, for example, within the Wallaby clan, Yimasɨndam and Samblakmari do not marry with the Bird of Paradise and Eagle-2 clans, Imanmari do not marry with the Crocodile-1 and Eel clans, and Marikapinjam do not marry with the Bird of Paradise, Crocodile-2, Pig-1, and Eagle-1. One can see that in this respect some lineages behave as if they were clans in their own right. To make the situation even more complex, there are cases where men from the same clan who call themselves 'brothers' take wives from the same lineage of another clan but from different

TABLE 5. *Ambonwari marriages in 1991*

	Wife's clan same as father's father's wife's clan	Wife's clan different from father's father's wife's clan	Wife from other village	Considered improper marriage	Unable to trace	Total marriages
'One-way' marriage	18	10	3	3	11	45
Exchange marriage	7	6	–	–	6	19 (30%)
TOTAL	25 (39%)	16 (25%)	3 (4.5%)	3 (4.5%)	17 (27%)	64 (100%)

generational levels. For example, Elias from Yimasɨndam calls Linus from Imanmari 'brother'. Both lineages belong to the Wallaby clan. Elias can marry a woman from the Sɨmbanimbuk lineage of the Pig-1 clan, but only Linus's son can get a wife from the same lineage, not Linus himself. The marriage paths of the two lineages of the same clan are thus a generation apart.

Despite the difficulties, from the existing marriages in 1991 I was able to reach some conclusions. The results are summarized in Table 5. Almost 40 per cent of all marriages in Ambonwari were of the 'preferred' kind, that is, in which a man married his FMBSD. Almost 30 per cent were exchange marriages.

Relationships and Behaviour

In Figs. 3 and 4 I have presented the consanguineal terminology for male and female ego respectively. With a few exceptions (see below), terms of address do not differ from terms of reference. The Ambonwari use collective terms, which are actually plural forms, to cover categories of people including classificatory kin.[11] Thus a man calls:

his older brothers	*wasɨnarm*
his younger brothers	*wasanarm*
his father's father's brothers	*wasamarinji*
his sisters	*maymɨndi*
his sisters and daughters	*maymɨndi-ya-kyawɨndi*
his wife's F and FBs	*masokurarsindi*
his wife's brothers (following her terminology)	*masɨnɨm*
his mother's lineage	*yarɨmbawkum*

[11] Two terms can be joined with an additional *-a-* between the nouns: *anay-a-wasan* (father and son), *asɨm-a-kyawi* (mother and daughter), *aykaris-a-pɨsanma* (son's mother and his wife).

his married sister's lineage	*yimɨnɨnggi*
his sister's sons	*maypɨnggi*
his grandchildren	*lamɨn yarakɨmbi*

A woman calls:

her brothers	*masɨnɨm*
her father's father's brothers	*wasamarinji*
her sisters	*wasɨnmɨndi*
her husband's	
younger brothers	*wasanarm*
older, or all together	*wasɨnarm*
her husband's father's brothers (children)	*kɨmbi*
her mother's lineage	*yarumbawkum*
her married sister's lineage	*yimɨnɨnggi*
her brother's and father's lineage	*masɨnbanunggum*
her and her sister's daughters	*kyawɨndi*
her grandchildren	*lamɨn yarakɨmbi*

TABLE 6. *Reciprocal kin terms*

anay	*wasan*	(F–S)
anay	*kyawi*	(F–D)
asay	*wasan*	(M–S)
asay	*kyawi*	(M–D)
wasanar/kupan	*wasapuk*	(Be–By)
mamyang/kupanma/wasakɨnday	*masɨn*	(Z/Ze/Zy–B)
kupanma	*wasakɨnday*	(Ze–Zy)
mayn	*yarma*	(H–W)
yipɨnma	*yipɨnma*	(W1–W2)
wasamari	*wunyar*	(FF–SS)
wasamari	*wunyma*	(FF–SD)
arkay	*lamɨn wasan*	(FM, MM–SS, DS)
arkay	*lamɨn kyawi*	(FM, MM–SD, DD)
akay	*lamɨn wasan*	(MF–DS)
akay	*lamɨn kyawi*	(MF–DD)
aki	*anayn wasan*	(FZ–BS)
aki	*yimari*	(FZ–BD)
away	*maypɨnggɨnar (wasan)*	(MB–ZS)
away	*kyawi*	(MB–ZD)
awayma	*wasan*	(MBW–HZS)
awayma	*kyawi*	(MBW–HZD)
pɨsanma	*aykaris*	(SW–HM)
kwandamayn	*masɨnmɨnanma yarɨma*	(HZ–BW)
sakuri	*masɨn*	(ZH–WB)
And two avoidance (taboo) relationships:		
apuk	*masok*	(DH–WF, WM)
kasɨnma	*akimban*	(DHM–SWP, DSM–HMP, FZSW–HMBS)

Both men and women relate to their parents as *anayok asayok* (father and mother), or *anay-asɨm* (father-mothers), or *anunggum-b-awkum* (fathers-women). A couple calls its children *kɨmbi kapanaki* (children of two of us).

In Table 6 I present a sketch of reciprocal kin terms. I put in brackets only the main persons who are designated by these terms. In the following discussion we will look at the behaviour between individuals in particular kin relationships.

Anay *(F) and* asay *(M)*—wasan *(S) and* kyawi *(D)*

Obviously, in societies organized by patrilineal descent such as the Ambonwari, the relationship between father and son is a vital one for group continuity. Even before their initiation, sons in Ambonwari are looked upon as those who will care for their parents in their old age, take over the land and creeks, and continue to use the same hunting and fishing areas as their father did. Although 'the way of the village' (*imɨnggan kay*) implies the primacy of the oldest son, a father watches each of his sons carefully while they are growing up to judge which of them will be the most suitable to be entrusted with the secrets, myths, and stories about the most important spirits. If sons are few in number their father looks after them with special care and they become the centre of family attention. After initiation an unmarried son usually leaves his father's house to stay in the men's house. The main reason for this is said to be that his father wants him to grow up properly. But if nothing else, people say, it would be extremely shameful for a young man's mother and father to hear their son farting in his sleep. This is understandable when we consider that in Ambonwari there is a strict prohibition against women seeing, hearing, or smelling anything concerning male buttocks.

After initiation sons and grown-up daughters are not allowed to eat their parents' leftovers. People say that this would drain the strength from their parents, who would soon become old and die. On the other hand, parents can eat the leftovers of their children. Such observances are found in individual family houses as well as in the men's houses, where young men eat before the old.

After a son marries, his *warɨmbar* (partner) takes his mosquito net and his stool from the men's house and brings them back to the father's house. The young couple make a separate cooking-hearth near the entrance where they also sleep. The parents stay at the back of the house. In this way the son blocks the entrance through which the spirits of his father and mother could leave and never return.

People say that in the past all the sons, after they married, stayed in one 'long house' together with their parents. Persistent disputes changed this custom and most sons began to build their own houses. Unless he is judged by his father to be unworthy, however, the first-born son, *kanapanggɨnar* or *kanapanggɨn wasan* ('head son'), still stays with his parents, and looks after them until their death.

When he marries, his father transfers the name of the lineage to him, though he is still not considered *yamɨn anay*, 'father of the house', but *yamɨn wasan*, 'son of the house'. When he becomes a father himself and thus 'the father of the house', he begins to refer to his younger brothers as *yam-bɨna-ki*, 'children of the house'. The younger brothers are referred to by their father as *mapinjinggɨnar* or *mapinjinggɨn wasan*, 'middle sons' (lit. 'other sons'). After marriage they can move back into their father's house from the men's house, where they slept as bachelors, but only while they build houses of their own. This takes place in the same ward, usually just next to their father's and their brothers' houses. Only the last-born son, *sambukwarinyar* or *sambukwarin wasan*, 'tail son', can stay permanently (even after he marries) with his parents and his oldest brother.[12] Whenever this occurs the youngest brother will take his place at the front of the house, the oldest at the back, while their parents stay protected in the middle. It sometimes happens that parents prefer to stay at the back, leaving their sons to take each side at the front of the house.[13]

A father can transfer the name of the lineage to his youngest son if the first-born does not 'sit' well, is unreliable, or remains unmarried. A father explains his decision loudly so that the ancestral spirits of the house can hear him. If the youngest son takes over the lineage's name, he still remains the 'tail son'. When the eldest son has his own family and his sons get married, the youngest brother can again transfer the lineage name to the eldest son of his eldest brother, or he can give it to his own eldest son.

The one who takes the lineage's name becomes the one who has a priority and obligations in those social institutions which pertain to his position as the 'father of the lineage'. He inherits from his father the role of *sɨmɨrinyar*, relationships with spirits, the role in the men's house, responsibility at death, initiation, and other rituals, and control over names, myths, and stories. His 'house' is called when people want to use his lineage's land for group fishing or hunting. He gives the payment which he gets in the name of his lineage to his brothers. Often he does not even take his own share but waits for his brothers to give him a share later.

Caring about the future of his children, a father divides coconut and betel nut palms and banana plants and piper betel vines between his sons and daughters, and shows them the places on creeks where they can build barricades for fishing. The land and creeks are there to be used by all of them. When there is land which does not actually belong to the lineage but was given for use

[12] Daughters are referred to in the same manner: *kanapanggɨnma* (head daughter), *mapinjinggɨnma* (middle daughter), and *sambukwarinma* (tail daughter). While this may seem irrelevant for the continuation of the lineage, it is not so when considering marriage order. The oldest daughter should marry first.

[13] Frequently an old man makes his own exit at the back of the house so as not to disturb his sons' families when he or his wife go to the toilet at night. Everyone would feel ashamed if the parents saw their sons having sexual intercourse. For this reason people often cough when passing near the sleeping places of others.

in the past by a lineage of some other clan, the father decides which of his sons will manage it. This person has to compensate the original owner on major occasions such as death. Brothers can help a manager with their own contributions of money but often do not, even though they may use the land.

If the oldest son marries while his father's father is still alive, he makes himself a new house. The lineage name is transferred to this new house while his father and paternal grandfather stay in the house without a name. When the father's parents die, father and mother move to the back of their first-born son's house. Only then can the father's youngest brother build his own house.

As I have mentioned above, a father can call his son 'father', and his father, 'son'. For example, when Tony's 5-year-old son (not the first-born) had an attack of malaria, Tony held him in his arms crying over him: 'Oh, my father, please do not die!' This practice may be extended to a man's classificatory sons, especially to those who bear his father's names.

Daughters marry away from the clan (and place) where they were born and are thus, as people say, unable to 'hold the post of the men's house'. If a man has daughters only he has to adopt a boy, either from his siblings or from elsewhere (see below). If a daughter does not marry she stays with her parents and brothers. With regard to marriage, people say that daughters are more unpredictable and 'flighty' than sons, and may suddenly leave the house. Because of this their father 'shows' them their fruit trees, banana plants, and piper betel vines even before he 'shows' them to his sons.[14]

People say that a mother never ceases to take care of her children, even when they are married. Mothers protect their daughters from their father in domestic disputes but can nevertheless also fight with them. When daughters get married and leave the house and the ward, their mother is the one who cries piteously. The link between a mother and her daughters remains strong throughout their lives.

Brothers—sisters and kwandamayn *(HZ)*—masɨnmɨnanma yarma *(BW)*

In every family younger children take priority among siblings and are carried around and nurtured by the older ones. It seems to be almost a rule that when two siblings fight a third will hit the one who started the fight without considering whose fault it was.

People say that when there are only two brothers in the family there is never

[14] When a young unmarried child dies, the father cuts down in his grief all the palms he had already given to the deceased child. On such an occasion the heartbroken father may also cut down the posts of the house or even burn down the whole house, as happened during my stay. In this way, by destroying the objects with which a dead child was in intimate contact, and which were thus extensions of her or his body, a man destroys the child's body in its totality. When mourners hold the objects of a deceased person, they hold them not only because they represent a dead person but because they are part of the person's body, her or his *kay*. Holding her or his plate, basket, cup, or paddle is equivalent to holding a hand.

any problem. They usually live in the same house and support each other. Of course this is not always the case. One of my friends will never forgive his older brother because he married a woman from another village, a woman whom he thinks is useless. Gossip, quarrels, and fights between two brothers are often instigated by their own wives. When two grown brothers quarrel it is hard for the third one to decide whose side he should take. If brothers fight frequently during their childhood their father distributes palm trees and other property between them before they become adults. On the other hand, there are larger groups of brothers who cooperate, help each other, and settle their arguments quickly.

If a man gets involved in a fight with someone from another group, his brothers will help him regardless of whether he was right or wrong. In everyday life it is not consanguineal but classificatory brothers who are good friends, those who usually come first to a brother's aid and leap into a fight without thought. If they were in the bush when the fight broke out, they start it again immediately they return to the village. They shake in anger, screaming about who asked for a fight, who is the *bladi fakin bastet* ('bloody fucking bastard' in Tok Pisin). While the villagers gather around to see what is going on, the man comes up to the house of those who fought with his brother shouting: *amangok yay anjangara*, 'give me the blood too'.

When classificatory brothers and sisters refer to one another they can use terms relating not to their own age but to which of their parents is younger or older. Thus, for example, if a man's mother's sister (or father's brother) is younger than his mother (or father), a man can call his mother's sister's child (or father's brother's child) *wasapuk* ('little brother') or *wasakɨnday* ('little sister'), regardless of whether they are actually older than him. In a longer version he can address them with *wasapukɨn kupan*, 'the big brother of the little brother' or *wasakɨndayn kupanma*, 'the big sister of the little sister', while they respond with *kupan wasapuk*, 'the little brother of the big brother' or *kupanman wasakɨnday*, 'the little sister of the big sister'. This is interesting when we look at namesakes living in different villages. Father's and mother's namesakes are simply called 'father' and 'mother'. If two women or two men are namesakes they may use the terms *wunyma* (SD) or *wunyar* (SS) for the younger one and *kupanma* (Ze, FM) or *wasamari* (FF) for the older one. But this is the case only when the older person is from the clan (or lineage) from which the name originates. If the younger is from the clan (or lineage) of the name's origin she or he can call the other *wasakɨndayn kupanma* or *wasapukɨn kupan*, 'the big sister/brother of the little', while the older from some other ward or place reciprocates with *kupanman wasakɨnday* or *kupan wasapuk*, 'the little sister/brother of the big'. This becomes significant when one looks at the relationships between different totemic clans throughout the Sepik, and it sheds light on which is the original clan (or lineage).

A brother (or classificatory father) arranges a woman's marriage. Her 'brothers' and 'fathers' provide a kind of dowry which she takes with her: a clay hearth, a clay plate, a paddle, a pot, a mosquito net, a lamp. When a brother gets married he should take care of his sister and their children to the end of his life. His wife ideally becomes her friend. When women go fishing or gathering the most usual combination of partners is *kwandamayn* (HZ) and *masɨn-mɨnanma yarma* (BW). A man's sister and his wife do not ask each other for anything but use their husband and brother as the mediator.

The relationship between sisters is usually very close. When *kupanma*, 'older sister', gets married, her *wasakɨnday*, 'younger sister', tries to stay with her. After the marriage the younger sister carries her older sister's dowry to the new house. She can even stay with her in the new house and may sleep with her under the same mosquito net. People say that the newly married girl is shy to be alone with her husband and needs someone for support. After a month or so, when the husband and wife have come to know one another better, the younger sister moves back to her parents' house. During all this time the younger sister serves as a messenger between the two houses, carrying not only news but also food, betel nuts, kerosene, salt, newspaper for rolling cigarettes, batteries, and other goods of daily importance. When a girl falls pregnant her younger sister helps her with her work. If the relationship between two sisters is still as close when the younger sister gets married, the latter can call her older sister's husband *kupanma asɨkɨn*, 'the man who married the older sister', while the older sister can call the husband of her younger sister *wasakɨnday asɨkɨn*, 'the man who married the younger sister'. These terms of address are more intimate than the usual *sakuri*, 'sister's husband'.

Mayn *(H)*—yarma *(W) and co-wives*

Among the advice that a woman receives just before getting married is that her husband will not tell her if he is hungry. She is supposed to be the one who senses his needs. When in the house, she should be around the clay hearth and not gossiping near the entrance. This, as well as all the other normative behavioural patterns described in this section, depends on the personal characteristics of the individual. Those women who are most effective in their work and at the same time very communicative are most admired, even if they do sit at the house entrance and laugh and talk a lot.

A wife takes care of her husband's father, whom she calls *wasan* (son) while he calls her *asay* (mother, son's wife). If she argues with some other woman he will eagerly support her. If her husband argues with her or beats her his father takes her side. People explain that in such a relationship the man's father acts like an eldest son who helps his mother. For the same reason a woman rarely beats her own children in the presence of her husband's father, who acts as her children's elder brother. If her son does something wrong his father's father can scold him.

I once asked a man who had had two wives how people in such situations behave. His reply illustrates the importance of care in the relationship between spouses and co-wives. The events described in his story might easily have taken a different course. They are presented from a male point of view; nevertheless, this sequence shows the expected behaviour under comparable circumstances.

If a man wants to marry a second woman he has to be tactful and careful. During his courtship of the second wife, he takes pains to be particularly kind and generous towards the first. He allows her to use all his belongings, even those personal things in his woven bag which are otherwise considered his individual property. In this way the second wife-to-be sees that the first wife is happy and expects that she will be nice to her as well. When the new relationship is publicly disclosed, the village councillor calls a meeting at which the man's wife finds out about the relationship for the first time. But her husband does not admit that he wants to marry this woman. There is no mention of marriage. After paying compensation for adultery (the husband, the woman involved, or both pay this to the man's wife) the case appears to be over. But the second wife-to-be continues to give presents (mainly fish that she has caught) to the man, while his wife is alert and wary. Even if she becomes angry and screams at him, or if her brothers joke at his expense, the man remains quiet. Thus his wife sees that her husband is worrying about something.

The wife begins to push her husband around, urging him to do work (get firewood, fetch water, go to the garden, cut sago palm, and so on). Finally the husband resists her, saying no. His wife gets angry, telling him that he thinks only about the other woman. The man beats her. Her brothers ask him if the other woman has given him something again. The man admits it and shows them a rubber bracelet (the relationship between two lovers can begin only with the exchange of small gifts as signs of 'feeling Heart'; if no gifts have been exchanged no one will discuss the possibility of marriage). The village court is again convened. The councillor asks the brothers of the wife-to-be if they are prepared to give her as a second wife to this man. If they agree the first wife is asked. She has to give her consent, too, otherwise there will be no marriage.

Before she agrees she wants to know more about her husband's thoughts and feelings as well as more about the woman. She tries to persuade her husband to tell her more. He talks about general things, but offers no details. She urges him to sleep with her and then asks him: 'Do you still think about her? I think that you saw her vulva and because of this you still think about her.' He replies that men marry two or three women, and women do not marry two or three men. Then she realizes that her husband is unhappy and that he really wants to marry the other woman. She decides to help him. When the women next go fishing she gives the other woman a basket, a net, or a child to carry. Thus the relationship between the two women starts. Other women will say: 'Yes, you two are sisters now.' The second wife gives all her fish catch to her prospective co-wife.

The marriage is arranged and everyone knows that the second wife will be under the control of the first one. But the man now has to relate equally to both. If he brawls with one she will pull the other inside so both can be beaten. The man cannot leave the food prepared by one but has to eat what each has prepared. When they go together to make sago flour a man has to carry back two baskets, one for each wife. A man cannot separate one wife from another; he must wait until one goes to see her parents, for example, before he can sleep with the other.

The wives behave as sisters. One wife calls the brothers and fathers of the other brothers and fathers, and so on. The brothers of one wife (her children's mother's brothers) give priority to the children of the other. Children thus get two maternal lineages.

Wasamari *(FF)*—wunyar *(SS) and* wunyma *(SD)*

In a society with patrilineal descent and a naming system as described in the previous chapter it is understandable that the relationship between father's father and son's son is one of some importance. On the other hand, *wunyma* (SD) leaves her *wasamari* (FF) when she marries and takes up her new obligations and new relationships. These prevent her from having closer relations with her paternal grandfather.

The term for a person's father's father, *wasamari*, is derived from the word stem *wasa-* (little), as are other terms for brothers and sisters: *wasa-puk* (little brother), *wasa-nar* (brother), *wasa-kɨnday* (little sister), *wasa-narma* (sister), or, for a son, *wasa-n*. *Wasa-narin* means 'slowly' or 'easy, easy', 'little by little'. But *wasamari* is alternatively called *kupan*, 'big brother'. His main obligation is to bring his *wunyar* (SS) or his *wasapuk* (little brother) to the men's house and take care of him during the initiation period. As in all other similar relationships the real paternal grandfather usually steps aside and allows the primary tasks in rituals to be performed by his classificatory brothers and namesakes (and their brothers) from some other clan or even from some other village. *Warɨmbar* (companion, partner) can also assume the role of protector of his 'younger brother' during initiation.

Arkay *(FM, MM)*—lamɨn wasan *(SS, DS) and* lamɨn kyawi *(SD, DD)*

The relationship between grandmothers and their grandchildren is extremely warm and caring. One can easily witness the pleasure of grandmothers when they see that their grandchildren think about them. There is always some meat for a grandchild hidden in its grandmother's house.

Aki *(FZ)*—anayn wasan *(BS)*

Aki is a 'female father'. As a man can relate to his father as 'son', so father's sister becomes *kyawi* (daughter) or *akin kyawi* (daughter as father's sister).

Reciprocally she can call him *anay* (father). Everyone expects that *anayn wasan*, 'father's son' (i.e. brother's son), will remember his father's sister whenever he kills large game or has a surplus of fish. To put it concisely, the man more or less follows his father who is *away* (mother's brother) of his *aki*'s children. When a man gives something (food, clothes, other goods) to his father's sister, she pays him back with money.

Pɨsanma *(SW)*—aykaris *(HM)*

In this relationship, where both women are married into the same lineage, they are expected to cooperate. It often happens that the husband's mother lays down her own rules in the house. As I have already mentioned, a new wife is supposed to take care of her husband's parents.

Relationships of Alliance

Of the relationships produced by marriage (in ego's own and ascending generations) the most important are: *akay* (MF)—*lamɨn wasan* (DS), *away* (MB)—*maypɨnggɨnar* (ZS), *sakuri* (ZH)—*masɨn* (WB). A man can always ask his mother's or his wife's lineage for things: a paddle, a canoe, a clay hearth or plate, a pig, sago grubs, clothes, a lamp, and so forth. But when he receives something from them he has to repay with cash (shells in the past). Among all these relatives the man's *masok* (WF) is asked the least often because, as people say, he gives things according to his own wishes and expects to be paid.

Once married, a man would only reluctantly move into the house of his younger brother and would never ask him for help. Nor does he eat in his younger brother's house. If his own house has been demolished and he has not yet built a new one, or if there is persistent sickness in his household, he would rather move to the house of his wife's father or brother. During the time that he lives uxorilocally he regularly visits his own clan's ward and men's house. After anything between a month or a year, when a man leaves the house of his wife's lineage he has to 'compensate' them for his long stay there. As usually happens, he complains of not having any money and leaves his personal wooden stool there. When he comes into possession of a small amount of money, perhaps after several months, he returns to his wife's house and pays the sum to the father of the house. Then he can take his stool away.

Staying with the wife's lineage is quite a common arrangement. The first two months after my arrival in Ambonwari I lived in the house of Tony Simiwaria Andi. Tony's sister's husband's house had fallen apart and his family had moved into Tony's house. The two families had separate households, that is, cooking hearths, within the one house. They shared their food, and betel nut whenever they had some. During periods of plenty, when they caught lots of fish and small animals, there was no problem. The complaints began during times of scarcity.

Any large request, such as for lamps or mosquito nets, has to be announced in advance so that mother's or wife's lineages can collect money, plan a trip, and buy the required goods in Angoram or Wewak. When they return they present the gift with some food to those who asked for it (sister's son, sister's husband, daughter's husband). The recipient is expected to reciprocate the gift with money. As he often does not, the others complain about how much money they have spent, how far they had to travel, and how they have not been properly repaid for all their troubles. Such gifts carry extra value and are never given to anyone else, even if someone asks for it. People do not ask their mother's or wife's lineages for money. They borrow it and usually 'forget' to return it. If a person receives too many gifts from his affines and does not have enough money to pay for them, he can ask his 'brothers' to help him. Later he will distribute the goods between them. He may also ask his mother's and wife's lineages to stop giving him things as he has no money to pay them. This nowadays happens between a lineage in which someone has a permanent income and another lineage in which nobody works for money.

The most important figure in one's maternal lineage is the 'male mother' *away*, mother's brother. He is the one who always thinks about his *mayp*ɨ*ngg*ɨ*nar*, sister's son.[15] During initiation the mother's brother acts as an 'enemy', displaying anger because a young boy is leaving his mother. He may even hit him while at the same time bringing him the best food. The boy eats this food together with his *wasamari* (FF) who, being his 'elder brother', protects him. Later in life his *away* will fight to protect him and, people say, even die for him. If a man's mother's brother comes to the house his wife is expected to cook him a meal. If she shows reluctance her husband can beat her. *Away* can then feel guilty, thinking that the fight started because of him. He may later return to the house with a gift such as a clay hearth, a pot, or something of similar size. Afterwards the man will have to repay his mother's brother with money. In all important rituals, such as those of death, *away* and *akay* present him with gifts. They can show him a creek, a patch of bush, a coconut, or betel nut palm which he can use. He, of course, has to pay them back with money on every occasion. His mother's lineage is also important during the hair-cutting ceremony which is a culmination of a long period of mourning. Mother's brother's wife, *awayma*, in many ways acts as a real mother towards her husband's sister's children. People say that if, for instance, *awayma* forgets to give something to her HZS or HZD her husband can beat her. If he does, her HZS or HZD will later go to her house and pay, as she was beaten because of them. People say that *akay* is the only one among all maternal relatives from whom a person cannot get a meal without paying for it.

There are many celebrations and rituals where one's maternal lineage is very important. One such event is the *kurang* ceremony, in which all the members of

[15] The term *mayp*ɨ*ngg*ɨ*nar* is a masculine form of *mayp*ɨ*ngg*ɨ*nma*, a small mouse with a short snout which lives in a bush.

a maternal clan dress as women and all the members of one's own clan as men. The dance lasts the whole night. These celebrations are organized in recognition of someone's notable achievement, such as the successful hunt of his first pig, the building of his first house or of a large canoe, his return from a long trip, and in recent years a child's completion of six years of schooling. The last can be performed for both men and women. Half a century ago the *kurang* ceremony was also performed after the successful killing of an enemy. The ceremony is known as *naven* among the down-river Iatmul (Bateson 1958).

The relationship between *sakuri* (sister's husband) and *masɨn* (wife's brother) is the source of joking relationship between two groups of people from different clans. Metaphors referring to genitalia are used in joking relations, mostly between a man and his sister's husband's classificatory brothers in adult male company and between young boys when alone, but also between children who are imitating adults, and sometimes when young boys tease girls. Women also use the same 'jokes', though only on special occasions and when no man is present.

Men who marry women of your lineage will say to you 'you are my wife' (*mi yarɨma amananma*), or in the plural, 'you are women, you are wives' (*ipa awkum, ipa yarɨmɨndi*). Jokes refer to all women's work: 'woman, you fry sago' (*awkunma sipi mi kasin*) or 'go and stir sago pudding' (*karis mi wusɨra*). You will reply that you are not a woman, that they are women, 'you not, you are my wife' (*mi kaya mi yarɨma amananma*). While still 'polite' within the village, conversation in the bush (while making a new canoe, for example) is more direct and has more sexual overtones. Men make gestures of copulation, touching and 'kissing', saying who has 'male genitals' (*panyangusɨm*), 'penis'-polite (*usɨng*), 'penis' (*aruwa*), 'testicles' (*panyang*), and who 'vulva' (*yawi*). Such conversations are not considered to be 'bad talk'. Your classificatory sister's husband's brothers will say 'Come, I will copulate with you' or less directly: 'Do you women want me to give you betel nut so that you will dream about me?' Such conversations are accompanied by loud laughter.[16] Similar talk may occur among the women when they are alone, especially when half of them are dressed as men and half as women in the *kurang* ceremony. Ambonwari women do not have their own dancing performances as do neighbouring Konmei and Iatmul (Bateson 1958: 148–51). Nevertheless, they have plenty of time on their own while decorating themselves for a public performance.

Taboo Relationships

People can generally address each other by personal names, though there are some relationships in which not only the use of personal names but also any

[16] Many metaphors for genitalia may be used in such a conversation: 'a tree' (*yuwan*), 'a blunt adze for pulverizing sago' (*pandɨpɨnam*), 'a root of a Pandanus' (*ingrayn pamɨng*) are used for penis. There are more metaphors for a female vulva because men use these expressions more than women do, wishing,

kind of physical contact is forbidden. These relations are *pan kupambi*, literally 'very ancestral', the term implying secrecy and prohibition. Such institutionalized relationships are between *kasɨnma* (DHM, DSW, FZSW) and *akimban* (SWP, HMP, HMBS) and between *apuk* (DH) and *masok* (WP). Like *warɨmbar* and *sɨmɨrinyar* discussed in the previous chapter, taboo relationships are strongest between those linked by an 'ancestral path' of marriage. If the actual marriage is not on this 'path', affines sometimes address each other by personal names and even visit one another's houses. What I describe below pertains mainly to those who are in taboo relationships based on people's understanding of their ancestral past, regardless of the actual marriages in the present.

Prohibitions on names are extended to everything with names which resemble those connected with the person with whom someone is in a taboo relationship. Thus, for example, *kasɨnma* does not speak the names of the forest, land, or creeks of her *akimban* but refers to them by the general term *akimbaym* (plural of *akimban*). She does not utter the names of his men's house, lineage, and clan, his canoe, his spirits, or his dogs and she does not even speak his Christian name, which otherwise carries no importance.[17] On the other hand, a man can more freely pronounce the names of his *kasɨnma*'s land or clan, for instance, as the names in any case belong to her or to her husband's patriline.

Ambonwari, as they say themselves, love to 'turn around' words and play with them (which becomes quite a problem for an ethnographer). For example, my friend Elias Wapun has, among others, the name Moon. All those who are not allowed to say his name call him 'the one who sleeps above' (*maun parar*). On the other hand, Elias cannot, for example, say: *Mi ingana kɨndɨngaykɨr* '(I suspect) you will trick me', as his *masok*'s name is *Kɨndɨngay*. Elias also has to be careful not to say: *Mɨn makuraywariar*, 'he went around and came back', as his *masok*'s other name is *Kuraywaria*. One can imagine, on the village level, how many avoidances of words are involved in everyday communication. To take another example: a small broom for cleaning a frying-pan is called *sapasa*. As Andrew's wife's name is *Sapasay*, everyone in a taboo relationship with her calls a broom *sapikrang*, a word which is otherwise hardly used. In the same

in their joking relationship, symbolically to feminize those men whose sisters they marry, and, regardless of sex, liken their whole lineage (and clan) to the feminine aspects of life in which both the men and the women act in a womanlike way. Thus, for a vulva they use the following expressions: 'a turtle's tail' (*pandɨm*), 'a mountain' (*yimbuk*), a fish called 'big mouth' in Tok Pisin (*samɨndɨ*), 'a little fish basket' (*injing*), 'back of a canoe' (*kayapi*, thus *kayapi yamban karingan* means 'I will carve the back of a canoe'), a pool in a sago bush (*kayam*), a sort of Malay apple fruit (*wimɨnɨng*), a small pool (*usɨm*), or the shells *wimɨnɨng*, *king*, *sandam*, and *pari*, the first 3 having holes in the middle.

[17] As an example we can see how *kasɨnma* refers to the men's house of her *akimban*. If he is from the Crocodile-1 clan she will call it *Wakiarandi* ('garden-land'). If he is from the Bird of Paradise clan she calls it *Kupaiman* ('big men's house'); if he is from the Wallaby clan she calls it *Awyanggmay* ('jump into canoe'—female suffix). She calls the men's house of the Pig-1 clan *Sakurismay* ('to make a noise as pigs do'—female suffix) and the men's house of the Cassowary clan *Pambangguri iman* ('the men's house from the back'). She can refer to the men's house of the Eagle-1 clan as *Akupin-soa* ('at the back—?') and to the men's house of the Pig-2 and Eagle-2 clans as *Kupa Suwanmas* ('the big Sepik men').

way, people from the Cassowary clan rarely use the ordinary term for a broom, *sɨkay*, as this is also a name of a young boy in a taboo relationship. They call it *krasɨngndi*.

The most restrictive taboo is between *kasɨnma* and *akimban*. The former is not allowed to enter the house of her *akimban* and even keeps away from the ward of his clan. She cannot hold his belongings such as a paddle, lime-container, or shirt. A man can go to the house of his *kasɨnma* but does not approach her cooking area nor touch her personal belongings. The strongest taboo is against touching her paddle and mosquito net, which are, as part of her *kay*, in the closest contact with her most of the time. If they must give each other anything, they would place it on the ground between them.[18]

If anyone sees *kasɨnma* and *akimban* touching each other, the whole village soon knows. She will have to decorate a sago stalk with money and, in the company of other *kasɨnma*, bring it to the main men's house, make a circle around the coconut palm near the entrance, and then proceed to her *akimban*'s house. As she has never entered his house before, he will wash her feet with water (the only time they are permitted to touch). He will present her with a basket full of betel nut, a new paddle, clay hearth, bowls, and plates, and even a spear or bow and arrows for her husband. He may also give her purchased goods such as a mat, bedsheet, clothes, pot, and so forth. He will then offer her a plate of sago pudding. She and her husband will estimate the value of the goods and repay him with money.

Such a taboo relationship is extended to the lineage and clan, and *kasɨnma* and *akimban* have to be careful not to enter the house of the same name in some other village, even in a different language group such as Yimas, Imanmeri, or Iatmul. During my stay a young married woman was unaware that the house she entered in Kunggriambun village actually belonged to the Sapɨndam lineage, the one to which her *akimban* belonged. She was suddenly presented with a large number of gifts and subsequently her husband had to collect money and repay them.[19]

Similar though less powerful restrictions apply to the relationship between *apuk* and *masok*. They are not allowed to speak each other's names, the names of their dogs, spirits, or land. A man will refer to the bush of his *masok* simply as *masok*. *Masok*, however, is allowed to speak the name of his *apuk*'s bush. Similar restrictions apply to the female *apukyawi*.

Because of the remembered relationships, present relationships, and those

[18] If one looks at this relationship from the perspective of identity it will be obvious that any close contact between DHM and SWF, for instance, would mean that the two had embodied each other, as husband and wife do. This would indicate that their children cannot be married, as they would be like brother and sister. This is also why a father advises his son not to marry a woman with whose mother he once had sexual intercourse.

[19] Despite all these restrictions, Ambonwari songs sung during a 'house dance' recapitulate events in which *kasɨnma* and *akimban* not only touched each other but had sex together. When they were discovered their families were ashamed, while many were amused.

anticipated in the future, and because of marriages deviating from the preferred paths or between people who usually do not marry, the terms of address and reference can become a sort of combination of several terms. For example, take a young woman from the same lineage as ego's mother. She marries ego's classificatory *lamɨn wasan* (daughter's son). Before marriage she called him *wasan* (son) but after marriage he became her *akimban*. Thus she calls him *wasan akimban* and he calls her *asay kasɨnma*, using 'mother' before the restrictive term. If the marriage had followed the 'ancestral path', as the Ambonwari would point out, there would be no need for such dual terminology.

Another example is from Kumbranggawi lineage of the Bird of Paradise clan. If Tony Simiwariya had followed the preferred marriage path he would have wed Godfried Yangus' sister from the Eagle-1 clan and Godfried would then have become mother's brother of Tony's children. Instead, Godfried married Tony's sister. Thus, Tony's children call Godfried *away apuk* ('mother's brother-father's sister's husband'). In terms of his marriage Godfried is *apuk* (FZH), but in view of the traditional marriage between two lineages he is *away* (MB). For the same reason Godfried's children call Tony *away apuk*. After the actual marriage Tony became their *away* (MB), while in regard to their traditional relationship between two lineages he was and has remained *apuk* (FZH).

Adoption

Adoption within the clan as well as between different clans is common, one might almost say a day-to-day characteristic of life in Ambonwari. It is an important device for maintaining the continuity of descent and other relationships and for later marital arrangements between different lineages. If a lineage has no boy to continue its name its elders may look around to see who has a suitable child, which woman is pregnant, or even which one is expected to become pregnant. Of course, people do not like to give their children away, especially to those whose ward is relatively distant and with whom they do not communicate much anyway. Thus consanguineal and classificatory brothers and sisters are normally the first to help their brother. In the same clan it happens all the time, even when there is no need for it, that a man bestows a name on his brother's child instead of the father. Children are also adopted from one clan to another. A girl can be adopted to solidify good relations between two groups, for the purpose of exchange, or simply for the purpose of arranging her marriage and sharing her bride-wealth. The Ambonwari do not adopt those children whom they call 'brothers' or 'sisters' but only those whom they already call 'sons' or 'fathers', that is, those born in the next or the previous generation.

If the above explanation seems clear, the reality is more complicated. A baby or child given in adoption immediately creates the obligation of a future return. It represents a debt that has to be returned in the form of a child of the same sex, not in the same or even the next generation, but in the second one. Such a 'gift' or 'debt' is called *kunaypa*, 'a splinter of sago leaf's stem'. When the splinter is nicked, it sticks out of the stem. This notion is metaphorically used to mark a debt. When a 'gift' is returned in the form of some other child, people say that they discarded *kunaypa*. There is often no other reason for the adoption of a child than the desire of people for their 'gift' to be reciprocated. *Kunaypa* thus links the past to the future, 'before' to 'after'. It refers to the common time shared between two groups, its beginning when *kunaypa* 'sticks out' and its end when it is removed. This tempo between the gift of a child and its return two generations later takes its meaning from anticipation of the future. While the other group has a sense of obligation, it may be reluctant to return a child if its members have only a few children. This results in a delay between the initial gift and its return. The actual return of a child is acknowledged and expected, and it is in collective terms recognized as proper, that is, as the way things should be done. Because of the hope, uncertainty, and doubt of the child-receiving party, because of the possible delay, the 'intragenerational exchange' receives its characteristic tempo and rhythm.

Adoption procedure is quite simple. A man who has no daughters, for example, goes to his sister and asks her if she can give him a daughter for adoption. If she and her husband agree, when she gives birth her brother comes to her house and bestows the personal name from his lineage on the baby, *yamɨn wi yan kandɨ mɨnaki*. When a girl is between 4 and 6 years old she moves to the house of her adopted parents, in this case her MB and MBW. She refers to these new parents as mother and father, while calling her true parents by their names.[20] She is fully aware that she has been given away, but soon comes to regard those who gave her her personal name and take care of her as real parents. They feed her, caress her, and usually treat her with as much attention, if not more, than any child of their own. If a child is adopted by another clan it becomes a member of its new clan immediately after receiving its personal name.

Adoption at birth is another, more radical way of obtaining a child. It is also kept secret between those involved. The adoptive mother appears at the back of or underneath the house at the time of birth and takes the baby immediately after its birth. She may even be present at the birth, and may cut the umbilical cord and bury it herself. She takes the baby to her house, where she washes it and covers it with clay. Both real and adoptive mother then observe all the prohibitions and behaviour connected with childbirth. They stay isolated inside their houses for five (in the case of girls) or six (in the case of boys) days. After a period of seclusion the real mother washes in a creek and continues her

[20] Ambonwari may indicate their classificatory parents by terms such as *sandukwan anay* ('sit-watch father') and *sandukwan asay* ('sit-watch mother'). Nevertheless, they do not use these terms for parents who adopted them. To give some idea of how often adoptions occur, in 1991 47 living people in Ambonwari had been adopted, 24 of them female.

life, trying to forget the loss of her baby. This may be quite hard for her if she had been told to do this by her husband. She may lament: *amanan wasan (amananma kyawi) mɨn yan yar*, 'she took my son (daughter)', *ama kambra ama san*, 'I stay/sit nothing/empty'. But she may receive ten kina to compensate her for the 'hard work' of birth and her grief.

If an adoptive mother already has a small baby, but has adopted another because of its sex or just as a return of a gift from the past, she can immediately breastfeed the adopted child. If not, as usually happens, she has to prepare appropriate food: squeezed sugar cane, coconut milk, sago pudding, cleaned, boneless fish, sago grubs. Adoptive parents may tell the child that he or she was adopted, but often they do not. It is prohibited to talk about this matter in the village. If a child hears about it from someone else, its adoptive parents could demand from this person as much as 1000 kina in compensation. This is hypothetical, as there has never been such a court hearing in Ambonwari.

Conclusion

In this chapter I have surveyed formal structural and behavioural aspects of kinship and marriage. Although I presented a largely normative picture, stressing formal aspects of kin terminology, 'ancestrally' sanctioned institutions, prescribed kinship behaviour, and so on, I do not wish to obscure the interrelationship between 'way' (*kay*) and Heart (*wambung*, understanding) analysed in earlier chapters. For the most part 'the way of the village' (*imɨnggan kay*) and those esteemed institutions brought under the term 'the way of the ancestors' (*kupambɨn kay*) 'go without saying', as Bourdieu might put it. But Heart is an aspect of *kay* in its most fundamental sense, and when circumstances dictate new departures are undertaken. On the other hand, it would be wrong to view *kupambɨn kay* as lacking real meaning for the Ambonwari, as an ideological cloak under which the real business of social life is hidden. On the contrary, *kupambɨn kay* is the locus of those values and orientations that give life its shape, and to the extent that people through *wambung* can objectify and set aside the dictates of *kupambɨn kay* in particular circumstances, those same values and aims rationalize their departures from it.

So, for example, a 'path' (*konggong*, 'marriage') as a dimension of *imɨnggan kay*, even in its 'ancestral' form, *kupambɨn konggong*, is always comprehended through understanding (Heart) and thus lived anew; conversely, *kay* and the life-world it embodies manifest themselves in new marriage relations just as much as the most perfect reproduction of 'ancestral path'.

In this connection it is interesting to consider certain issues in the literature on Crow/Omaha terminologies. Lévi-Strauss (1968: 74, 301) proposed that the Hopi kinship system incorporated three different models of time: 'empty' time, stable and reversible; progressive, non-reversible time; and undulating, cyclical, reversible time. Barnes disagrees:

Lévi-Strauss has not shown that there really is any question of different kinds of time; nor has he given us any good reason to suppose that there is more than one attitude towards time implied throughout the terminology. In any case, the use of the same word for successive positions in a descent line indicates neither an empty nor a frozen time . . . (1976: 392)

Barnes denies that Crow/Omaha kin terminologies 'abolish' time or ignore the succession of generations; on the contrary, he maintains that they emphasize it (see also Harrison 1984: 398). Williamson, in a survey of Omaha terminologies in the Sepik, suggests that they indicate people's concern for their children's future marriage arrangements (1980: 542). The Ambonwari ethnography suggests that both views are correct as far as they go, but that each is strong where the other is weak. In Ambonwari kin terminology, the emphasis upon 'ancestral paths' and the putatively eternal nature of 'exchange partner' and 'ritual inaugurator' links between descent groups all suggest both the anticipation of the future and conformity with a perduring past. The duality is nowhere more clear than in the perpetual reproduction of alternate generational identities, which seems both to negate time by collapsing the past and the future and yet, through the intervening, adjacent generation, to separate and guarantee both. The duality is itself an illusion, produced only on the condition that we forget about *kay* which as habit, manner, custom provides the context within which the distinction between sameness and difference can have meaning.

I would argue in a similar manner to Annette Weiner (despite the fact that she dealt with a matrilineal society) that the model for differentiation between male and female practices in Ambonwari is the relationship between brother and sister (Weiner 1977; 1982: 59–62). Ambonwari people say that the problems of brothers become the problems of their sisters and vice versa: 'My skin is hers, her skin is mine', a saying which applies only to brothers and sisters. No wonder that (after siblings of the opposite sex) the most important figures in a person's life are mother's brother and father's sister. No wonder that the most elaborate joking relationship occurs between a man and his sister's husband's lineage. The roles of a sister and a brother as 'female/male symmetrical and complementary oppositions constitute elements basic to a conceptualization of the social order' (Weiner 1977: 11). If we see the essence of male and female gender dichotomy in the relationship between brother and sister, then we can see that through exchange between the two a brother helps his sister (by giving her food, clothes, material objects, and so on) to achieve a reputation as a nurturing person who makes people around her obliged to her. This contributes also to her husband's prosperity and fame. She, on the other hand, helps her brother (by paying him with money or shells) to gain a reputation (along with his wife) as a wealthy person who shares his wealth with people and makes them obliged. If mother's brother has no family of his own, the feminine aspect of his *kay* will take over. He will think of and give to his sisters' children. To balance his 'femininity' his 'brothers' (or someone else) may give him a child to adopt, though he can still remain single. His new relationship to this child will appropriate him as a man, as a 'father', and will bind him to his own clan. If a woman remains single she stays with her clan. If she has children outside marriage by an unknown father they will become members of her clan—as if she were a man. In this instance, she does not bind together two separate clans, which is normally characteristic of women. Her *kay* is that of a man, regardless of being a woman. While the fathers of her husband-to-be call her 'mother', she does not act in such a manner.

Unmarried and childless women prevent their brothers from becoming mother's brothers, 'male mothers'. They cannot take up their 'female' roles which are the sign of expanded social relationships. Unmarried and childless men prevent their sisters from becoming father's sisters, 'female fathers', and from taking up their 'male' roles. An Ambonwari person, to be complete, has to be both 'father' and 'mother', male and female. To incorporate gendered characteristics of Ambonwari life-world in one and the same person is to become a person in all his or her social complexity. This is possible only if the central concept of human existence is *kay* (being) without sexual difference and without physical bodies. As *kay* is not a physical body, the sexualization of it is impossible. In Ambonwari cosmology, male and female *kay* are social categories attached to differences on the skin, but not, however, the consequences of these differences.

A complete Ambonwari person merges both male and female *kay* and acts accordingly, depending on the context of situation and on his or her relation to those involved. One may pay gifts with money as fathers do, or one may give food and useful things as mothers do. The identities of individuals and groups are continually formed and reformed through the processes of giving, receiving, sharing; through the processes of obligation and expectation; through greed, conflict, jealousy, resentment, happiness, and hope. One might say that male and female *kay* are the result of, and are expressed through, multiple relations manifested in Ambonwari kinship. The dimensions of Ambonwari cosmology as well as the complexity of social life, however, cannot be materialized simply through the relationship between brother and sister (the central aim is to unite them in illusionary 'oneness'). A husband has to replace her brother and a wife his sister. By the exchange of his sister for a wife, he materializes the perpetuation of his own lineage and clan as well as those of his sister's husband. The bond between their lineages becomes strong as both men, through their children, secure their progeny (future) as well as their ancestry (past). The 'same' brother and sister then 'reappear' in his paternal grandchildren. Tempo, rhythm, and time are inseparable from these processes and feelings, and from the continuous identification and differentiation of individuals and groups.

It remains to explain the ethnographic significance of the final concept of the set introduced in Chapter 2, namely *mariawk* (speech, story, myth). For as Ricoeur says: 'time becomes human to the extent that it is articulated through a narrative mode, and narrative attains its full meaning when it becomes a condition of temporal existence' (1984: 52). It is *kupambɨn mariawk*, 'mythology', which provides the narrative setting within which Ambonwari institutions and identity exist and to which I now turn.

5

Myth and Identity

Mariawk (talk, discourse, story) refers to all forms of discourse including myth, everyday speech, and those important occasions such as public moots, discussions in a men's house, and disputes when oratory comes to the fore. Every event is told and retold several times, from the different perspectives of individuals and groups. Such reconstructions should be considered as extensions of the events themselves, as their continuation in a verbal medium—as parts, that is, of collective and personal *kay*. Continuous communication, as well as people's relation to their past and future through the recollection and narration of events, constitutes the narrative time of the village. Such public time 'is not the anonymous time of ordinary representation but the time of interaction. In this sense, narrative time is, from the outset, time of being-with-others' (Ricoeur 1980: 188). *Mariawk* might be seen as constituting identity and temporality on two different axes: vertical, relating people who lived in the past with those who live in the present; and horizontal, between consociates.

Myth is the most obvious example of what *mariawk* (speech) represents in the constitution of identity in Ambonwari. *Kupambɨn mariawk* (ancestral story, myth) directly narrates the mythic-historical past which for the Ambonwari is the core of identity. More broadly, since *mariawk* is implicated in understanding and thought it is intimately connected with *kay* and *konggong*. Of course, myth itself is influenced and constructed by these broader processes. In other words, myth is 'reconstructed through lived experience which mediates culture; and culture is reconstructed through lived experience which mediates myth' (Young 1983: 35).[1] For the Ambonwari, myth is considered to be inherited and to a large extent a sedimented aspect of their collectivity in terms of past knowledge and knowledge of the past. But in practice, when retold, reinterpreted, and reinvented myth reflects present relationships between people. As such, it reveals the significance of narrativity in the process of identification. As an aspect of collective identity, and with its connection to thinking and speaking, myth does not represent any kind of abstract thought separated from lived experience, but represents what we might call 'Ambonwari-ness' as lived, thought about, and spoken about. In a real sense this thinking itself has its origin in the process of identification, in the process of being-with-others. Just as *konggong* is an aspect of *kay* articulated in

[1] Young's approach to the lived myths of Kalauna is 'through narrative and the contexts in which narrative is constructed and construed' (1983: 16). Mythology and narratives in New Guinea have been a major focus in the works of only a few other anthropologists, e.g. van Baal (1966), Burridge (1969), Gillison (1993), LeRoy (1985a; 1985b), Wagner (1978), and Weiner (1988).

institutional practices conditioning the identity of clans and lineages *vis-à-vis* one another, *mariawk* is an aspect of *kay* that provides a narrative context for those practices and a means for expressing people's experience of them.

An individual is born into a social past which forms and shapes her or his cosmology and identity. The many hows and whys of life are answered through numerous collectively owned myths which 'confer collective identity and influence personal destiny' (Young 1983: 9). An individual should not be seen to be passively conditioned by his or her 'social context' or 'collective past'. Each person brings *mariawk* into being anew, transforms ideas that others have already entertained, uses his or her own past experiences to answer the hows and whys of life. All those aspects of life which are evoked by the word *kupambɨn* (ancestral) are seen as traditional, even when they are not in fact ancient. In this chapter I want to emphasize *mariawk* (talk, story, discourse) and, especially, *kupambɨn mariawk* (ancestral stories, myths, legends).

In contrast to many areas of Papua New Guinea where important myths are very secret, in Ambonwari they are a mundane dimension of collective life. Moreover, Ambonwari mythology is very rich. My aim in this chapter, however, is simply to demonstrate the way *mariawk* articulates with *kay*, not to analyse speech or myths in themselves. Accordingly here (and in Appendix B) I present only three myths to serve two purposes. First, I want to discuss myths of origin which tell of the beginning of the world and the beginning of Ambonwari village. The 'heroes' of myths of origin represent the collective identity of, and collective differences between, Ambonwari groups. Second, I want to explore the lives of ancestors who are 'caught' in the process of creation, in a state of becoming, without a definite identity. Mythological beings are without familiar *kay*, constantly on the move through an unknown space without fixed paths (*konggong*), and existing in a temporal horizon that contrasts markedly with that in which their descendants live. Their being is pure becoming.

Moon and Sun

While the myths of origin of individual clans are important in terms of people's clan identity, the myth about the origin of the moon and the sun is not. However, insofar as it explains the creation of the world, it encapsulates the dual nature of Ambonwari collective life.[2]

Once, the story goes, when our 'fathers' and 'brothers' lived, when there was no white people's knowledge or the knowledge of the Bible, people said: 'It seems that the sun and the moon made all the places.' When the stories of the Bible became part of our knowledge, we said: 'O yes, perhaps it is true that God made all these "ways". God

[2] The myth was told to me by Bob Kanjik Anjapi, Sapɨndam lineage, Crocodile-1 clan, and recorded in Tok Pisin.

made the earth, God made people. He sent humans to the earth.' The earth, I think, existed earlier. Before there was no other belief. Only the belief of Moon and Sun. Two men only. And their story goes as follows.

A man and a woman made their place here, before, long ago, at the time of the ancestors, when there were no men on this earth. Only ancestors who were born out of the earth, I do not know, half-human, lived here. It was the same mother who gave birth to the two. It is true that *Sunggwi* (Moon) was born first, and *Sɨmari* (Sun) was the second son. Their mother's name was Karingganma. 'Fathers' and 'brothers' who lived before said so. The father of the two children was Karinggan. They had the same names. The bush where they lived is called Sanggur.

Bob explains that he is the father of this forest. Ambonwari people do not collect and eat eggs of wildfowl and cassowary in this area, nor do they touch their nests. They say that if they gathered eggs from the place of origin of Moon and Sun many villages would be destroyed. If Bob collected eggs in this forest, put them into a basket, and went home, they would change into red earth. If people saw this red earth in the basket they would know that someone had broken the taboo. They would tell everyone in the village and send messages to other places. People would know that the eggs gathered from the nest of a wildfowl were actually the placentas of the two men (Moon and Sun). People believe that their mother dug the earth to bury the placentas. This earth turned into the nests of a wild fowl. The placentas of Moon and Sun which their mother had cut off and buried had turned into the eggs. Bob continued:

Moon is the first child and he stayed with his mother all the time. There was no sun. It was dark, really dark. The man who wanted to find food searched with his hands for dry leafstalks in order to light them. When dogs chased a pig the man followed them and thrust the torch into the ground while he shot the pig. He did not know that his son was born. Moon was with his mother. When the man went on a hunting trip, the mother got up and took the basket from the house corner where she had hidden her child. She held Moon with his back towards the forest so that his light did not reach her husband. Only she could see his light. She was happy, she had powers. Whenever she expected her husband to return she turned off his light [Bob attributes to Moon a knob as found on kerosene lamps] and covered it with the scrapings of dried banana leaves. She hid Moon again in the basket. The poor man had to walk in the dark.

So they stayed. The woman got pregnant again. This was her pregnancy with the second 'man who lights'. She gave birth to *kriak*, a large frog which croaks at night during rainy periods. Afraid of her husband, she hid the frog inside the grass used for mats and baskets. But her husband became suspicious.

'I know that she was pregnant. Now her breasts are black and her nipples are erect. Who is the one that she feeds?'

The man began to think, 'What did she give birth to?' This was her second pregnancy and for the second time her husband did not see their baby. Where were his baby boys or baby girls? He did not ask her. He continued to go to the forest in search of meat. During the day Moon took the shape of a human body and sat near his mother. Once when the father returned he found the boy and asked:

'Hey, who is this?'

'It is Moon whom I hide from you.'

'What power does he have that he has to hide? Why did you not show him to me when he was a baby?'

'You know, he had his own routine. Now he is a bit bigger he came back to us.'

She lied to her husband. When the light began to emanate from his body, his mother wrapped him again, afraid that her husband would see his power. She made a sago pancake and with the flour which stuck to her hand covered his face so that his head would not shine too strongly. She was afraid that her husband would discover her secret.

Ambonwari explain shadows on the moon with this myth. They say that they are traces of flour from his mother's fingers. The darkness during the absence of the moon is explained by the mother's hand covering Moon's whole face.

One day her husband asked her to go and fence the creek to catch some fish. She left her second child, the frog, in the house. While she was away, the spirit-woman Kandinggrimay came to the house. She knew about Moon's power. She was angry with Karingganma because she held her son for herself while everyone else had to walk in the dark. She wanted to take the 'boy who lights'. Searching through the corner of the house she found a frog. Thinking of a quick meal, she took it and broke its legs. She cut red shoots of *aibika* (TP) and the leaves of red banana. She put the leaves with *aibika* onto the fire for a while and then wrapped the frog in them. Then she put the bundle into the fire. She placed a baking pan made of clay onto three stones. She got some sago flour, rolled a pancake with a coconut shell, and baked it.

If someone from the village were to go to this forest and see the baking pan on the fire or the old house, they would know that they were going to die. If this happened during fighting with neighbouring villages a man would be certain that he was going to be killed.

While Kandinggrimay was cooking the frog its mother was still catching fish in the creek together with Moon. She hid his light so that only she was able to see. She hung Moon on the tree near the grassland so that his light was directed towards the creek where she was fishing.

The spirit-woman began to eat the *aibika* which covered the frog. Suddenly a strong light came out of the bundle. She covered it quickly. She said:

'Hey, what did I do? What does this light mean? I saw everything around the house. I could see everywhere.' She uncovered just a bit again, the light came out, and she covered it again.

'Hey that's it. I wanted to eat it but now it makes such a light.' She got scared. She took a large clay plate and placed it over the bundle.

Near the creek the mother wanted to return to her house. She called her son:

'*Sunggwi, Sunggwi* [Moon, Moon], what are you doing? Come! Hurry!'

'Mother, come and get me! You put me on the tree, I don't have legs to climb down.' She climbed the tree and brought him down. She turned his light down a bit, covered him and walked home. His light showed her the path. Near the house she turned the light off completely. She covered him and put him into the basket. She heard that someone was in the house. She said:

'Who is making this noise in the house? This is not a place where many people live. Just my husband and I.'

The spirit-woman appeared at the entrance. She thought that the angry mother was going to kill her. The mother entered the house. The spirit-woman said:

'I did not come without reason. I am thinking of something you hide all the time. What did I find? Now the light will be for everyone. How many of us walked around and died in darkness. That is not right.'

The mother was worried about her baby. She went to the corner, searched and found her half-cooked son. The mother learnt about the power of her second son, *Sɨmari* (Sun). He cried:

'Mother! Mother! Mother! Breast! Breast!'

The spirit-woman said:

'What are you waiting for. Take him to your breast. You talked about a frog. He is not a frog.'

The mother took him and gave him her breast to drink. The light was everywhere. Karinggan who was in the forest jumped in surprise when the light came up. He hid at first and then ran home. He heard the voices from the house, he came in and was told what had happened. He said to the spirit-woman:

'What am I going to give you? I got scared and I left the pig which I had caught in the forest. When I saw the light I ran home.'

Karinggan went back to the forest and brought the pig to the house. He was pleased with the way of his children. Next day he went again to catch another pig. When he returned the spirit-woman told them that she was a spirit and not a human and that she could not stay with them. The only concern which brought her here was that she wanted to make use of these powerful children. She thought that the two should be in a clear spot so that everyone could use their light. Karinggan and Karingganma gave her the pigs. Leaving them, the spirit-woman said:

'We do not know who made the earth but we were born together with it. I am leaving you now with these two children of yours, you will know what to do with them.'

The parents decided where Sun and Moon were going to be. They told their children:

'You, the elder son, were born into the darkness. You began with "time". You are the one who started this way. Sun, he came after you, he is the little one. So you, Moon, will look after night. You will go far up. If you stay closer everything will die. Your light will burn everything. All people will die. So you Moon, you will go far up and you, Sun, the young one, will be closer down.'

So the younger brother Sun is lower and the elder Moon is higher up. And because of this the light of Moon is not as strong and hot as the light of Sun. Moon is far up.

This story was, according to people's accounts, brought to Ambonwari by their first ancestors. It is not clear how Moon and Sun were related to Kapi and others from that first group of ancestors. They are all said to come to life together with the earth, thus sharing the same 'becoming' period. Nevertheless, it is Kapi who is regarded as the father of this myth. In present-day Ambonwari village, the Crocodile-2 clan claims a totemic relationship to *Sɨmari* (Sun), and its men are called *Sɨmari yanmas*, 'the ones of Sun'. This clan is also the holder of the personal name *Sɨmari* (Sun). People say that when Kapi took Ambrɨpi, the ancestor of the Wallaby clan, to his canoe, he looked at him through the smoke of his tobacco and said to him: 'You will hold the name *Sunggwi* and Moon will guide you.' From then on the men of the Wallaby clan were called *Sunggwi yanmas*, 'the ones of Moon'. They use *Sunggwi* as their personal name. Though this totemic relationship with Moon and Sun is not associated with leadership or hierarchy *per se*, it is my understanding that the 'gift' of two such powerful beings was made mainly to moderate the precedence of Kapi (Crocodile-1) and Akɨmbrɨkupan (Bird of Paradise) which they gained as founders of the village. If we observe that the gifts given in myths of origin include such important things as men's houses, carved spirit-crocodiles, names, and also women and children, thereby forming stronger ties between groups, we can also understand the generous gift of Moon and Sun.

As I have already mentioned, it seems that Ambonwari once had two moieties, the Sun and the Moon, the first being the moiety of Kapi and the latter the moiety of Akɨmbrɨkupan. The arrival of the third brother, and third father of the village, is offered as an explanation of why the moieties became less important. However, a principle of duality is evident in many aspects of Ambonwari life.

Another aspect of the myth relates to *kay*. We can see that the visible body does not truly represent the human body. While behaving like a human, Sun looks like a frog, for example, yet his *kay* is to shine brightly.[3] This is what determines who/what he really is. His particular *kay* gives him power which others (except his brother) do not have. The unusual *kay* of ancestors is most explicit in Ambonwari myths where beings with human characters can change their physical bodies according to the activity they perform. This gives some insight into the intimate relationship between body and practice for the Ambonwari. Indeed, the distinction between what or how a being is and how it actualizes itself in its transactions of the world barely seems to exist.

'I Am Kapi'

Ambonwari myths of origin are significant for the structural and institutional order of the community. More pragmatically, they are also crucial in regard to the ownership of land and are often recalled during disputes over fishing-places, over gathering and hunting in different areas, over ownership of betel nut and coconut palms, and over the ownership of trees used for house posts and canoes. These myths of origin not only serve the purpose of identity inside

[3] The world formed from humans or from parts of their bodies is familiar in 'mythical thinking', which 'seems to negate and suspend spatial distance' (Cassirer 1955: 91). 'The objective world becomes intelligible to the mythical consciousness and divides into determinate spheres of existence only when it is thus analogically "copied" in terms of the human body' (p. 90).

the village and privileges and prerogatives concerning land, names, spirit-things, and roles in rituals but also relate their owners to other Sepik societies. The names of lineages, clans, and men's houses relate groups of people from different language groups in the Middle Sepik area and are most important for an understanding of relationships and behaviour between them.[4] Thus, Ambonwari people often cite the fact of sharing a mythic origin in explaining why they should not see the blood of people in other villages and groups. If the Ambonwari see a mosquito on such a person (from Yimas, for example), they would tell him or her to slap it, while turning their own heads so as not to see the blood. With these groups Ambonwari do not marry, lest a man see his wife's menstrual blood or the blood resulting from a fight within the family. These groups were the most reliable allies when fighting other villages. They did not fight 'shoulder to shoulder', as they might see each other's blood, but they took revenge on behalf of one another. Origin myths, then, are involved in processes underlying the politics of identity in relations between disparate groups.[5]

I now turn to the myth of Kapi, who is recognized as the founder of Ambonwari village. The essential characteristics of this myth of origin pertain also to the myths of other clans and to their own conceptualizations of identity.

The arrival of the first ancestors (three brothers, two in particular) of the Crocodile-1 and 2 and Bird of Paradise clans to the no man's land illustrates the beginning of Ambonwari village (see Maps 4 and 5). The father of the first ancestors was Angguru. He had four sons. He was able to transform his body: if he encountered strangers he changed into a crocodile and swam under water to hide. Following his extraordinary *kay* he was able to change his body accordingly—his skin became the skin of a crocodile.

Angguru arrived at Chambri Lakes from the area called Kumbranggawi. People say that the hills suddenly opened 'throwing up' people who fell on different places around. Thus, the land near the hill Aibom became the starting-point for Angguru and his four sons. Akɨmbrɨkupan [the eldest son who was already married with two wives] and Kapi [the third son] took two paddles and two clay hearths and went to Kamindimbit. Then they followed the Karawari and Korosameri Rivers up-stream and arrived in Kabriman. In Kabriman they left their canoe and hearths, and took with them only their paddles. The two brothers walked from Kabriman to Marukamari Mountain [near today's Amongabi], where they built a house. The brothers quarrelled because Akɨmbrɨkupan had two wives while Kapi had none. They both visited Yimas, where they separated.[6] They were discouraged from staying there because too many people already lived there.

[4] 'Shared totemic clanship' (Harrison 1993: 41) relates Ambonwari descent groups to others from linguistically different villages, and not the villages as wholes. Harrison properly argues that 'the important point about totemic clanship in the middle Sepik is that it is a system wholly of links, of channels of communication and sociability and for the flow of goods' (p. 49).

[5] However, I could not reconcile these statements with all the facts concerning marriage and warfare between groups linked in this way.

[6] Even the name Akɨmbrɨkupan reflects his marriage situation. It is a compound of *awkum-b(ungg)ri* (two women) and *kupan* (elder brother).

Kapi alone paddled down the Karawari, arriving at a place where Konmei Creek joins it. The place was blocked with bush vines, showing that no one had entered the creek yet. Kapi cut the vines with freshwater shells. The entrance to the creek was guarded by two spirit-crocodiles. Kapi killed them both and took their jaws, which he put in his hair to change his face. With his canoe he looked like a crocodile. Meanwhile, Akɨmbrɨkupan noticed that his younger brother had taken some sago leaves from their men's house, Yanbonman, obviously with the intention of building a new shelter for their spirits.[7] So Akɨmbrɨkupan and his wives tried to follow Kapi. They noticed freshly cut vines at Konmei Creek and followed the trail up-stream. They found Kapi at Aranggmai [also called Wakiarandi] near Yaramun Lake, where they all stayed. They were followed to this place by their father and another brother, Mamanggamai. The youngest brother, Kwarisa, did not enter Konmei Creek but started a new place by himself, known today as Manjamai. Their father died in Aranggmai.

The brothers took their father's head, washed it, decorated its nose, made its eyes, and wrapped vines around it. They placed the skull on a carved piece of wood (with a special hole made for it) and put it into their men's house. The brothers worshipped their father's spirit and gave him the name of a forest spirit, Kɨmbyaman. Whenever they travelled they carried the skull in a basket. People recall that the skull was kept in the Auriman men's house of the Wallaby clan, their maternal clan, until the baptism of the village. It was then left to decay with other secret objects in a small forest house. It has never been seen since. A piece of what is supposed to be Maran's (Kapi's son's) skull is kept by the Singgayjam lineage of the Bird of Paradise clan. This piece of skull used to be worn on a necklace during dancing.

In Aranggmai the brothers met Pandi, a wanderer from Manjamai. He tried to follow the brothers but Kapi, claiming that the surrounding forest was theirs, killed him [today a small creek there is called Pandimban].

More and more people from the settlement which is today called Manjamai came to fish at Yaramun Lake. A man called Yambɨn noticed that because of the long trip the fish always stank by the time he returned to Manjamai. So he built a bush camp near Konmei Creek where he could smoke the fish before returning to his village. Many people followed him and the camp expanded into a village, today known by the name Konmei.

Akɨmbrɨkupan gave his son to Kapi but he died after being bitten by a poisonous snake. Because of the death of his 'son', Kapi left, following the creek upstream. He arrived, alone and the first in that land which is today Ambonwari.

Again, Akɨmbrɨkupan and his two wives and a daughter, Mamanggra, followed his younger brother. Kapi saw their arrival from his canoe. He asked his older brother why he had followed him. Akɨmbrɨkupan answered that they were brothers, the sons of one mother and one father, and that they should not quarrel. When the two canoes came close enough to each other Akɨmbrɨkupan put his daughter into Kapi's canoe, saying: 'Here is your new wife.' They all stayed.

[7] Whenever the Ambonwari build a new men's house they take a bundle of leaves from the roof of the old one and incorporate it into the roof of the new one. In this way the *kay* of the men's house is preserved.

During one of many further quarrels between Kapi and Akɨmbrɨkupan, Kapi angrily decided that they could not share the same men's house any more. He wanted to keep Yanbonman for himself and told Akɨmbrɨkupan to start a new group: Bird of Paradise. Kapi thus got his own men's house, where he put all his secret and sacred objects: his father's skull and decoration, parts of the crocodile's jawbone, the stone breast from the time Kapi was at Aibom, a slit-drum, flutes, and the carved spirit-crocodile Aymakan. He and his brother worshipped Ibrismari, the spirit from a nearby creek, and Kɨmbyaman, a spirit embodied in their father's skull.[8] Their third brother, Mamanggamai, followed Kapi and Akɨmbrɨkupan. He married a woman from the Wallaby clan. This was the first marriage of the Crocodile-2 clan, a type still practised today. The three brothers thus founded Ambonwari and became fathers of the village.

It is necessary to acknowledge how Kapi received his special powers and how he is related to the Dog clan, which is not found in Ambonwari. It seems that Kapi met Mambi (Dog clan) on one of his exploratory trips, after he had already settled in Ambonwari. Mambi came from the Sepik River. He also wandered around alone, sometimes as a man and sometimes as a dog. He stopped at Tambunum and then entered Kangramai Creek at a place called Yipmari. Wherever he went there were already settlements, and the land had been marked with clay on tree trunks. He moved on and on. He arrived at Kungriambun and Kaiwaria (two Karawari-speaking villages) and always stayed with the people from the Crocodile clan (the men's house, Mambiman, in Kungriambun is named after him). When Mambi met Kapi, he, not owning any land himself, decided to give all his knowledge and power to Kapi. Mambi said: 'If you go hunting for your family, you will be able to change into a dog and hunt as a dog. When you go to the bush you must carry a basket and a spear. Cover the basket with leaves and call the name of Sapɨn in a magical spell. Call Sapɨn and you will be a dog. You can carry your spear with you. Once you hold a pig or a cassowary, you can transform back into a man and spear the animal.' When Kapi returned to Kanggramai Creek Mambi had already disappeared. Kapi named his first lineage Mambinjam after Mambi, and the second one Sapɨndam after Sapɨn.[9]

We see from the myth that the well-known Iatmul area of the Middle Sepik called Kumbranggawi is the place of origin of the two Crocodile and Bird of Paradise clans, as it is also for Aibom people. During my short visit to Aibom, the Iatmul people settled there acknowledged that other people speaking a foreign language were living there before them. 'Kumbranggawi represents the southern of the two world hemispheres resulting from the breakthrough of the Sepik in the primeval past' (Schuster 1990: 14–15). Aibom people say that the very first men's house which appeared together with the earth was called

[8] Today Kapi's decorated paddle, carrying its owner's name, remains permanently in the main men's house, together with the carved crocodiles. People from the Bird of Paradise clan keep Akɨmbrɨkupan's paddle in their houses. Whenever a woman in such a house falls pregnant, the paddle has to be moved to some other house.

[9] It might be mentioned that *mambi* is also the term for a stone adze used for cutting down sago palms; and *sapɨn* is white clay used for decoration. It is still used as a dog's name.

Wereman and did not belong to them (pp. 16–18). This name strongly resembles Wariman, the name of the men's house of the Bird of Paradise clan in Ambonwari, and Kumbranggawi is one of this clan's main lineages. Regarding the significance of Kumbranggawi hills, Bob Kanjik commented:

Kumbranggawi broke and everyone came out of these broken hills. When there were no people on this land and there were no villages, that was when Kumbranggawi broke. The hills broke and Kapi fell near Chambri Lake. He 'made' Kamindimbit [Iatmul village], and he brought with him one 'sister', whose name was Kapimbonma. He left Kapimbonma with Kawkapiyapan [Iatmul man]. Kapi told him: 'You are the man who will marry my sister.' Kapi also gave him presents including a plate made of clay. Kapi said: 'I made this place, you people stay.' They asked him: 'Where are you going now?' He said: 'I want to see these Karawari waters. I want to see if people already live there or not.' And he left for Karawari. But there were already places with people in Karawari. At the place which is today known as Koki camp of Mindimbit village, people were already living there when Kapi arrived. Yes, that was the first place of Yimas ancestors. Kapi then went to Kabriman; he slept in Kaningara, found a path in Siriakai and arrived at Marukamari. He stayed there to make a canoe. He put everything inside and travelled at night. He entered Yimas waters and he saw a mark. Someone had already marked *pitpit* near the river. He asked the sign: 'Whose sign are you?' And the sign answered: 'I belong to Karanggapi.' 'All right, you stay.'

Kapi went down-river and entered our Konmei Creek. There was nobody here. His brother Akɨmbrɨkupan arrived and two of them made this place of ours. Kumbranggawi was still breaking up. Some people entered Karawari, others followed their own paths. Kwarisa arrived. Kapi asked him: 'How did you come?' 'Oh, Kumbranggawi is still breaking. We are many now.' Kapi said: 'Come and sleep with us. We will take you to the middle of Karawari River.' Next day Kapi and Akɨmbrɨkupan took Kwarisa to Karɨnggupa land. This is how Manjamai village began. Kapi and Akɨmbrɨkupan stayed here [Ambonwari] when another man, Mandɨm, arrived. The two brothers asked him: 'How did you come?' Mandɨm answered: 'Oh, Kumbranggawi is still breaking.' 'We thought that it had already stopped.' 'No, Kumbranggawi is still breaking up, many people will come.' The two brothers took Mandɨm inside Kanggramai Creek and Mandɨm made Kaiwaria village. You see, the place where you and I sit [Ambonwari] is the main place of origin. Kapi and Akɨmbrɨkupan directed all ancestors of other villages and distributed land between them. There were other ancestors too. All these places say that Ambonwari is their real place of origin. Then people from Yerem [Upper Yuat River] arrived, and their ancestor was Pamɨngapi [Eel clan]. Kapi asked him: 'Where did you come from?' Pamɨngapi answered: 'I came from afar. Do not think that only Kumbranggawi broke. Other hills broke up as well. I fell a long way and so I came to find you.' And Kapi replied: 'You will stay with us, you who came from afar.' Do not think that we are all the same here in the village, we are all sorts of people here in Ambonwari. But our language is the language of two brothers. These two said to others: 'Leave and forget your own language.' Kapi and Akɨmbrɨkupan were the first to enter these waters and we sit with the language of these two men.[10]

[10] To comprehend better the arrival of the Eel clan, I enquired about its origin also among Imanmeri, Imboin, and Meakambut people. The last maintain that their mountain, Kom, is the place of origin of all the people down-river and even, they claim, of all white people and all spirits. They say that the first ancestors turned to rocks and hills. From these rocks and hills the myths of origin of different groups

As both the myth and Bob Kanjik tell us, Kapi, the primordial ancestor of the Crocodile-1 clan, moved from place to place, made relationships with people whom he met on his way, exchanged with them goods, objects, secrets, 'sisters', and 'daughters'. Kapi contributed towards the formation of different settlements in the area. By remembering the names of places where he left his 'footprints' and by remembering the sequential order of these places, Kapi formed his physical and social paths. However, we can view the myth from a slightly different perspective, from that of the identity of his descendants. First, what Kapi's agnatic descendants share is collective path (*konggong*). This is defined in terms of their shared residential movements rather than merely sharing a place of origin. In other words, it is the specificity of journey which lies at the basis of their *konggong*, as a shared process of identification. Kapi's descendants thus share a common path. Second, Kapi, as a narrative identity, appears not to be simply an individual who lived through all these events in all these places, but rather signifies a social person as well as a whole group of patrilineal descendants. The name Kapi stands for this group. This mythic history collapses the agency underlying significant events of the past into a single narrative figure. Regardless of who was responsible for new relationships or for the formation of a new settlement, he/they did it in the name of Kapi. Still today Kapi's descendants in the Crocodile-1 clan identify with him, for example by using the first person when telling of his life. In short, Kapi's life-world seems in reality to represent the collective life-world of the members of the Crocodile-1 clan before they settled in Ambonwari. Person and context, individual and society are collapsed into one another, though every person is inevitably situated somewhat differently from any other in his or her social environment.

In disputes or during important speeches, a man from the Crocodile-1 clan often says, 'I, I, I, I am Kapi.'[11] If the dispute, for example, is within the clan, the other man may reply, 'Me too, I am also Kapi.' By saying, 'I am Kapi' a man says 'I am the one who came here first', 'I am the one who belongs here', 'I am the one who owns this land', 'I am the one who is father of this place', 'I am the one who dances in the front', 'I am the one who is father of the most important spirits of the village', 'I am the father of the main men's house', 'I am the father of the village'. His identity has its roots in the primordial past to which he is related through *mariawk*, *kay*, and *konggong*. As Kapi represents both personal and collective entity, this 'I' should be understood in a similar way. Through this use of 'I' the man presents himself as embodying both personal and

in the wider area began. Because of this, Meakambut people guard the 'stone vagina', as they believe that if anything destroyed it everyone on earth would die. Among the 41 living Meakambut people in 1991, I am especially grateful to Papielam and the young boy Kwairara Ayaw who took me to the rocks of 'human creation', told their stories, and looked after me during my stay in their rock shelters.

[11] As the living person whose actual name is Kapi was quite young, he never joined these vigorous disputes.

collective identity. Intersubjective and personal identities are aspects of one another. They enable a living man to assert his own sense of self, because he embodies what it is 'to be Kapi'. Being the first ancestor, Kapi is not only himself; he is at the same time every man who was and will be from the Crocodile-1 clan. His descendants are 'instances of his own personage' (Mimica 1988: 80).

The meaning of 'I am Kapi' is most powerfully expressed in the appearance of one of the two principal performers in the all-night dances, who is dressed as Kapi. With all Kapi's decoration the dancer's 'skin' becomes the 'skin' of Kapi. Moreover, considering that the body and the way it acts are identified in *kay*, then the man who plays this role is both Kapi, the ancestor, as well as himself. Both *kay* (way of doing things) and *wambung* (understanding), as well as the past and the present, are functions of the whole person in social relations with others who themselves incorporate their own past and present social relations. Since Kapi is the founder of the village as a whole in such a context (see Chapter 7), Kapi and the man who embodies 'his ways' represent not only the Crocodile-1 clan but the village itself.

The notion of Kapi attaches to itself several other signifiers (often with their own names), such as his decoration (plumes, feathers, shells), his decorated paddle, a slit-drum, a crocodile jawbone, a canoe. All these things are either 'extensions' of his *kay* or, when they have their own 'personal' names, beings in their own right (his canoe, slit-drum, a mask on his men's house, his carved spirit-crocodile).

If we can agree that Kapi represents the collective identity of the Crocodile-1 clan and that temporality is an inherent dimension of identity (being a practice itself, identity—i.e. identification—is in process of continuous preservation, formation, or re-production), then we see that the myth of origin historicizes and temporalizes the Crocodile-1 clan's collective identity and the identity of each of its members. At the same time, however, this mythic history of Kapi's arrival derives its significance from the future. This means that people themselves, through their own present process of identification, are those who choose this myth and its hero, and then mythologize, historicize, and temporalize it.

Identity, not only as sameness but also as oneness, is most explicitly present in the cosmogony of Iqwaye, in which the creator Omalyce made five men from lumps of earth. He is at once not only himself but 'every single one of them, regardless of the fact that the mud-men are his progeny' (Mimica 1988: 82). Though the people of the Crocodile-1 clan are, through their *kay*, embodiments of Kapi to whose *kay* their own *kay* owes its existence, they are nevertheless also Kapi's successors who live in the present.[12] Young argues that

[12] Identification with mythological beings is well documented among Melanesians (Harrison 1990a: 91; Malinowski 1922: 412; Schwimmer 1973: 63; Williams 1932: 161, 163–6). It is self-evident, as Young pointed out, that 'such identifications and imitations do not gainsay the everyday identity of an

[t]here is no mass delusion . . . that would make a conspiracy of society by viewing its members as mythical incarnations, eternally enacting their given destinies through the bodies of the living. This is as far as the 'impersonation' or 'mythical mimesis' view can be taken, and it patently slides into absurdity. One may presume that no human society has ever adopted the literal assumption that all or most of its members 'are' other than they seem. (1983: 18)

This, of course, is applicable to Ambonwari. Durkheim (1961 [1912]: 101) pointed out that it is not intuition of natural phenomena but society which is the true model of myth (cf. Cassirer 1972: 79). 'Mythical incarnation' or 'mythical mimesis' (i.e. identification with mythological heroes) is an analytical artefact rather than the reality of Ambonwari existence. For a man 'to be Kapi' does not mean simply an unreflective identification with his mythical ancestor, nor with a 'hero' as a collective entity and collective identity which can be portrayed by a single individual. It means, above all, an assertion of his sense of self through his understanding of the originary acts, i.e. *kay* of Kapi, and the temporal relations with others that were, and still are, made possible by these acts. This means that by 'being Kapi' one can hold an attitude towards the future in which his *kay* will still 'be Kapi', regardless of its transformation.

It is in this way that the ancestral past and its apparent 'sameness' in the present have to be understood. Kapi's *konggong* is followed by movements between the 'same' named places and the 'same' marriages. Kapi's *kay* is reproduced by using the 'same' named men's house, slit-drums, canoe, and other objects such as a paddle or a spear, and by retaining the 'same' relationships with spirits. Kapi's *mariawk* is reproduced by knowing and telling the 'same' stories.

None of this is 'available' to someone from the Wallaby clan. He cannot claim that he is Kapi's descendant, that he is the father of the main men's house, that he has same stories, names, and secret knowledge, nor can he carve a crocodile head on the prow of his canoe (or, to be more precise, not without permission from Crocodile-1). He is not allowed to use the same land as Kapi did, nor are his marriage paths the same. However, the final episode in the myth of origin of the Wallaby clan, the arrival of Wallabies in Ambonwari, is also part of Kapi's myth (*kupambɨn mariawk*). Accordingly, the Wallaby clan through its connections with Kapi is also part of Ambonwari *kay* in its totality.

Kapi and his myth from the period of formation of Ambonwari village have an illusory, eternal character for those who can identify with him. Following Ricoeur, we might say that this 'permanence in time' means identity in the sense of 'sameness' and not 'selfhood' (1991: 190; 1992: 117). Of course, Kapi as a person had his own character and his own Heart. These constituted his

individual; nor, we may presume, do they abrogate his sense of himself as distinct from the hero or ancestor with whom (aided by imagination and the mild dissociation induced by areca nut narcosis) he seeks a temporary identification' (1983: 17).

selfhood in his own epoch—a selfhood which, for Ricoeur (1991: 192–3), does not belong to the category of events and facts and does not have 'permanence in time'. But from an Ambonwari perspective one could question the very dichotomy between 'sameness' and 'selfhood' as well as attribute permanence to the former. The sense of self, which is always the producer and product of intersubjectivity, constructs 'sameness' (as a part of selfhood). This endures only as long as one's understanding (*wambung*) of the world and one's 'ways of doing things' (*kay*) choose such a 'sameness' (and 'selfhood' of which it is a part) to exist.

The past in the present and the present in the past may be seen also in the connection between other myths and everyday life in the village. One of the most prominent examples is the one concerning the stone breast. While Ambonwari ancestors still lived in Aibom a young man was tricked into the horrifying act of cutting off his mother's breast (see Appendix B). The Ambonwari keep this stone breast hidden in the house of Kapi's descendant. When I held the stone breast (which reminded me of a stone club I had seen in Menyamya in Morobe Province) people said that from then on I would eat a lot. They explained that the breast is hidden in two coconut shells; a natural hole in one half of the shell enables the breast to look around the village. The breast is only shown to small children when they refuse to eat to increase their appetites. During the first months of my stay I was surprised by the number of adult women whose right breast had shrunk while their left one was large and hanging down. I assumed that this was due to their breastfeeding habits: holding a child with the left hand while working, so that the child hung onto the left breast. But women denied my explanation, claiming that it was because in the past a man had cut off his mother's breast. This is simply another example of how *kay* through *wambung* (understanding) makes its connection with the collective past.

The Beginning of Time

Kapi is not the creator of the world, nor is his father. The creator could be traced either to the Kumbranggawi hills or to the rocks near Meakanbut. Nevertheless he, as well as the other primordial ancestors, is historicized as a creator. Regardless of the actual chronological time of his arrival, he is placed in the 'organic time of epoch' (Wagner 1986: 81–95) between the creation of humanity and the creation of identity. Although Kapi is not the creator of the world, I maintain that he has certain similarities with the Iqwaye creator:

> the first human being, the creator, is himself in the process of creation . . . As yet there is no time. However, there is an originary temporality since this primordial state of being is actually a process of becoming . . . The originary temporality is a paradoxical flow which perpetrates only duration in the same state of being. The cosmos subsists in the mode of everlasting *now*, the eternity. Further, the primordial condition . . .

represents a peculiar intuition of chaos—chaos as a potentiality of all forms, the principal of which is human corporeality. (Mimica 1988: 78)

Kapi, likewise, has no significant identity. His identity is in a process of becoming. Since his time is one of becoming, from the point of view of his successors real, lived time has no existence.[13] When Kumbranggawi broke there was a chaotic succession of events which is only retrospectively ordered in relation to a way of life that emerged from that chaos. Their chaotic and unpredictable existence does not temporalize mythological beings but preserves them in an eternal process of creation that only ends with the institution of the world as it is now known.

The beginning has nothing whatsoever to hold on to; it is as if it came out of nowhere. For a moment of beginning, it is as if the beginners had abolished the sequence of temporality itself and were thrown out of continuity of the temporal order. (Connerton 1989: 6)

Between the breaking of Kumbranggawi and the beginning of Ambonwari the world as it is known today was in the process of being created: there was no period of moon or period of sun, nor were there any other periods *(pɨsɨnɨm)* with which the Ambonwari are familiar today. As such, this period represents an eternity of the sort so characteristic of mythological beings.

So mythological beings live outside the boundaries of Ambonwari time. Accordingly, the *kay* which characterizes Ambonwari is not yet constructed. The extraordinary *kay* of those mythological beings is responsible for *imɨnggan kay*. Thus even the appearance of mythological beings changes and they can take the forms of animals—those which will later become the totems of their descendants. In many other myths one finds that even houses are built by themselves, sago processes itself, all the beings have extraordinary powers. They lack the habitual activities which order time in terms of remembering and anticipation, in terms of familiar sequences of 'before' and 'after' which today structure the tempo of life. The beginning of habits, being the beginning of familiar bodies, is also the beginning of time as humans know it. The non-habitual practices of these first ancestors precipitated an unpredictable future. They were constantly confronted by new situations, new places, new beings. They were constantly on the move, having neither a familiar home nor familiar paths.[14]

[13] 'For myth there is no time "as such", no perpetual duration and no regular recurrence or succession; there are only configurations of particular content which in turn reveal a certain temporal *gestalt*, a coming and going, a rhythmical being and becoming. Thus, time as a whole is divided by certain boundaries akin to musical bars. But at first its "beats" are not measured or counted but immediately felt' (Cassirer 1955: 108).

[14] When discussing Sophocles' Antigone, the intrinsic meaning of the poem and the authentic Greek definition of human being as 'the strangest of all' that is strange, Heidegger says: 'it is necessary to insert a note calculated to ward off a misinterpretation of the whole poem—a misinterpretation to which modern man readily inclines and which is indeed frequent. We have already pointed out that this is no description and exposition of the activities and fields of activity of man, an essent among other essents,

Mythological beings inaugurate relationships and reciprocity. The landscape through which they travel changes and as yet has no names; groups are in process of being formed. Consequently, their *konggong* paths are still being constructed. There are no identifiers or 'markers' of identity which could be ordered in any kind of sequence. Only Kapi's descendants, by knowing that Ambonwari village became his final destination, 'reordered' Kapi's life by retrospectively conferring on him his identity as their ancestor. Thus Kapi receives the necessary 'after' and becomes temporalized. But during his lifetime, or the lifetime of collective Kapi(s), he had only a beginning which was an absolute one that had no 'before'. He lived in chaos, where the becoming was most uncertain and the outcome of forms and actions most unpredictable: it could lead either to the continuation of chaos or to the formation of the world. Kapi was unable to anticipate the future, something that is essential for those with identity. Only those who act habitually to determine their future, those with a collective *kay*, can be conscious of temporality and make their own past and history. If mythological beings appear to have time which is continuous with that of human beings, it is only because later human generations, by virtue of narrative logic, attribute familiar characteristics to them. Only the people living through time, after identity and temporality have been created, confer identities on their mythological heroes, and in turn make these heroes inform people's own identities. Only humans living later will temporalize them, while the heroes themselves are in process of creation.

It is during this period of becoming that the moon and the sun come into existence (from darkness to light), that plants such as tobacco and bananas begin to grow, and that beings inaugurate social life.[15] They still lack relations and they search for places—home and familiarity (the appearance of first names). After they settle down and form first marriage paths among themselves, people acquire human *kay* while simultaneously losing the special powers of eternal beings. Ambonwari village comes into existence.

Ambonwari myths must be looked at from several different perspectives if

but a poetic outline of his being, drawn from its extreme possibilities and limits. This in itself precludes the interpretation of this chorus as a narrative of man's development from the savage hunter and primitive sailor to the civilized builder of cities. Such a notion is the product of ethnology and psychological anthropology. It stems from the unwarranted application of a natural science—and a false one at that—to man's being. The basic fallacy underlying such modes of thought consists in the belief that history begins with the primitive and backward, the weak and helpless. The opposite is true. The beginning is the strangest and mightiest. What comes afterward is not development but the flattening that results from mere spreading out; it is inability to retain the beginning; the beginning is emasculated and exaggerated into caricature of greatness taken as purely numerical and quantitative size and extension. That strangest of all beings *is* what he is *because* he harbors such a beginning in which everything all at once burst from superabundance into the overpowering and strove to master it . . . The knowledge of primordial history is not a ferreting out of primitive lore or a collecting of bones. It is neither half nor whole natural science but, if it is anything at all, mythology' (Heidegger 1987: 154–5).

[15] The origin of these and other culturally significant equipment (in Heidegger's sense) are recounted in other fragments of the extended myth of Kapi.

their temporalizing effect in the present is to be fully understood. As everyday *mariawk* concerning discourse referring to the recent past is under continuous construction, the final story and its outcome are not yet formed. On the other hand, *kupambɨn mariawk*, 'ancestral stories', are supposedly fixed and not negotiable. They are narrated frequently for people to recognize their identities and differences, thus enabling them to constitute their present and future relationships. However, as *kay* itself unfolds, a process mediated by *wambung* (Heart), *kupambɨn mariawk* is transformed. For *kupambɨn mariawk* is not merely a text but is an aspect of and responds to life itself, which as always is a matter of *kay*. While I do not have any concrete data on the point, it is hard to resist the thought that the mythic insignificance of Kapi's third brother, the ancestor of Crocodile-2, is a reflection of the clan's political decline. Although this insignificance strikes an outsider who is inclined to treat the myth primarily as a text, it is not an issue for the Ambonwari (except perhaps for the members of the Crocodile-2 clan), just because *kupambɨn mariawk* is an expression of Ambonwari *kay* including its political processes. Let me now examine the temporal aspect of myth as a narrative.

Ricoeur (1980: 178–9; 1984: 66–8) proposes two temporal dimensions of narrative: one is its episodic dimension, which draws narrative time in the direction of the linear representation of time; the other is a configurational dimension, which brings the succession of events into one meaningful whole. Ricoeur's main argument is formed around the mediating role of a plot (which transforms the events into a story) between time and narrative. 'The composition of the plot is grounded in a pre-understanding of the world of action, its meaningful structures, its symbolic resources, and its temporal character' (1984: 54). The plot serves the purpose of ordering temporality and historicity, and thus brings narrative time into existence.

While Ambonwari myths of origin are sequentially interrelated—for example, as the myth of origin of the Wallaby clan incorporates Kapi (Crocodile-1)—other myths stand alone, unrelated to other myths, but still in the eternal 'epoch' of creation. Every myth can be examined from an internal perspective, i.e. the events themselves (chronology, episodic dimension), or from the perspective of its totality and its significance for collective life.

Myths, or most usually parts of myths, are told and retold on many occasions in everyday life: when people pass by a particular place where the ancestor stopped, marked the land, or fought; during leisure hours, when younger people are most interested; at moments of death, sickness, misfortune, when people search for causes; during land and other disputes; and during rituals such as initiation. *Kupambɨn mariawk* ('stories of ancestors') are intervowen with the present as are other aspects of identity. A myth is not just a story from the past. Whenever it is told it is perceived and explained by the listeners in their 'now'.

I will look at the episodic structure from the perspective of such acts of narration. The actors and events in the story undergo temporalization as the narrator takes his or her listeners through the process themselves. Only in this way, 'afters' following 'befores', is a story comprehensible. In myths of origin the names of the first ancestors serve as collective 'we's' for every group. What is important is who came 'before' and who 'after', who helped the other upon the latter's arrival, who owned what land where, who gave things away and how he was reciprocated, who married whom at that historical moment—all those things which contribute towards the constitution of the village social structure, its institutions and relationships between individuals and groups.

It can be seen from the myth of the stone breast (Appendix B), as well as from the other myths, that linearity of time is achieved in several ways. The Ambonwari use different affixes to emphasize that something comes after: the verbal suffix *-bɨn* (just after) indicates that something has just finished and that something else is happening or is about to happen. The word *mba* (enough, so) is used after the completion of any minor act. This word is often replaced by 'all right', *olsem* (TP) (in this way, thus, and so, therefore), *em nau* (TP) (so). Words such as *sayn* (again, but), *sambin* (after, later), along with other temporals, as presented in Chapter 2, are sometimes used in narrative, though more often in everyday conversation. The duration of an event or activity or the lengthy period between events is indicated either by imperfect suffixes, by nominalization, or by the repetition of a verb. *Kay-* (to be, to stay) and *sa-* (to sit, to hold) are the most often repeated in this way. After the direction of movements is established, listeners can anticipate the succession of events also by the use of prefixes *ma-* and *ka-*, with their spatial meanings of 'up-river' and 'down-river' respectively. With the additional spatial distinction between right side (*ipɨndamban*) and left side (*yakrapaynjan*), and up (*mawun*), where Moon and Sun live, and under (*isinar*, 'down'), where the spirits of the land dwell, the six sides of a collective space are formed, with the village at its centre. Sometimes the narrator can change the tense of his narration, jumping from the irrealis to the remote past; at other times the narrator brings the past into the present by the frequent use of dialogue, aspects, modalities, and moods, and by her or his own techniques of narration. The last includes mimicry, theatrical gestures, non-verbal sounds and noises, the use of objects, and so on.

The plot of the story, along with temporals, principally determine time in the sense of 'earlier' and 'later', and in this way orders temporality and brings narrative time into existence. With this order, 'narrative time' can be reckoned. But such an 'ordinary representation of time as a linear series of "nows" hides the true constitution of time' (Ricoeur 1980: 170, following Heidegger).

In myths and legends the main aspects of identity are presented through situations in which the habitual or taken-for-granted is suddenly transformed into something strange, unpredictable, extraordinary, often terrifying. There are always emotional and moral aspects to these stories. Resentment, shame, fear, anger, and other emotional states colour the stories, motivating the actors, as they do in real life. Everyone in the village can, for example, understand and sympathize with the spirit who ate only plain earth and did not know about the sweetness of sago. When Kapi offered him baked sago filled with sago grubs he ran away, ashamed of his ignorance, and turned into earth himself. To give another example, everyone in the village can understand and sympathize with Kapi's act of leaving Aranggmai after the death of his 'son'. *Wambung*, which here equates with Ricoeur's notion of pre-understanding, shapes the responses of the audience to the actions of mythological heroes.

The difference between the not yet formed identity of beings in myths and the fully formed identity of people in the present is made plain through the contrast between the way mythic heroes and ordinary people do things. While some beings are admired for their special powers (ancestors can change their appearance, for example), others can be ridiculed because of their ignorance of how to do things (the spirit who does not know how to make a canoe or how to bake sago). Other aspects of mythic *kay* are also constituted in ways that are not familiar in the present and are therefore strange and mysterious. Their *konggong*, for example, entails travel through an unknown land, marriage inside the family, the continuous search for home and for the relational aspect of identity. Again, they are involved in relationships where the normal expectations of trust are violated by their extraordinary lies and trickery (see Appendix B). People often remark upon the language of the ancestors, their secret stories, spells, and chants, which are often incomprehensible to those living in the present. Women in myths, who otherwise represent home and familiar place, are presented as wandering around without their distinctive attribute of stability, while men have no patriline, no alliances, and no past. Most of them are solitary wanderers.

The ways of doing things within this mythopoetic period are different from those of the present. But as soon as the wanderers settle down, their lives assume the 'same' identity as that with which present-day villagers are familiar. This includes not only rituals, such as those of initiation or first menstruation, and dancing as the highlight of their 'way of the village', but also daily practices.

Myths of origin, like other stories from the past, contain a large number of place and personal names. Sequential order is marked by such names, and people who are unfamiliar with them have difficulty putting different events, especially from different stories, in some sort of chronological order.[16] 'The

[16] In *Naven*, Bateson (1958) recognized the importance of the Iatmul naming system but found it almost impossible to penetrate. This became a starting-point for detailed research by Harrison (1990a) and Wassmann (1990; 1991). The relationships between movements through space and corresponding place-names, and the experience of temporality through oral performances, can be found also in other Papua New Guinean societies (Feld 1990: 150–6; Schieffelin 1976: 184; Weiner 1991: 93, 105–6) as well as elsewhere (Basso 1988; Rosaldo 1980: 56; Thornton 1980: 170). The problem arises when movement through space becomes equated to movement in time as the only way of experiencing temporality, thus neglecting its complexity. In regard to place-names, these 'form the immanent order of the "field of visual sensation"' (Husserl 1964: 25). If we consider the interrogative 'where' we can see that the Ambonwari term *sanggwana* ('where?') is a homonym of the verb *sanggwa-* with the meaning 'to see, to look'.

order in time is given as the sequence of localities associated with events, and even genealogy has its temporal aspect—already contained in the sequence of persons—enhanced by the association of persons with events in designated places' (Schuster 1990: 18). Chronology is thus importantly related to people's collective identity, especially to its relational aspect.[17] *Kupambɨn konggong* ('path of the ancestors') is one of the tools for memorizing the movements of ancestors, and takes the listeners through the events which follow the temporal order of successive places (as happens in songs figuring in the all-night dances: see Chapter 7). Though most of the settlements on this path do not exist any more, people still use the same names for those places. While the main place-names are well-remembered, people speculate about less important ones (and about personal names as well) in the way one might speculate about the year in which some historical event took place.

There is yet another important aspect of identity of every lineage and clan in Ambonwari: the relationship with their own spirits. This will be discussed in the next chapter.

[17] Although Ambonwari do not have the knotted cords so important to Iatmul, Wassmann's observations are apt: 'The past becomes articulated through the spatial pattern of the migration. The beginning is indicated by the creation of the world and the end by the founding of the present-day villages; in between there are the migrations as the successive occupations of places of settlement' (1990: 30).

6

Spirits

Until recently many ethnographers have tended to look at items of material culture as mere objects, either as art objects or in terms of an interest in their production, the materials used and decorations applied. In his discussion of Trobriand canoes Malinowski (1922: 105–23) argued that such descriptive accounts do not tell us much about the function of these objects, about people's emotional attitudes towards them, or about their social implications and cultural importance. It seems obvious today that to separate the objects of material culture from their social and cultural contexts and from the particular historical moment in which they were fashioned constitutes a misinterpretation of their essential aspects. Such things belong to the world which is determined by people through their imagination and cosmology. At the same time they reciprocally influence people's perception and experience of them. In Ambonwari such 'objects' can be classified into two groups: first, those things which are used in daily life such as tools, weapons, and all other things which are handled daily by people, helping them to carry out their activities; second, those objects which represent the embodiment of spirits and are therefore regarded as beings themselves. The second group, as I will show, is the province of men. I argue that by manipulating the visibility of these spirits Ambonwari men seek control over their own existence.

I have mentioned several times in previous chapters that the human body is seen through its activity. It is defined by *kay* (the way of doing things) and extended to the things which are used in particular activities. The 'life' of things such as small canoes, stools, paddles, spears, adzes for pulverizing sago, axes, bush knives, hand drums, mats, baskets, body decorations, and so on is derivative in that it depends on the life of humans. Most of this equipment is handled only by those who own them, and the identities of these things are determined by the people who use them. A thing becomes 'something' only when a person uses it. The Ambonwari employ the same verb, *sa-*, with the meanings of 'to use' and 'to hold'. A typical round stool, for example, becomes a part of the person who sits on it, absorbs his 'dirt', and is not supposed to be used by anyone else. While being used, a thing 'gives something' to its user. At the same time, its usefulness and the way people use it gives a thing its character: thus the term for 'something' is *kanggɨnɨng* or *k(a)-anggɨn-ɨng* ('should-give-ness'), which could also be translated as something which produces *kay*. A paddle, for instance, gives its user the ability to move a canoe on the river; its usefulness gives existence, identity, and value to the paddle. We could say that the essence of *kay* (habit) 'which is not performed but "acquired" by mastering an activity . . . lies in its use-value' (Ricoeur 1966: 281, 283). A paddle and a person become one in terms of *kay*, the way of paddling a canoe. It would be interesting to compare *kay* as way, habit, ritual, custom with *kay* as canoe, for although in different nominal classes, the same word is used for both. It might be just a coincidence, but I prefer to think that there is a more significant relation involved. *Kay* (canoe) enables people to float firmly, securely, and soundly, and to move easily between different places. It does not sink. It is something good. A bridge is called *kwarisi-kay*, 'across-canoe', it connects two sides, making them one. All these meanings and values are familiar to, for example, *imɨnggan kay* (the way of the village; tradition) which by definition, among other things, is something good. It enables people to keep their lives firm, secure, and sound by integrating the different village wards. To put it another way, in using a canoe a person is stable on the water as if she or he is stabilized by custom, habit, and knowing the ways of doing things.[1] A canoe is 'something' because it 'gives' this stability.

It is tempting now to take a different tack and look at those things which are not used, either because they are not yet finished (they are still 'becoming') or because they are old, broken, or abandoned. People would say that such things are *kambra* ('nothing'), from *k(a)-ambr-a* ('should-ripen yet' or 'should-make fire'), the translation well suited to something unfinished (not 'ripe') or something only fit to be burned. 'Nothing' is related to that thing which destroys habit and 'burns' life, or it cannot yet be recognized as a 'something'. To summarize: in everyday life the difference between 'something' and 'nothing' is that a thing becomes 'something' by being held and used whereas 'nothing' is not held and not used. While 'something' gives, 'nothing' either has to 'ripen' before it can give by being used or it has nothing to give, it is not used, and it can be burnt.

Most of the small things used in daily life have carved designs with patterns resembling those of face decoration or those found on slit-drums, posts, and other things in a men's house. They represent the spirits of dead women, pig's intestines, shells, birds, or male and female figures. These designs carry the mark of those who carved them and differ sufficiently for people to know their owner (the Ambonwari use the word 'father'). People recognize those who are skilled carvers, and if their kin relationship is favourable they ask such individuals to make things for them. Many of these things (paddles, stools, and hand drums, for example) are also valuable gifts, usually given to people by their maternal relatives. But to overemphasize the appearance of things in terms of people's aesthetic perception would be to disregard their main feature: that is, their usefulness.

[1] Using something *y* to do *x*, as I have already mentioned in Ch. 2, is expressed by the construction *x*-ing-with-*y*. Such a construction, involving the 'postposition' (Foley 1991: 107) *ngandɨk*, 'with, together', suggests an intimacy between a person and a thing used which is absent from English constructions.

Opposed to the small things of quotidian life in Ambonwari are spirit-things: masks, flutes, posts, slit-drums, and carved crocodiles of the men's house, as well as important things of daily life such as large canoes, all of which carry spirit-names.[2] The relationship between people and their spirits influences their perception of these things. Manipulation of their visibility (by hiding them) adds another dimension to their meanings. By considering this class of things we will better understand Ambonwari cosmology and existence, of which these things represent a significant part. In the first chapter I introduced the main spirit of the village, *imɨnggan saki*, referred to as the 'ridgepole' (*kil* in Tok Pisin) of the village. In this chapter I will look at those objects which are most important for understanding Ambonwari relationships with their spirits and are kept in the men's houses. But let me first present the spirits as they are conceptualized in Ambonwari cosmology.

Spirits

At birth people inquire about the sex of the child: 'Is it *panmari* (male, man, lit. 'very-male suffix' or 'very-mortal') or *awkunma* (female, woman)?' Both terms are used mainly to express the sex difference between humans and can, albeit rarely, be applied also to the gender of spirits. An encounter in the bush with an unidentified being leads people to ask first, is the being a human or is it a spirit? second, if it is a spirit, is it a bush-spirit or a spirit of a dead person? third, what sex is the spirit? Suspecting an encounter with a spirit, people would ask: 'Was it *saki* (male bush-spirit) or *wunduma* (spirit of a dead woman)?' If the being was human, the answer would not be *panmari* or *awkunma*, but *yarmasɨnar* (male human being) or *yarmasɨnma* (female human being), terms meaning 'the one who cries' or 'the one with blood', thereby distinguishing humans from bloodless spirits. The questioner would not ask whether the spirit was *sakima* (female bush-spirit) or *wundumbunar* (spirit of a dead man), because they are too dangerous. Distinctions between humans, spirits of the dead, and spirits of the bush, and between the sexes of these beings, constitute a two-by-three classification, as shown in Table 7.

The above distinction between two kinds of gendered spirit would seem to be based on people's experience of succour or care. *Wunduma* (spirit of a dead woman) represents the 'sister' whom brothers always take care of, giving her presents of food, especially fresh meat. *Wunduma* also represents 'the mother' who never turns her back on her children and always 'sits firmly'. In contrast, *wundumbunar*, the spirit of a man, a 'father' or a 'brother', is considered wrathful and capable of striking or harming even his own children or brothers, either because of their wrongs or because of his own spite. Indeed, *only* a man's immediate household is under threat from his ghost.

[2] Large non-traditional items of contemporary significance in Ambonwari, such as outboard motors or guns, are also given spirit names.

TABLE 7. *Humans and spirits in Ambonwari*

	Female *awkunma*	Male *panmari*
Humans: distinction from spirits	*yarmasɨnma*	*yarmasɨnar*
Spirits of the dead	*wunduma* (good)	*wundumbunar* (bad)
Spirits of the bush	*sakima* (bad)	*saki* (good)

Saki are powerful 'big' spirits of the bush. Those living around Ambonwari share the land with the villagers and are generous and protective. Especially important are those—such as spirit-crocodiles—who came to Ambonwari with Kapi or other ancestors and share *kupambɨn konggong*. Thus, male bush spirits are not to be equated simply with wild spirits. While some of them could aptly be described as hostile or indifferent to humans, several such spirits are sustainers of Ambonwari life and central to their cosmological preoccupations. In contrast, *sakima* are alien female bush-spirits who do not care for anyone and are particularly dangerous to other women's husbands. They are unpredictable wanderers who do not 'sit firmly' in the house looking after children.

Though they may harm people (especially in unknown territory), male bush-spirits and female spirits of the dead are those who help and care for people, and are worshipped by them. Female bush-spirits and male spirits of the dead, on the other hand, are malevolent, full of envy and resentment, and are feared. Contacts with female bush-spirits, which occur mostly in prohibited (taboo) areas, are considered deadly, as many terrifying stories reveal. But there are also stories about people being able to overpower these malevolent female spirits, as in the following example.

This story comes from Wakriyamarimbuk hill (the place of origin of the Pig-2 and Eagle-2 clans).[3] The waters were low and one of two sisters saw an opportunity for a fishing trip that night. She told her younger sister she would wake her when the time came. But she was overheard by a female bush-spirit outside the house, who promptly made the younger sister deaf. When the elder called at night her sister could not hear her. But the *sakima* answered using the voice of the younger sister. They took fish-traps and went off. In the middle of the grassland the elder sister gave her baby to her younger sister and told her to wait while she placed fishing-traps in the water. As soon as she left, the *sakima* began to eat the baby. As she swallowed its leg the baby screamed. Hearing her baby's cries the mother called out in concern, but the *sakima* assured her that everything was fine. The *sakima* ate the whole baby and threw its liver into the creek. When the woman saw it floating past she knew that something was terribly wrong. She took off her headband and attached it to a fishing-trap. She told the headband (in the past things were able to act and talk) to make a noise in response to the *sakima*. Then she ran away. Whenever the *sakima* asked the woman about her fishing, the headband

[3] This story was told to me by Raymond Anggurapi Yakopia and again by Elias Wapun Kandangway.

answered. Finally the *sakima* discovered she had been deceived and ran after the woman. The woman did not return straight to her house, but climbed a palm and broke some sharp dried pieces from the crown. When the *sakima* came and saw her up the palm, she nodded in delight at the discovery of her meal. Then she asked the woman how to climb the palm. The woman answered that she climbed it legs first. The *sakima* followed this instruction many times, slipping off and trying again. When she finally succeeded in nearing the top, the woman stabbed her between the legs. The *sakima* fell down and the woman ran back to her place. After recovering her breath she told her brothers what had happened. They took their spears, bows and arrows, and stone axes and found the terrible female spirit. They shot her from every possible angle. They collected firewood, laid it over her body, and burnt her. When they searched through the ashes they found a figure made of earth. It looked like a rock but it was earth. It is still there today and people call it *andi sakima* ('the earth female spirit').

In this account we can see that a woman was able to resist the *sakima* and easily deceive her. Ambonwari explain that if this kind of female spirit attacks women they can fight back, as they are all women. But if a *sakima* attacks a man he is lost. Similarly, although not inclined to do so without provocation, a male bush-spirit (*saki*) can simply hit a woman with red betel nuts or 'shoot' her with the seeds of a tree, taking her *anggɨndarkwanma* (spirit) away. But men's personal spirits, who are also under the protection of their clan-spirits, can fight other male bush-spirits, though with only a slender chance of winning.

The other kind of bad spirits, those of dead men, who usually operate inside the village, are vengeful and bring sickness and misfortune. Because people fear them, the words *sakima* and *wundumbunar* are seldom pronounced, not even when playfully 'scaring-to-comfort' small children. *Wundumbunar* shoots his human relatives with tiny invisible spears and implants small destructive objects (such as teeth, bones, thorns, shells, stones, nails) in their bodies to make them sick. Mostly this happens because the angry spirit wants to reveal and correct the wrongs of the living. Consider the following:

Godfried Yanggus from Eagle-1 clan and his wife boiled some ginger plants in a pot and sprayed and treated their sick daughter with the aromatic mixture. As Kambianma, who was approximately 6 years of age, did not recover that day, Godfried asked Tobias Yangi, one of the two village healers, to visit them.[4] Late that evening Tobias entered the house. He held six or seven ginger plants with leaves. He chewed ginger with betel nut and rubbed ginger over the forehead and chest of the sick child. He made sure to spray her under her finger- and toenails, and inside her ears and nose. Every orifice is a place where spirits can enter the sick person and carry the spirit from the body. While doing this Tobias inquired about any wrongs in Godfried's family and lineage. Godfried confessed that he had been to a sago forest which was not part of his clan's land. A few years ago two children had died after their father, Godfried's classificatory brother, had cut some palms in the same forest. They had decided then to leave the

[4] I am not interested here in presenting the healing activity of Tobias so much as in considering the social aspects of sickness (but see Telban 1997). Nevertheless, it is worth noticing that 50 years ago Ambonwari people did not have these kinds of healer, but rather a kind of 'shaman'.

bush unused for five years. Tobias said that it was Godfried's mistake to cut sago palms in this forest, and that this was also the reason for his own previous sickness. Tobias also commented that the Eagles had arrived in the village following others, and that the village spirits were angry. But the sickness was given to both of them by the spirits of the house (*wundumbunar*) because he had used land which was not his. After massaging her for some ten minutes Tobias removed a stone from Kambianma's belly. Everyone chewed betel nut and Tobias was given sago pudding to eat. They praised Tobias, saying that he did not ask for money immediately as Bonifacio, the other healer, supposedly did (though this was not true). Kambianma went to bed and Tobias left after a couple of hours of chatting and chewing betel nut.

But the two healers are not always called to attend the sick. When their children or grandchildren are ill, some women hold a thin vine and twist it. They call the name of the sick child and, as they say, tie its spirit (*anggɨndarkwanar*, *anggɨndarkwanma*) so that the *wundumbunar* cannot take it away. Then they take a thick vine and twist it, first next to the child's head then moving down to its toes. In the morning, without going to the toilet and without having anything to eat, they take the thick vine to a huge tree, saying to it: 'You are a tall man, your skin is fine but this child is sick. Take his (her) sickness from me and send it far away.' Then the woman puts the twisted vine on the tree trunk and cuts it with a bush knife. In this way the tree will get sores and absorb into them the sickness from the vine. Its bark will heal over and the sickness will climb high up into the crown. The woman returns to the house and untwists the thin vine over the forehead of the sick child, returning his or her spirit back to the body. The woman will cut the skin of the sick child with a razor blade and wash it with a herbal concoction.[5] She will cook food and give it to the sick child.

Twisting a bush vine without tying knots in it, in the manner of people when building a house, is the way that spirits of the dead and bush-spirits are thought to kill people. They strangle the person's 'watchman'. Quite often in healing rituals this 'watchman' is symbolically represented by a small stick around which a vine is twisted. The stick should not be broken; if it is, the Ambonwari say that the person will die. It is kept in a container, a bamboo tube, a bundle made of bark, or any kind of box until the person recovers his or her health. Similarly, when someone leaves the village for distant places someone else will 'tie' his or her spirit with a vine or a *Cordyline* leaf, and give it to the person leaving. The person should not bend or crush it, in order not to harm his or her own spirit.[6]

The main sign that someone is sick is that her or his *kay* is not as it should be. A person's breathing is different from usual, their skin is hot or cold, she or

[5] Plants in general have symbolic powers rather than pharmacological uses, as I have argued for other parts of Papua New Guinea (Telban 1988a; 1988b).

[6] My 'brother' Tony told me that before I left the village he would go to the place on the river where I washed, and he would tie my spirit and give it to me to take it with me to Australia. Because of the excitement on the day of my departure, Tony forgot to do so.

he sleeps more, and does not perform their usual activities. A person is not *mar* (dead) but *mari* (sick). People say that a person is 'with sickness' (*mɨn mari ngandɨkɨn*), that is, she or he 'has sickness'. Sickness 'has taken hold of a person' (*mɨn mari yan sarinyan*). By holding and 'using' a person, the sickness and person become one; they share the same body. To get better, a person has to remove 'sickness' as a part of the body. Only in this way will she or he rehabilitate their previous *kay*. The thorns, stones, teeth, and shells extracted from the person by a healer are not 'nothing'. They are 'something' because they 'give something', i.e. sickness, to the person. These intrusive objects implanted into the skin have to be pulled out. People expel the spirits that are responsible by talking to them and telling them to leave the person, and by twisting them in the vine and carrying them away. In the same way a healer recovers a person's spirit.

In everyday discourse the words for good spirits are often heard. The huge round masks, called *saria* (faces), at the front and back of each men's house represent female spirits, *wunduma*, each having the name of an ancestress, a 'mother' or a 'sister' of the 'father of the men's house'. Such a mask becomes a female spirit following the ritual installation of a stone, which is the spirit's Heart, behind the mask. Such rituals are performed whenever the mask, which is the spirit's *arɨm* (skin), needs to be replaced. When, fifty or so years ago, the Ambonwari fought their enemies (mostly Imanmeri), they painted their bodies to resemble *wunduma-nga* (i.e. female spirits). They say that *wundumanga* came into their bodies and helped them during the fight. Through this 'transformation' they became *wundumanga*. If the prow of a canoe, especially a large motorized canoe, is in the shape of a crocodile's head, the Ambonwari carve a special face on top of it. The face is called *wunduma* and bears the name of the worshipped female ancestor from the house of the owner. The same is true of the carved faces on the posts in the men's house, on paddles, shields, and drums. The *wunduma* design is also the name of one of the types of face-painting done just before a dance. All these female spirits are believed to protect and support the individual houses and the village in their 'standing up'.

As I discussed in Chapter 1, the huge snake which 'lies' under the village with its tail up-river (east) and its head down-river (west) is the main *saki* (male bush-spirit) of the village. The snake, or 'spirit of the village', *imɨnggan saki*, is the most important being in Ambonwari: it is the guardian of the village and, if something goes wrong, its destroyer. The women of the village, the carved faces representing spirits of dead women *(wunduma)* on the men's houses (see Plate 7), and most notably the three mothers of the village, by 'sitting firmly' anchor this snake and preserve the well-being, literally 'well-standing', of the village. Women secure the firmness of the place. Second in importance to the snake are the spirits which occupied the land and creeks around the village before humans appeared. They formed close relationships with, and are controlled by, the three main village clans: the Ibrismari spirit by the Crocodile-1

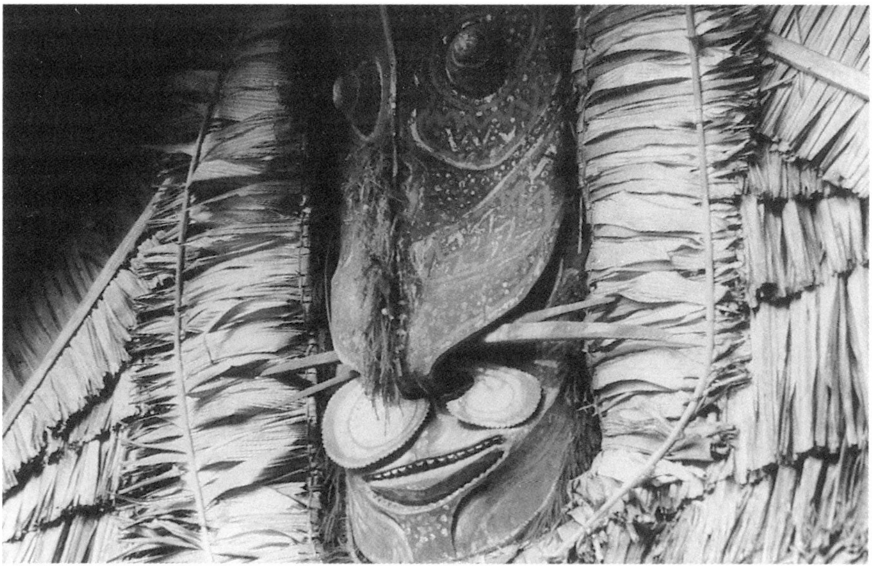

PLATE 7. *Wunduma*: the carved face of a female spirit on the men's house of the Cassowary clan

clan, the Yawundaki spirit by the Bird of Paradise clan, and the Amburmari spirit by the Crocodile-2 clan. People do not carve figures of them, but use their names for large canoes. The third group of powerful bush-spirits is the one brought to the village by the first settlers. These spirits appear in the form of crocodiles with individual names, and are represented by the carved totems stored in the men's houses (see Plate 8). These carved spirit-crocodiles are worshipped as protectors who guided and cared for the Ambonwari people, especially during periods of fighting. They also have an important role in initiation, and all presentations of food in the men's house are addressed to them. While land-spirits and those living in the creeks stayed at their own places, the crocodile-spirits moved in the past with Ambonwari ancestors wherever they went. Even today, whenever a person from a particular clan goes far away (women are not expected to leave the village), people say the spirit-crocodile follows, helps, and guides him.

Wherever ancestors settled on their way to Ambonwari they encountered land- and creek-spirits who already lived there. But their crocodiles, which they carried with them, were always spirits of their own. They are carved by men, hidden by men, and worshipped by men. They secure male command over the environment and give men power over their enemies. If women and children saw these carved spirits the spirits might leave the village. Then men would lose their power, authority, and mastery of the living environment. A

PLATE 8. *Saki*: spirit-crocodiles inside Yanbonman men's house: Aymakan (front) belonging to the Crocodile-1 clan; Kasmari (back) belonging to the Hornbill clan

consequence would be the destabilization of the village and even its destruction. It is interesting to recall an event which occurred during my stay, when an old woman suddenly entered the men's house of the Wallaby clan.

Magun is an old woman from the Pig-1 clan who had married into the Wallaby clan. Just four months before, her daughter Sanggaymay had died. Unable to overcome her grief, she directed her anger towards her son, and one day while he was sitting in a men's house she suddenly appeared, telling him it was his fault that Sanggaymay died, and that she wanted to die too. She deliberately entered the men's house wanting to commit suicide. Her angry and confused son immediately pulled her out while word about the event spread throughout the village. First of all they had to find two *sɨmɨrinyar* (ritual inaugurators) from the Pig-1 clan, one of whom was responsible for the front and the other responsible for the back of the Wallaby men's house. They soon arrived and pushed a stick decorated with shells into the ground at the front of the men's house. With the assistance of two younger members of their clan, the *sɨmɨrinyar* circled the men's house and sprinkled the roof with coconut milk. They were 'washing' the men's house while emitting sounds resembling *uuuuu*, *uuuuuu*, *uuuuu*. They placed a stool at the front of the men's house and put two kina on it. In the mean time women from the Wallaby clan cooked sago pudding and garnished it with sago grubs. Together with a cluster of betel nuts, the food was brought into the men's house for the *sɨmɨrinyar*. The men discussed how they would have had to pay much more if the woman had entered the main men's house, where she would have seen all the carved spirits (in this one she only saw a slit-drum). They would probably need to go hunting and present the spirits with a pig, hanging one of its back legs at the front of the men's house and the other at the back. They also commented that because Magun was an old woman she was unlikely to die. But if a young woman or one who was sexually polluted in any way entered the men's house she would most certainly die.

There was once also a fourth group of *saki* spirits, comprising those which belonged to the individual lineages and were kept in their main houses. Thus, in the past, every lineage had its own *saki* (male bush-spirit); it animated a stone or a carved object, and acted as a protector of that 'house' group. I was told that in the past a big man would find such a stone on a neighbouring hill and carry it to his house. The bush-spirit from that land followed its stone and, on finding it in a house, began to inhabit the stone. The man would talk to the stone, call the spirit by name, and present it with food. The spirit would thus become the protector of the house and warn people of the lineage through their dreams whenever something was wrong. Men hid these spirit-stones from women and children in decorated baskets, taking them with them only when they went on dangerous missions (such as killing someone) or when they intended to stay in the bush for a long period. Today, with the exception of the Kunbaynjam lineage of the Crocodile-2 clan, not a single lineage keeps its spirit-thing any more. When the village was baptized in the 1950s these were abandoned and never revived. The 'father' of Kunbaynjam lineage keeps the spirit-thing permanently hidden in a woven basket. Every twenty years or so, he says, when the basket needs replacing, the oldest woman from his lineage will be asked to make a new basket, and the oldest man of the lineage will put the spirit-thing from the old basket into a new one. He will not tell anyone what it looks like.[7]

Other bush-spirits inhabit the surrounding land and creeks and are known by individual names. Bush-spirits are considered much more powerful than spirits of the dead. One category of extremely powerful spirits is that of the sky. The Ambonwari call them *maun saki* ('above spirits'). Dark clouds are called *krambri*, from *kra-*, 'to fight' or 'to cut', which means that the spirits above are fighting; they come down to earth with the lightning. They 'fall' down with, and in, stones. Some stones enter the earth, others stay on the surface. The latter become those important stones seen at the front of the main men's house and those representing spirits in individual houses. People say: 'The rain are spears of the "above spirits". If the rain drenches us our skin becomes heavy and we can get sick. Later the wind takes all the spirits back above. We cannot see the wind. We just hear the noise in the bush, we see how the grass and trees move.' In Ambonwari cosmology there are no other spirits or beings.

The majority of Ambonwari people are relieved to say that they have never seen any spirits except the carved ones. But as both the seen and the unseen world are apprehended with the same Heart, whatever is felt is known and therefore real as well. This 'reality' is reflected in myths, stories, and healing

[7] Just before I left the village my friend Elias Wapun gave me such a spirit-stone which was once used as a house or lineage spirit. It is black, smooth, and about 12 cm long with a few incisions on it.

rituals, in which the superhuman powers of spirits are most evident. Such spirits are said to appear without heads and with grotesquely long fingernails, which makes people wonder if they would ever want to see them. Almost any unusual event may be attributed to spirits, such as the creaking of a house floor, the sound of unknown voices or unfamiliar noises, the shaking of a canoe, or a glimmer of light in the bush at night. If men are fishing at night and a torch bulb burns out they will say it is because a spirit looked at the torch. When 5-year-old Sangɨrmari was suffering a malarial attack, he called out fearfully: '*Saki*, *saki* (male bush-spirit)!' During their initiation young boys are warned that they should never copulate in the bush where spirits dwell. The carvings on two posts represent the punishment: on one a huge snake is biting a man's penis, on the other a snake is entering a woman's vagina.

During the divination following a death, when a long bamboo pole is placed on the floor (usually in the hand of the deceased), people await with nervous anticipation the responses of both spirits of the dead and bush spirits from their land. As the whole procedure is performed in total darkness (even smoking is forbidden) people generate a state of anxiety, afraid that spirits will cause the bamboo to hit them. The person who is hit will die.

Whenever anyone died, people from the area near the cemetery would leave their houses to stay with their wives' relatives, at least on the night following the burial. People fear the nightly visitations of ghosts, especially male ones, of the recently deceased. This is because a new ghost does not know its way around and, seeking to return to its home, it wanders. Such misadventures are made more frightening for the Ambonwari because new ghosts are invariably accompanied by other spirits.

As 'caring' is all-pervasive in the Ambonwari's social sentiment and ideology (Telban 1993), spirits are defined with respect to it as well. Thus, the familiar *wunduma* and *saki* are regarded as those on whom people can count while *wundumbunar* and *sakima* are unreliable, even if familiar. The spirit of a dead woman (*wunduma*) representing a 'mother' or a 'sister' is made publicly visible in carvings which are not hidden from women and children. Male bush-spirits (*saki*) are those who own the land where people dwell. They are particularly powerful; and just as they hide in the creeks and in the bush, their carved representations are hidden within the men's houses. Both classes of spirit, *wunduma* and *saki*, have ongoing relationships with people. While they protect people they can also punish their wrongdoings. Their names are invoked to scare children, and they spring to mind when one is frightened in the bush. By contrast, the spirits of dead men (*wundumbunar*) and the female bush-spirits (*sakima*) are those with whom people avoid any relationship whatever: they are alien and malign. As people have no relations with them they are never represented on carved objects, nor are they offered any food. They are those hardly visible beings which, because of their evil nature, people are most fearful of seeing.

'Spirit-Things': Introduction

An Ambonwari saying, *arɨm sambɨs ngandɨkɨm kwandɨkas ngandɨkɨm* ('skin has eyes, has ears'), specifies the main communicative aspects of human existence: the ability to see (and be seen) and the ability to hear (and be heard). Anger amongst the Ambonwari can be expressed by screaming and swearing and by jumping, shaking fists, 'shooting' imaginary arrows towards the opponent, and actual fighting. After the death of a person, the Ambonwari perceive that everyone present is sad and grieving, that everyone cares even though their external behaviour might suggest otherwise. In the morning many people would say that they did not sleep at all (even if they did), because they grieved so much. What they say is accepted, and mourners appreciate this. The articulated expression along with visible bodily signs—including simply physical presence—tell people how someone feels. Thus, both articulation and visual expression represent a communicative aspect of people's lives, and as such are most important for the formation of people's thoughts and understanding of their own existence.

In a previous chapter I examined some aspects of Ambonwari discourse about their past. By repeating their myths and stories people confirm their 'heritable property', reshape their collective identity (Young 1983: 9), and consolidate the way people think. When narrative identity is augmented by visual representation the experience of familiarity, of belonging, of being at home, is deepened to an even greater extent. The revelation, in the appropriate narrative context, of forms, patterns, symbols, and colours on carved things embodies these narrative meanings for those who are unfamiliar with them. While both hearing and seeing are 'tools' for apprehending the world, they are simultaneously constructed by it. But 'world is never an object that stands before us and can be seen' (Heidegger 1971: 44). The perception of situations, people, or things is not restricted simply to 'seeing them' or 'understanding them' but is determined by the standpoint from which they are apprehended (Sartre 1956: 593).[8]

Being aware of the immense power of interaction with spirit-things of the men's house and the possible harmful consequences of, for instance, rumour and gossip, Ambonwari men guard those secrets which they believe are important for the preservation and well-being of society. Not only do they keep certain secret stories to themselves but they also manipulate the visibility of particular spirits. By making the spirits visible by carving them, then hiding them inside the men's houses out of the sight of women and children, men not

[8] 'Thus we find, that for man in so far as *"there are" things*, there is in things a perpetual appeal toward the integration which makes us apprehend things by descending from the total integration which is immediately realized down to this particular structure which is interpreted only in relation to this totality' (Sartre 1956: 593). Similarly, we could say that there is in the different aspects of temporality and identity 'a perpetual appeal toward the integration which makes us apprehend' and interpret these aspects from the integrated perspective.

only claim to establish more balanced relations with things which have a fearful dimension but obtain command over them and over their own existence. As the witnesses on behalf of the whole village, they contribute towards village stability as women do by 'sitting firmly' and anchoring the village. Men, being more involved than women in the public discursive aspects of village life, gain control over something that, at another level and from their own perspective, is a strategy for control itself: the body of restricted male knowledge.

But there is much more to the carvings than the representation of spirit-things: they are beings themselves. Once completed and installed, they are no longer an extension of an individual's *kay*; they are no longer part of the body of the person who created them. By having their own personal names, by having their own bodies (or being embodied in objects), and by having their own Hearts ('insideness'), these spirit-things themselves become a part of the *imɨnggan kay*. They become attuned to and an extension of 'the way of the village'. These beings are conceived to have their own lives, including the main aspects of identity and temporality: they have their own ways of doing things (*kay*); their own thoughts, stories, and discourse with other spirits and people (*mariawk*); they have their own places and paths between them (*konggong*).

Take, for example, the large named canoes. These are few in number and categorically different from ordinary canoes. They were formerly used only for special occasions, such as warfare, and were the prerogative of the first lineage of the clan. The names of these main canoes arrived together with the ancestors, thus sharing with them 'ancestral paths' (*kupambɨn konggong*) and 'ancestral stories' (*kupambɨn mariawk*). Consider the canoe Simbraynja, belonging to the Kanjimarpinjing lineage of the Hornbill clan. The female name it carries is also the name of one of the posts in the main men's house.[9] By carving a canoe, by a ceremony in which 'ritual inaugurator' (*sɨmɨrinyar*) puts shells and money on it while the owner's lineage provides food, by bestowing a name on it, a canoe, as a physical artefact, becomes a being in itself. The being of a canoe is transferred from the old one to the new one along with the name after the ceremony. It has its own body and its own *kay*. It communicates with people by rocking or shaking. When in the past the Ambonwari went to fight with their neighbours, the rocking of a canoe would induce the men to return to the village. It was interpreted as a form of advice from the spirit. It often happened that the men asked a canoe about their chances before they paddled off towards their enemy. A man thinking about his sick wife may turn his canoe homeward if it is shaking and perhaps telling him that she may be dying. As I mentioned in the first chapter, a canoe's female character is an inherent part of its exist-

[9] Though canoes, like men's houses, do have feminine characteristics, people usually do not bestow female names on them. Other names used for canoes include Ibris (Mambinjam lineage, Crocodile-1 clan), which is the name of one of the three main land/creek spirits, Ibrismari; Ambarmari (Mɨndapmay, Eagle-1), which is a name of another bush-spirit; Aranggmay (Sapɨndam, Crocodile-1), which in Ambonwari is the only female name used for a men's house.

ence: both women and canoes move in space, connect different places, and both in their *kay* should embody stability.[10]

The life and therefore temporality of a spirit-thing exists through its name, its place in ritual, its relationships with humans, and so on. Spirit-things are in this way personified or, we might say, spiritualized. If humans want to share their identity and time with them (which of course they do), they have to implement similar practices to those in their own human relationships.[11] So people talk to these spirits, present them with sago pudding, pig's heads, and clusters of betel nut, preserving in this way a continuous relationship of reciprocity. Through this reciprocity, humans and spirits coexist and share the world. These spirits had relationships with those who lived in the past, they have relationships with those who live in the present, and are expected to have relationships with those to come in the future. Relationships with these spirits are traced back to the mythological period, to the period of becoming. Most of them were present at the formation of the village. The spirit-things with their eternal character are thus 'symbols of permanence' or 'symbols of eternity' (Adam 1990: 127–8). Their existence for people is guaranteed as long as they are 'still produced in continuous, newly appearing primal impressions' (Husserl 1964: 61). Whenever people replace an old carving with a new one the old one becomes useless, or as the Ambonwari say, it becomes *kambra arɨm* (only skin, 'nothing skin'), *kambra arɨng* (skin removed from the rest of the body) or *pinjingɨnar* (a half-man or a half-wood). While the spirit enters a new carving after the proper ritual is performed, the old carved object is left to decay or is used for firewood. As long as the name is remembered, the spirit lives even out of its skin. But once the names are forgotten the spirit ceases to exist. This happened, for example, to the spirit-things of individual lineages.

Men's Houses

Paradoxically, men's houses (*iman*), recognized as beings in their own right, are entities with pronounced female characteristics. As Bob said during the short initiation of a young boy, they 'open their legs' to let the men in; while the

[10] The importance of canoes for the Ambonwari invites comparison with the role of canoes in Gawan life as described by Nancy Munn (1986: 127–47). The comparison is all the more inviting in that she too is concerned with cultural construction of space and time. However, my approach differs significantly from hers, and Ambonwari and Gawa are very different. Accordingly, a worthwhile comparison would be a substantial piece of work which I cannot undertake here. In Gawa, however, it is also true that 'canoes have paths', are identified with the human body, and are associated with the female principle. More broadly, Munn examines kula transactions, especially their movement, that is 'the path (*keda*) of a shell' (1986: 57). She identifies 'spacetime' as the major dimension of this exchange, which is then subsumed by a wider sociocultural spacetime. For her, practice produces spatiotemporal beings for whom '*bodily spacetime* serves as a condensed sign of the wider spacetime of which it is a part' (1986: 17, emphasis in original).

[11] Forge (1973: p. xviii) argued that 'in primitive art, art objects are rarely representations *of* anything, rather they seem to be *about* relationships' (emphasis in original).

initiates secluded in a men's house are like the young inside a 'possum's pouch'.[12] At first glance it seems odd that, as among the Kwoma (Bowden 1990: 483), all men's houses except one have male names. From a different perspective men's houses could also be characterized as incorporating both male and female qualities with the latter dominating. As individual beings with all the characteristics of a person, they should be seen as 'male mothers'. This may well be related to those aspects of kinship and social organization where the influence of the maternal lineage and clan seems to hold greater sway over individuals, especially during their childhood and youth. By providing protection to the members of a patrilineal clan, along with their secrets, men's houses are like mothers' brothers to their sisters' sons. This justifies giving male names to such beings, whose male and female characteristics are most evident.

This blend of male and female characteristics can also be discerned in other carved things or by the way these things are grouped together inside the men's house. Posts in a men's house, as I will show below, have male and female figures carved on each side, while a *wunduma* (female spirit) 'sits' on the top of every carved figure. The secret flutes which are played during initiation ceremonies are organized in two lines: on the left side are four flutes with female names and on the right side four with male names (see also Tuzin 1980: 30).[13] The right side is male because, as people explain, men hold spears with their right hands. The position of the flutes in array is based on the order of arrival of the mythological ancestors of individual clans. Thus, the most prominent (Crocodile-1, Bird of Paradise) are at the front, while second in importance are furthest back (Crocodile-2, Cassowary). Also, the classification of flutes as male and female reflects the importance of brothers and sisters within the lineage or clan, and is extended to paternal and maternal clans in Ambonwari social organization. As seen in Fig. 8, the right side, called *kupangwar* (from *kupan* 'elder brother, ancestor'), is occupied by those flutes owned by those clans (lineage in one case) whose members are considered to be real or classificatory elder brothers of those on the left side. Thus we have the following moiety-type oppositions:

[12] 'Belly', 'womb', and 'net bag' are the expressions used by the Abelam and the Arapesh (Tuzin 1980: 156 n.). While Forge (1966: 26) saw Abelam men's houses as feminine, Tuzin did not find this perception among the Ilahita Arapesh (Tuzin 1980: 156).

[13] The story goes that in the distant past there were no flutes in a men's house but only birds which sang during the initiation rituals. But on one occasion these birds pecked the boys to death, ate them, and left only skeletons. The men were angry, so they burnt the men's house together with the birds. After that there was no singing during initiation. But the spirit of the flutes gave dreams to a man, telling him how men could make these sounds themselves. A man planted bamboo, and tied a vine at two places so that the bamboo developed knotted links. The first time he tied the vine too close but the second time he succeeded. He cut the bamboo, made a hole in it, and blew. When he played it in a men's house his wife came and she played it as well. But the man killed his wife and kept the secret for himself. From then on, people say, whenever men play flutes the women and children wonder what is going on inside the men's house, and fear that the birds have returned and will eat their boys who are inside.

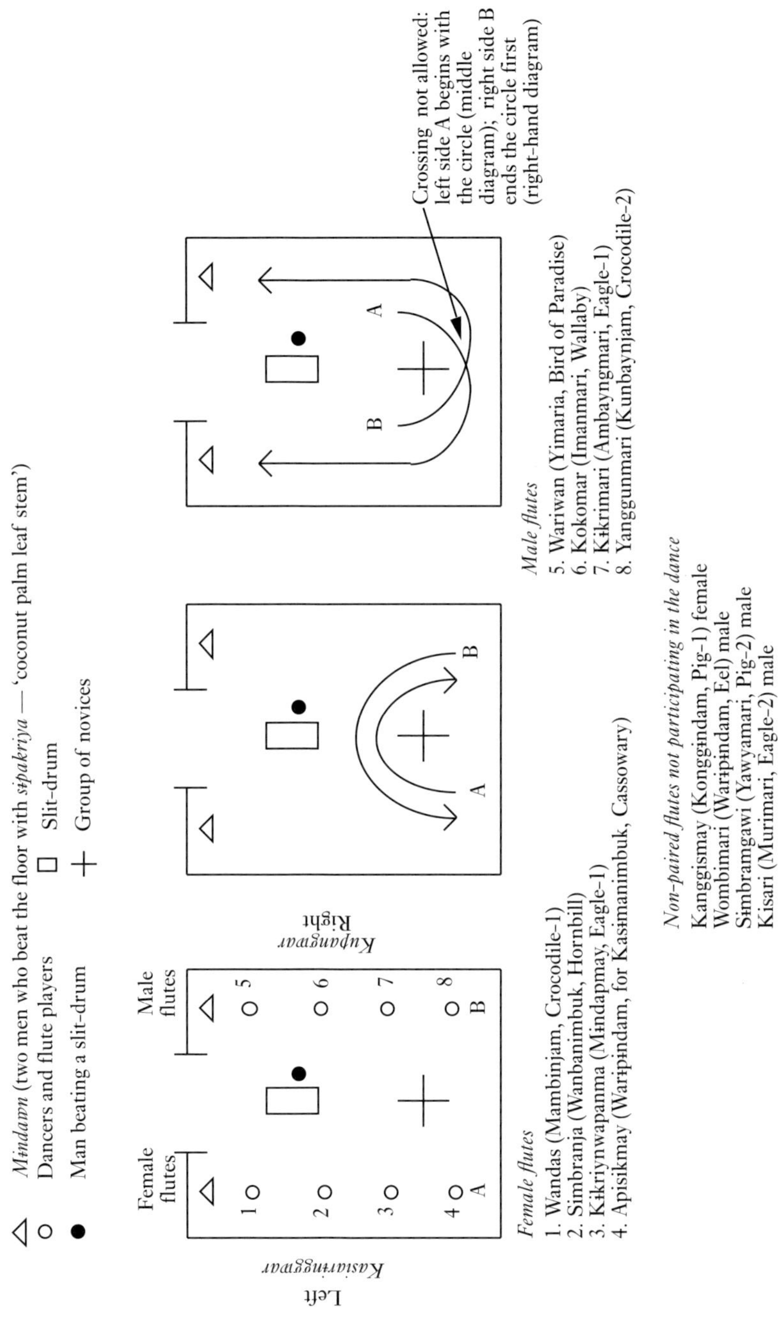

FIG. 8. Flutes and the movement of dancers in the main men's house during initiation

Left	*Right*
female flute (*sanggurma*)	male flute (*sanggur*)
younger brother	elder brother
Crocodile-1	Bird of Paradise
Hornbill	Wallaby
Mɨndapmay (Eagle-1 clan)	Ambayngmari (Eagle-1 clan)
Cassowary	Crocodile-2

The members of the Pig-1 clan participate in a dance as 'honorary' members of the Bird of Paradise clan, as they relate to each other as 'brothers', 'fathers', and 'sons'.[14] The few members of the Eel clan join the Cassowaries, and the one Cockatoo member joins the Wallabies. The only two groups which are not yet fully incorporated into the village structure are the Pig-2 and Eagle-2 clans, and they are the only ones which do not actively participate in the initiation dance.

Ambonwari men's houses, including the main one, are far less colourful and spectacular than those built by the Abelam and the Arapesh. Only the main men's house is raised from the ground on posts and beams and has walls, while all other men's houses are open and without an elevated floor. Every men's house has many stools which are used by men during their meetings. Every men's house also has at least one slit-drum, while only some have carved spirit-crocodiles. The names of all men's houses and spirit-things are given in Table 8. I will discuss here only the central men's house, Yanbonman, the most recent structure of which was built in 1982 on the site of its 'fathers' from the Crocodile-1 clan (see Plate 1). Though it is not the most popular meeting-place (men use the Cassowary men's house for village meetings), it is, in terms of initiation and other ceremonies, the most important for Ambonwari men. It is also the home of some of the most eminent spirit-carvings.

Long ago, as the story goes, a man called Uringgay from Kapi's clan went to wash in the creek. His namesake, a water spirit, pulled him down under the water and, though Uringgay was already an adult man, initiated him. The spirit showed him how to build a men's house and gave him names: Wandas for a post, and Kapaymbi and Maprɨkmay for the two 'spirit-faces' (masks). After the two brothers Kapi and Akɨmbrɨkupan separated, Kapi remained the sole owner of Yanbonman.

Over time different groups arrived in the village, and some of them were granted space to live in the ward of the Crocodile-1 clan. These groups were also given posts on which they could bestow their own names. In this way they received their own place in the men's house and in the village. Their identity was thus recognized. It is common in the Sepik for several clans to share a

[14] Just to remind the reader, Akɨmbrɨkupan (Bird of Paradise) was the older brother of Kapi (Crocodile-1); Wallaby is the older 'brother' of the Hornbill clan; Ambayngmari is the first lineage of the Eagle-1 clan and Mɨndapmay the second; Mamanggamay (Crocodile-2) was the brother of Akɨmbrɨkupan and Kapi who offered his land for the Cassowary clan to live on so that they became 'brothers'.

TABLE 8. *Names and owners of important spirit-things in Ambonwari*[a]

Clan	Lineage of father of men's house	Men's house	Parts inside men's house			Masks on men's house		Spirit-crocodiles	Slit-drums
			Front	Middle	Back	Front	Back		
Yamara (Bird of Paradise)	Yimaria	Wariman	Wariman (Bird of Paradise)	–	Maskwanbi (Cassowary)	–	–	Kongguniapan (Kanggris)	Maramundaun
Manbo (Crocodile-1)	Mambinjam	Yanbonman	Manggunburman (Crocodile-1)	–	Ambarsoa (Eagle-1)	Kapaymbi	Mapr‡kmay	Aymakan (K‡mbiyaman; (Crocodile-1) Kasmari (Hornbill)	Sangankumban
Warakapwi (Wallaby)	Imanmari	Auriman	Auriman (Imanmari-1, Wallaby)	no name (Imanmari-2, Wallaby)	Apinja (Yimas‡ndam, Wallaby)	Yangg‡nmay	Apinja	Warim‡ndan	Sakak‡sa
Awa (Cassowary)	Wayanggara	Yanjanman	Yanjanman (Cassowary)	no name (Crocodile-2)	Kundiman (Eel)	Kanj‡nggrimay	Kund‡ng (Eel)	–	Primari
Imbian (Pig-1)	Simbanimbuk	Yanggraman	Yanggraman (Simbanimbuk, Pig-1)	–	Sukurismay (Kongg‡ndam, Pig-1)	–	–	together with Bird of Paradise clan	Awunjapia
Yambari (Eagle-2)	Murimari	Kabriman	–	–	–	–	–	–	–

a This figure shows only those men's houses and those spirit-things which existed during my stay in the village.

men's house while the owners remain the 'fathers' by occupying its front portion. In Yanbonman the front portion, together with the clay hearth, belongs to the Crocodile-1 clan, while the 'fathers' of the rear of the men's house and its clay hearth are men from the Eagle-1 clan.

Yanbonman has no paintings on the front façade. The walls are made of a patterned thatch of sago leaves, while each gable is occupied by a huge carved face, with a wooden stick (denoting a cassowary quill) and two wooden circular plates (denoting shells) through its nose. Every *wunduma* (female spirit) which embodies an ancestress has its own name.[15] Once inside the men's house one can see the central horizontal beam under the roof, which is called Siar. Its carving depicts a male ancestor from the Crocodile-1 clan. There are also two female figures, with tails, facing outward: one above the entrance and one above the rear exit. They depict female spirits giving birth, representing the absolute beginning of newly initiated men. Following the division of the men's house interior, the front figure is owned by the Mambinjam lineage (Crocodile-1) and the rear figure by the Ambaynggmari lineage of the Eagle-1 clan. Each is the principal lineage of its own clan.

Posts

Yanbonman has fourteen posts supporting the structure (see Plate 9). Table 9 presents the arrangement of these posts and their names, which are the names of male and female ancestors. Each post has its 'father'.[16] Ownership of the posts does not reflect, on the village level, the mythological order of arrival of individual clans, the relationships between clans formed throughout the mythological past, and the resulting village structure. It reflects only those relationships which pertain to the Crocodile-1 clan. Other clans have their own men's houses.

Of the two posts at the front near the entrance, the one on the left (if looking from the inside) carries the name of Kapi, the founder of the village. I was at first puzzled when I learned that the post on the right has no name and no owner. Later, though no one wanted to say this explicitly, I figured out that the other should be called after Akɨmbrɨkupan, Kapi's elder brother and co-founder of the village. One might also expect that the elder brother would guard the rear of the main men's house and the younger the front. But as the myth tells us, the brothers separated. While Kapi kept the main men's house

[15] Along with the face as a female personification of the men's house, the Iatmul also had an additional carving embodying the great fish eagle, a symbol of the men (Bateson 1932: 259). This supports my contention concerning the male and female characteristics of a men's house.

[16] When Ambonwari fought their enemies they spread the blood of those they killed over the posts. They hung the belly of a victim on the post of the killer. During the vigorous nightly celebration someone would take this belly and hang it on the huge *Octomeles sumatrana* tree for Pambanggway, the 'fighting spirit'.

PLATE 9. Posts in Yanbonman men's house

the older brother was at first left without one. Later the ancestor of the Pig-1 clan gave one of his men's houses, Wariman, to Akɨmbrɨkupan's Bird of Paradise clan. Nevertheless, the descendants of Kapi never reordered and renamed this important post near the entrance, and it stayed there nameless and lifeless as a reminder of the split between the two brothers. Kapi kept his left side as appropriate to the younger brother in the same way as in the arrangement of flutes during initiation, or in the patterning of dancers in the ordinary dance inside a house.

As we move further inside the men's house we learn that posts have different owners. It is easy to unveil the pattern behind this. The owners of those posts which I have numbered 3–8 are all somehow related to the Crocodile-1 clan. The whole of the Hornbill clan (a 'branch' of the Wallaby clan), like the Warinjam lineage of the Pig-1 clan, lives on the land of the Crocodile-1 clan and has not built a men's house of its own. While the posts owned by Crocodile-1

TABLE 9. *Posts and names of carved spirits in Yanbonman men's house*

	No.	Name of post	Owner's lineage	Owner's clan	Name of female spirit on top of post	Name and sex of carving facing front	Name and sex of carving facing back
	1	–	–	–	–	–	–
Front	2	Kapi (M)	Mambinjam	Crocodile-1	–	–	–
2 1	3	Wandas (F)	Mambinjam	Crocodile-1	Wandas	Kapi (M)	Mambinja (F)
3 4	4	Simbraynja (F)	Kanjimarpinjing	Hornbill	Simbraynja	Konggundambɨn (M)	Apandi (F)
5 6	5	Kunggunmay (F)	Wanbanimbuk	Hornbill	Kunggunmay	Sunggwandɨk (M)	Awunday (F)
14	6	Kanggismay (F)	Warinjam	Pig-1	–	Wari (M)	Akamay (F)
7 8	7	Awinjari (M)	Sapɨndam	Crocodile-1	Aykmay (not carved)	Awinjari (M)	Andinggwanma (F)
9 13 10	8	Sɨmbramɨnggawi (M)	Warinjam	Pig-1	–	Sarɨpi (M)	Asawunja (F)
11 12	9	Wanjingmay (F)	Yuwaynjam	Eagle-1	Wanjingmay	Sukiyapmay (F)	Pmbɨnggay (M)
Back	10	Kɨkriynwapanma (F)	Ambayngmari	Eagle-1	–	Simbanmay (F)	Asarmari (M)
	11	Yakuras (F)	Mɨndapmay	Eagle-1	Yakuras	Manggumɨnda (F)	Kwanggɨndi (M)
	12	Kɨkrimari (M)	Ambayngmari	Eagle-1	–	Wapaymay (F)	Sandikin (M)
	13	Sɨmasa (M)	Yambarinjam	Eagle-1	Wanjing	Wanikupan (M)	Sɨmasa (M)
	14	Sapɨn (M)	Sapɨndam	Crocodile-1	Marɨminja	Sapɨn (M)	Sapɨnda (F)

are still kept on the left side, those on the right are owned by Hornbills and by the Warinjam lineage of Pig-1, the one adopted by Crocodile-1. And while the men from the main lineage of Crocodile-1 own the post called Kapi, the men from their second lineage own the front central post.

The five posts at the rear, including one of the two central posts, are owned by the Eagle-1 clan, the largest group living in the ward of the Crocodile-1 clan. The lineage of the older 'brother' keeps the right posts and the lineages of the 'younger' the left ones.

All the posts except the one called Kapi are carved on both sides, the figure of one sex facing the front and the figure of the other sex facing the back. The two figures on the single post are usually referred to as 'brother' and 'sister'.[17] The male figures on the front eight posts and the two central ones face the entrance, while the male figures on the four rear posts face the back exit. By way of contrast, the female carvings on these same posts all face inwards towards the centre of the men's house, where initiation is usually held. The reason for such an arrangement lies in the division of the men's house itself: the front belonging to Crocodile-1 and the rear to Eagle-1. Thus the male figures on the posts of Crocodile-1 and those associated with it face the front, and those of Eagle-1 face the back, all looking after the fireplace of their own descendants.

Most of the posts have an additional face of a *wunduma* (female spirit) on the head of every carved figure. These *wunduma*, as in all other carvings, 'press down' the other spirits, anchoring and securing the firmness of the whole men's house. Sometimes their names are those of the posts themselves. Though some posts have male names they are not *wundumbunar* (spirit of a dead man); they are all *saki* (male bush spirit). As the figures represented are not ordinary men, but those from mythological times when they had all kinds of powers, they are said to have been *saki* before becoming ordinary men after settling down in the village. Thus the first ancestor of Ambonwari, Kapi, is also called *saki* and never *wundumbunar*.

The majority of carved figures are not related to and do not depict mythological scenes, as Forge (1979: 279) has already pointed out for the sculptures of Papua New Guinea in general. Nevertheless, some of the carvings have their own 'stories' (see also Bowden 1984: 447; 1990: 485). I have chosen to describe more fully two of the posts in the men's house, both owned by the Eagle-1 clan. I do so for two reasons: first, by describing the main characteristics of these

[17] Analysing Poole's data on the Bimin-Kuskusmin, Annette Weiner writes: '[In] understanding the life cycle of women, we need to analyze more than women's reproductive abilities in the narrow biological sense. The role of women as sisters is essential to the life cycle of men who themselves are working throughout initiation to achieve their own regeneration through ancestors. If, however, we note that women are an integral part of the life cycle, we must understand that women also represent something essential about the society . . . The regeneration of agnatic identity can only be created within each generation through the work of men as brothers and fathers and the work of women as sisters and mothers. Agnatic identities must be attached actively to ancestral identities' (1982: 59–60; see also p. 62).

posts we can anticipate similar patterns on the others; second, both posts are used explicitly to give complementary moral messages during initiation and are carved with more detail than others.

If one looks at the rear central post called Sɨmasa (no. 13 in Table 9) from its back one can see that, like every other post, it is painted with red, white, black, and yellow ochre. Every pattern on the carving has an appropriate colour. There is a carved face, the *wunduma* Wanjing, on the top of the post with a short extension of her body. With her painted face and body, round mouth and eyes, and a spinal vertebra in the middle, the *wunduma* 'sits' like a headdress during a dance on the head of Sɨmasa. His face is elongated, his mouth open with his tongue stretching out and downward.[18] The carvings around his neck depict shells. His shoulders are emphasized and strong, lying just beneath the shells. His spinal vertebra, like the one on the *wunduma*, looks either like a central part of a skeleton or like a row of shells. This is why the shells of dancers also contain both meanings and thus identify the dancers even more with the ancestral spirits. Sɨmasa's hands hang down at both sides of his body. He stands on a belt made of shells in which he used to put *Cordyline* leaves and the dried skin of a flying fox. Because he is dead and a spirit, he is naked. More shells, representing his wealth, are carved on each side of the post. At one side there is a carved eagle, the 'totem' of the Eagle-1 clan. Below his belly a snake holds his flaccid penis (see Plate 10). Men explain that because Sɨmasa slept with Yakuras at the place of spirits, a *sakima* (a female bush-spirit) who lived there took the place of his lover. During intercourse her vulva changed into the head of a snake, bit his penis and killed him. Men explained that while usually snakes are *saki* they would never bite men's penises. They were sure that this must be a female spirit. This carving is used by Ambonwari elders as a warning to young men (and others) with its explicit instruction not to copulate in those places known to be the abode of spirits. Like most of the other posts, Sɨmasa is also carved beneath the house floor. The carving shows the lower part of a male body with bent knees. The hand holds a penis just near a bodyless vulva, which looks like a fish head (see Plate 11). The men explained that *yawin kwarɨk* ('split of vagina') is not yet opened, that *kasan yawng* ('the flesh/egg of the chin/shell') did not break yet. Her vagina will be 'closed' until her first menstruation—another message for young initiates.

Viewing the same post from the entrance of the men's house, there is another *wunduma* Wanjing on the top, this time with a more elaborated skel-

[18] I would agree with Leenhardt, who argued concerning New Caledonian carvings that 'when the carver wishes to honour the ancestor whose face he stylizes, he shows off the tongue, long and broad; he draws its outline. For him it is a concretization of the ancestor's power and wisdom when engaged in thought, speech, and action' (1979: 139). In contrast, Bowden associates the protruding tongue in Kwoma art with anger, aggression, and the penis (1984: 455). The two views, that of ancestral power and action and that of anger and aggression, however, are not so dissimilar if one considers the hardship and dangers that the ancestors had to endure. Ambonwari use an expression 'his tongue went outside' when they want to emphasize uncontrolled anger.

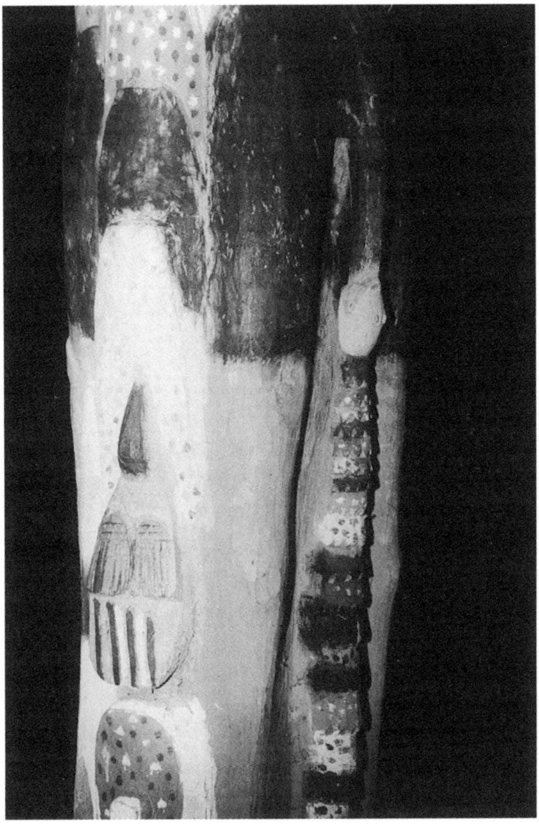

PLATE 10. Sɨmasa post: a carved eagle on one side and a snake holding a flaccid penis

eton, probably the work of a different carver. This time she sits on the head of Wanikupan, another of the mythological *saki* ancestors of the Eagle-1 clan. This post is the only one which has male figures on both sides. Wanikupan holds his penis just in front of an eagle's beak. Being his protector, the eagle does not harm him. Under the floor, the carving on the post portrays sexual intercourse. This time there is a headless female body with bent knees carved on the top, while the bodyless penis, with two circles representing the testicles, has entered her vagina.

The back of the post with its penis-biting snake on the top and a closed vagina at the bottom denotes those situations where sexual intercourse is forbidden. On the other hand, the front carvings show that if a man is under the protection of the eagle, and behaves properly, he will not be punished. It seems, though, that there is more to these messages. In the same way that spirit-things have both male and female characteristics, and thus embrace the life of Ambonwari in its totality, snakes are also androgynous and embrace 'otherness'

PLATE 11. Sɨmasa post beneath the house floor

in its totality. Thus boys, who go through initiation when their *kay* is under construction, are surrounded by their own existential world as well as by the threat of the world outside their own. The threatening strangeness of the unfamiliar other is presented in an extreme form—death. Before initiation Ambonwari boys are not yet beings in their own right. Initiation, then, makes them Ambonwari and provides them with both life and death. While before initiation they neither lived nor could die as Ambonwari, now both life and death become possible for them. Initiation in Ambonwari, as I argued elsewhere, should not be seen through symbolism of death and rebirth, but as an absolute beginning of both Ambonwari men and their cosmology (Telban 1997). To summarize, this experience centres on two propositions: first, the complementarity and integrity of maleness and femaleness in all aspects of existence; and second, the threat of 'otherness' and the familiarity and security of 'Ambonwari-ness'. This threat from the 'other' becomes even more explicit in the symbolism of the second post I discuss.

PLATE 12. Kɨkriynwapanma post: the figure facing the front depicts a pregnant woman

The danger of sexual intercourse in places where spirits dwell, this time for a woman, is shown on the other Eagle-1 post called Kɨkriynwapanma (no. 10 in Table 9). What is of interest here is that the figure facing the front indicates a pregnant woman (see Plate 12). It is best to quote Bob Kanjik's answer to my question about the meaning of this carving.

The bad snake came inside the woman's vagina and made its place there. It was that time when everyone was cutting wild sago palms and collecting sago grubs for the initiation party. She got this belly at that time. She emptied her basket filled with sago grubs on the ground in order to put them into the smoking baskets. While she sat there the smell of sago grubs entered her vagina. At the time she was apportioning sago grubs she did not sit properly. The way some women at some places sit is not the right way. She opened her legs and the smell permeated her vagina. And the snake smelled the scent of sago grubs. It got up from under the bed and came out. It ate all the grubs and slipped into her vagina. The woman took a piece of firewood and tried to hit the snake whenever it showed its head. But without success. She said: 'Oh, sorry all my brothers,

> my fathers, all the men from my family, all the men from my clan, all the men from my husband's clan, listen to me. I think I have a baby because this thing is inside me. This is the way that affects us women.'
>
> Her two brothers said: 'Let's go and get sago grubs.' Next day the two collected grubs and put them on the floor. They put one near the woman, very close. Another grub, and again another. They marked the length of the snake. The two discussed the situation: 'Don't put it too close. Leave some space in between.' The younger brother said: 'All right, now everyone move away.' He held a bamboo knife and his bow and waited. The woman said: 'Can you see me.' 'Yes.' The snake slowly advanced. One grub, another and again. Its tail slipped out. It moved away eating grubs. The man shot it and cut it. He killed the snake. He said: 'You, whoever you are, now I have held and fastened your power with my own. See, now you have a dead body here. Stop this behaviour of yours!' And so this way it ended. We mark this on the posts so that a new man can see it and say: 'That is true. Whichever bush you big men mark as being bad bush I must listen. I cannot sleep with a woman there. Nor can I sleep with my wife at such a place.' If the two have intercourse at such a place this is what would happen.

The majority of other posts are less flamboyant. Each post represents a man and a woman with *wunduma* on their heads. Small crocodiles, hornbills, and cockatoos are carved on the posts which belong to these clans. All these posts were carved either in the forest or in the village, with fences around to protect them from the view of women and children. The men painted the posts after the men's house was already completed.

There are four terms in Karawari language for what we would call colours, which are the colours of the clay found in surrounding creeks: black, white, red, and yellow-green. Most important is the dark/bright opposition. *Wurukupɨsan* is derived from *wur* (night, dark), *kupa* (big), and *san* (is, sit, hold), and people use it to express the dark appearance of something, whether it be black, dark blue, brown, or dark green. It is the colour of fighting. This term is also used when people talk about their own skins. For painting the posts the Ambonwari use 'dark ochre' (*wurukupɨsaki andi*) which they find deep under the water in creeks. They do not use soot or charcoal for painting the posts. *Angguringan* (white, bright) derives from *anggɨn* (a beam of light). People can say 'I open eyes' (*ama angguringan*) when they wake up in the morning. As a way of expressing the appearance of anything bright, white, or light, the term *angguringan* is used. A European is today called by this term. 'Bright ochre' (*angguringaki andi*), also called 'snake's ochre' (*wakɨn andi*), 'white bird's ochre' (*sawin andi*), or 'the ochre of white ants' (*amɨn andi*), is also found under the water of nearby creeks. People dry the clay in the sun.[19] The symbolic meaning of the white clay, especially when applied to the skins of mourners, is sadness, sorrow, and grief. The remaining two terms apply to appearances which might be called red and yellow-green. *Awiyan* (red) derives from *awi* (fire) and describes things such as the fire-like colour of the late sun, the dark

[19] For similar preparation of ochres among the Abelam, see Forge (1962: 12–13).

red and yellow betel nuts, brown leaves, and the reddish skin of young people. White people were once called by this term. People make 'red ochre' (*awiyar andi*) by heating yellow ochre over the fire. Ambonwari use the red oily substance from the fruit of *Bixa orellana* for body decoration, but not for painting the carvings. Red is symbolically the colour of anger (and therefore enemies), but even more it is the colour of celebration, especially following a victorious fight. While white can be metaphorically expressed simply by 'a white bird' (*saun*), red can be referred to by the term 'blood' (*yay*). Finally, *simbis* (yellow) is the name of a small shrub with bright yellow-green leaves. People find 'yellow ochre' (*sambisɨn andi*) inside the forest near Arkwas.

Slit-Drums

The Ambonwari have two sizes of drum, very large named ones, which always remain in a men's house, and smaller ones which do not have names, do not embody spirits, and can be moved from house to house when needed for ordinary house dances. The smaller ones are portable replicas of the larger ones.

The large, immobile slit-drums (*yimbung*) have their own names (see Table 8) and are regarded as beings in their own right. People say that these spirits used to live under the water with others, such as crocodile-spirits. They had their own villages under the lakes, their own men's houses; they performed all sorts of rituals, initiation being the most important one. Embodied in these 'heavy, hollowed logs' (Bateson 1932: 259), the *saki* were carried with the ancestors of the Ambonwari wherever they moved.

Drums have a body which is called *yimbung* ('slit-drum'). One end is designated the head (lit. 'chin'), the other end the 'tail' (sometimes called its 'penis'). Just under the 'chin' and emerging from the main body is a solid hanging piece of carved wood resembling a pouch (see Plate 13). This is called a 'fish-basket', where the spirit keeps its food. The slit at the top of the body of the drum is referred to as its 'mouth'. Midway along the slit of the mouth is a solid piece in the shape of a half-circle. This is called its 'fruit', which, people say, is important for the quality of the sound. Of course, there must be also a heavy stick to beat the drum, which when not in use pokes through the drum's mouth and rests in its empty inside, called its 'belly'. On either side of the drum the patterns depict its intestines and lungs, in appearance resembling a line of crescent moons or kina shells. Between the head and the main body is a 'belt' of carved shells with the face of a *wunduma* in the middle. On the top of the head the two 'breadfruit seeds' indicate the spirit's eyes. Other carved patterns depict shells, frogs, and birds: the spirit's wealth and those foods which new initiates are not permitted to eat. All the main slit-drums have incisions on their bodies denoting the number of people killed by members of the owning clan.

PLATE 13. Head, pouch ('fish-basket'), and part of main body of Sangankumban slit-drum of the Crocodile-1 clan

The Ambonwari use slit-drums during ceremonies and rites or whenever they wish to announce a message or to call someone from another ward or from the bush (see Plate 14). The drum signals announce that someone has just died or has just been buried, that someone has returned from a long trip, that a pig has been killed, that there is a village gathering, court or community work, or that there is a Sunday 'church' meeting. As people are familiar with constantly unfolding events in the village, these signals in many ways confirm their expectations, hopes, and fears. Slit-drum signals are used to inform a hunter who is still in the forest that his dogs have already returned to the village. In the past the drumming announced an imminent fight with enemies or the celebration of a victory.

Every totemic clan has its own signal. As the clans are patrilineal and residence is patri-virilocal the totemic signals which are sent from the men's house are generally the signals of men. Like the Kwoma but unlike the Sawos (Zemp and Kaufmann 1969; Aitken 1990), Ambonwari drummers, when calling someone, first beat the signal of the person's own clan followed by the signal of his mother's clan. Since two brothers both have the same signal, the drummer can substitute the call of the mother's clan by one representing the clans of the men's wives. He can also add the latter to the previous two. Such a threefold signal can take over five minutes to transmit.

PLATE 14. Beating a slit-drum during a re-enactment of initiation ritual

Whenever important events occur in the village such as an unexpected death, sickness, a serious dispute, or the arrival of visitors, people may decide to call those who have already departed for their gardens, to process sago, or to hunt and fish in the forest. These personal signals seem very impressive to a foreigner. The drummer (*kapuk yarar*) beats the slit-drum using rhythm and pitch to mimic the sound, the walk, the appearance or the behaviour of a totemic animal of a particular clan. People say *mɨn mariawk yan pasar yimbɨnunggɨna*, 'he sent/threw a talk/message with a slit-drum'. Though Aitken is firmly of the opinion that in the Sepik 'the melodic element is more important than the rhythmic one' (1990: 546–7), I regard them as inseparable: the melodic element may be emphasized in singing and dancing, but it is the rhythm which makes people understand to whom the message pertains and what it means. Fewer than ten men in the village know how to beat all drum signals, though the majority of villagers, including women, are able to recognize at least their own.

When the Ambonwari want to summon all the people from different wards a man beats the slit-drum, beginning with the signal pertaining to the Bird of Paradise clan as the first village ward down-river, and continuing 'up-river' with the signals of all those clans who have men's houses built. He does not beat the signal of his own men's house, so that everyone knows where the call is coming from. Omission here is the key to identity. Though the majority of people know only their own signals, they nevertheless recognize the different pitches of various slit-drums in the village. A single clan signal, which lasts between thirty and forty-five seconds, is repeated three times, with a short pause in between.

Besides those signals which identify the individual, additional signals usually accompany them. After the message the drummer can add a sequence of beats to urge a person to hurry up: it lasts for about fifteen seconds and sounds like *apra*, *apra*, *apra* ('come, come, come'). Once they hear the call people can reply by beating a large tree or a stump if they are in the bush, or by uniform strokes on a slit-drum if they are in the village. The drumming for 'church' (lasting over a minute) and other meetings (lasting thirty seconds) regularly begins and ends with fifteen to twenty seconds of rapid rattling with a stick in the drum's mouth to compel people's attention. Table 10 presents the Ambonwari terms for the signals of individual clans.

We can see from this table that there is no named distinction between the first and second Crocodile clans, or between different Pig and Eagle clans, nor is there a name for a signal of Hornbills and Cockatoos. However, these groups all have their own signals. The names of Ambonwari signals are obviously descriptive. For instance, the name of the Bird of Paradise clan signal refers to its intensely combative past; the name of the Pig's signal refers to the fact that they do not have their own flutes; and Eagles' signal name refers to the time of their arrival in Ambonwari.

Whenever someone dies in the village, the drummer will use one of the smaller slit-drums. Because of its higher pitch, its sound echoes over a greater area than the sound produced by large slit-drums. In the case of the death of a man, Ambonwari beat the signal first of his clan, then of his mother's clan, and lastly of his wife's clan. Then the drummer beats the special 'call of a spirit' informing people that he died. In the past, after only a few seconds of silence

TABLE 10. *The main signals of slit-drums*

Clan	Signal	Translation
Bird of Paradise	*kambɨn kapuk*	the call of a shield
Crocodile	*mambin kapuk*	the call of a crocodile
Wallaby	*kambansapɨr kapuk*	the call of a type of grass
Pig	*kambra sanggut kapuk*	the call of those without flutes
Eagle	*wi mɨnak kapuk*	the call of night
Cassowary	*awan kapuk*	the call of a cassowary
Eel	*masungun kapuk*	the call of an enemy
When someone dies	*kwin yimbung*	the slit-drum of a spirit
Fight	*kanggan ambɨr yimbung*	the slit-drum of the 'beginning of something'
Before the release of novices from a men's house	*kambɨn yimbung*	the slit-drum of young boys

the drummer would beat the number of victims that the dead man had killed during his life. These single strokes indicated the heads (*kanapanggi*) of his victims. In recent times the drummer has ended the death signal with a single stroke.

Spirit-Crocodiles

The names and characteristics of spirit-crocodiles travelled with the ancestors wherever they moved before settling down in Ambonwari. Like all other spirit-things they were always carved anew, and after a short ceremony their spirits became embodied in new objects. These *saki*, male bush-spirits, are simply referred to as *mambo*, crocodiles, and are kept in the men's houses of their owners ('fathers').[20] They protect the members of the clan who own them and they 'follow' them wherever they go. They were especially potent and helpful during periods of fighting. People recall stories from the past when a man who had killed an enemy would call the name of his clan's spirit-crocodile, announcing that it was the spirit itself and not he who actually performed the killing. One such story tells of how a young woman from a neighbouring village was captured and brought into the Ambonwari men's house. The men were furiously chewing betel nuts with ginger and aromatic bark. They held the spirit-crocodile in their hands, singing and moving according to the dictates of the spirit.[21] They killed the woman not with spears but by beating her to death with the snout of the carved crocodile. The rhythm of the drumming on the slit-drum was fast and loud and covered the screaming of the dying woman. When everything was over, the men came out of the men's house and said: 'We did not kill her, Aymakan killed her' (cf. Harrison 1993: 113). Men say that once they hold the spirit they cannot control it. It is *saki* who moves and pushes them around.

These carved spirits were once very secret. In the 1950s the other Karawari people from Manjamai village reported to the government that Ambonwari was still a 'bushmen' place where people continued to perform their rituals. The Roman Catholic Bishop Leo and Father Raymond came to Ambonwari with

[20] The Karawari-speaking group is the only one in the Sepik to have these large carved crocodiles as totems. The neighbouring Imanmeri had a few in the past, but they were given to them by Karawari people. The carved crocodiles from Manjamai in 1936 became the first objects in Papua New Guinea whose acquisition by outsiders was prohibited. Nevertheless, many carved crocodiles were illegally purchased and exported. Their place of origin is now difficult to ascertain (Craig 1992). Earlier records concerning Karawari people and carved crocodiles (Bühler 1961; Haberland 1975; Tischner 1965) are inaccurate. Previous researchers in the area paid only short visits.

[21] In a similar way the spirit of a dead person 'pushes around' those who carry the body out of the house. While young men carry the body in a kind of box made of wooden sticks, the spirit of the dead person guides the carriers towards members of the house and other relatives, pushing them out of the house. In this way, people say, the spirit shows her or his concern for the living and does not want people to sit in the house mourning for too long.

some men from Manjamai. After the prayers everyone gathered at the front of the main men's house. They brought out the carved crocodiles, bamboo flutes, and other objects while Father Raymond told everyone: 'This is not a spirit, it is just a stone; this is not a spirit-crocodile, it is just a piece of wood. You cannot worship these things.' Men were told to blow into the flutes in front of women and children. Father Raymond baptized the men's house and renamed the female masks using Christian names. The crocodiles were baptized as well.[22] Twenty years later, in the same year that Papua New Guinea became an independent nation, the Ambonwari rebuilt their men's houses and carved new spirit-crocodiles. As always, they were faithful reproductions of the previous ones. They put them into their men's houses and recommenced the initiation ceremony. The spirit-crocodiles became secret again.

Not every clan has a spirit-crocodile (see Table 8). In terms of village relationships it is important to know that the men from the Pig-1 clan, upon arrival in the village, gave not only the name and some sago leaves from their men's house, Wariman, to the Bird of Paradise clan, but that they also gave them the spirit-crocodile Kangris. The men from the Bird of Paradise clan renamed the spirit Kongguniapan, using a name from their own set. The exchange of these important objects for a place inside the village formed a sibling relationship between the two clans.

Most food presentations inside the men's house are to this category of spirits. They have an important role during initiation. The novices who are brought into the men's house are frightened by the men who hold the crocodiles and push the boys around. Also, the carvings on their backs depict all the food prohibited during and after the initiatory seclusion: bananas, little birds from the grassland, different kinds of Malay apple fruits, breadfruit seeds, all kinds of frog, a large hornbill. The tail has a small carved crocodile which represents one of the last foods to be eaten once the initiation ceremony is completed. Where the tail joins the body there is a carved figure. Having only arms but no legs it symbolizes Panggamari, the fighting spirit. This *saki*, which is said to live in the creeks, represents a dangerous spirit which drives people to fight, making them fearless and strong. The design on both sides of the crocodile represents the belly, intestines, and lungs. Four holes on each side allow people to hold the crocodile firmly.

At the tip of its snout and on the carved belt of shells around the neck of every crocodile, the carvers inscribed a face of a *wunduma* (female spirit). Through the middle of the head runs a snake, also prohibited in the novices' diet. On each side of its head are two circular patterns depicting a large frog, while the two carved breadfruit seeds represent the crocodile's eyes.

[22] It seems strange and contradictory that while Fr. Raymond first announced that these objects are 'just wood', later he actually baptized the objects and named them. The whole event of the revelation of these secrets is today viewed as the cause of the deaths of village fathers and their children shortly after the village was baptized.

Throughout this presentation we have seen that all those spirit-things which are called *saki* remain in the male cosmological domain. By giving them form and maintaining the social relations between spirits and Ambonwari men since the time of the ancestors, men not only establish more balanced relationships with fearful things but gain command over them and thus over their own existence. At the same time, by secluding the carvings in the men's houses and by keeping them secret, men avoid the unpredictable consequences which they believe would occur if women and children were to see these spirits. All those objects which have a straightforwardly female character are visible. This is because their identity is defined by female *kay*, that is to say, by their role in sustaining domestic existence (home, familiarity), in anchoring the village, and in relating wards and clans through their movements in marriage. In contrast, the male spirit-things are important in inter-clan and inter-village relationships. When both categories of spirits are placed together (as in the men's house) they represent the stability and harmony of the village as a totality, bound in social relations with the beings that control the world upon which the village depends. In this way spirit-things, Ambonwari persons, and the forces upon which they depend are drawn, to use Cassirer's phrase (1972: 83), into a single 'society of life'.

7

Transformation and Presentation

Throughout the book I have been concerned with *imɨnggan kay* (the way of the village) in its various manifestations and refractions. For Ambonwari the most salient feature of *imɨnggan kay* is that it embodies and reproduces *kupambɨn kay*, 'the way of the ancestors'. As noted in Chapter 2, however, this phrase is often used to refer specifically to those practices which anthropologists refer to as ritual, and amongst these the paradigms of this ancestral way are male initiation and female menstruation rituals.[1] And for the Ambonwari there is no doubt that it is the male initiation ritual that exemplifies *kupambɨn kay* in its purest form, being also referred to as 'the way we are/live/stay' (*kay apia kaykan*). As in Fiji, 'ritual is doing something in a Fijian way; its meaning arises from *how* things are done, as well as from *what* is done. Moreover, if one does something in a Fijian way then the doing of it becomes Fijian. This is a processual notion and so it can incorporate change' (Toren 1988: 712).

The initiation ritual involves more people, more time, and more resources than any other; it implicates all the most powerful spirits, the most important themes and values, and aims to achieve the most important of ends—the reproduction of ancestral *kay* both in individuals and in the village as a whole. Characteristically, the ritual focuses on males and, like their spirit-things, is kept secret. For this reason, and because I promised to send a copy of the book to Ambonwari, I cannot discuss this ritual here. The menstruation ritual is less secret, however, and because it aims to reproduce the way of the ancestors as it pertains to women, it will serve the purposes of this chapter. In addition I discuss the all-night dance ceremonies, which are the most important secular expression of *kupambɨn kay* and are in any case always the first part of male initiation in which the whole village participates.

Having stressed, as the Ambonwari do, *kupambɨn kay*, I would not want to lose sight of the fact that *kupambɨn kay* has its life in *kay*: the ancestral way is the crystallization of the life-world, its themes and orientations as they present themselves to those who live it, but for that reason it cannot contain within itself the grounds of its own existence. *Kay* represents those grounds. Accord-

[1] I propose to sidestep the difficult issue concerning the specificity of ritual. For my purposes I can rely upon the intuitions of the reader about what kind of behaviour does and does not constitute ritual. Insofar as I am concerned in this chapter with rites of passage, the classic writings of van Gennep (1960 [1908]) and Victor Turner (1980) are relevant, though with my reservation concerning symbolic death and rebirth (Telban 1997). More generally, Kapferer's suggestion that ritual transitions 'involve, in many instances, a transformation of context and a transformation of the identities of the participants' (1984: 5) is apt. However, I take Lewis's point about the inadequacy of claims that ritual constitutes communicative or expressive aspects of actions and behaviour (1980: 16–38; see also Gardner 1983).

ingly, I begin this discussion with ritual practices whose character as bodily *kay*, materialized through actions, are obvious.

> To understand ritual practice, to give it back both its reason and its *raison d'être*, without converting it into a logical construction or a spiritual excercise, means more than simply reconstituting its internal logic. It also means restoring its practical necessity by relating it to the real conditions of its genesis, that is, the conditions in which both the functions it fulfils and the means it uses to achieve them are defined. (Bourdieu 1990: 97)

Ambonwari ritual, like the Fijian one, however, 'is effective not because it denies the passage of time and the changes time brings (as Bloch suggests for Malagasy ritual) but because it incorporates change under the rubric of appropriate action' (Toren 1988: 713). Individual ritual practices in Ambonwari are like the large public rituals performed to influence the outcome of a process which can be influenced and needs to be facilitated. For example, a man who wants to stop rain can wrap a dry sago pancake with smoked sago grubs in the flower sheath of the betel nut and hang the bundle on a stick at the front of a men's house. He then enters the men's house and talks to the spirits, saying something like *Inja sipi. Ipa sanggwari. Manam pasi. Wi yapakupaki kayimɨnggiam*, 'Here is sago pancake for you. Look at it! Do not make rain. Let there be a clear night.' Ambonwari say that by presenting dry food a man reveals his wishes to the spirits. By acting in this particular way (*kay*), that is, through this 'practical mimesis' (Bourdieu 1977: 116), a man activates his relationships with spirits in order to set in motion the rain-stopping process. In such a practice a man's interaction with his spirits should be understood as an aspect of ordinary, daily relationships with spirits; they are relationships comparable to daily relationships between humans.

Given what has been said so far about the intimacy of *kay* and those things and persons identified with it, it comes as no surprise that sicknesses of children, for example, are attributed to the wrongdoings of their parents. Through their close bodily relationship the parents' *kay maman*, wrongdoing in this case (lit. 'bad habit'), is easily transmitted to children, who are an extension of their parents' *kay*. The sickness of either a child or an adult is then manifested through unhabitual practices, through the impairment of their usual *kay*. In healing rituals, the specialist tries to find the cause and remove the objects which are the embodiment of sickness. A healer tries to restore the *kay* to its previous condition. The sickness becomes a collective issue (of household or lineage, for example) including their own familiar spirits. The wrongdoers then have to 'cut themselves off' from the practice which caused the sickness by the payment of compensation, seclusion, washing, food presentation, and so on. Ambonwari people often say that they have to look after their skin properly, which includes their acts. In regard to sickness we can make some common observations: a healthy body is defined by its visibly healthy skin and its activity; a sick body is defined by unhealthy skin and its aberrant activity. Thus, a healing ritual is concerned with the extraction of invisible stones, teeth, bones, and thorns from beneath the unhealthy skin, the restitution of *kay* by identifying the cause of sickness in wrongdoing, and finally by addressing the spirits of the household, lineage, or clan who share collective identity with the sick person.

Unlike these rain-stopping and healing practices, large public rituals promote and further the attainment of collective aims, although they do not rest upon a different cosmology. However, they do mobilize the village as a whole and involve many secret and sacred elements of village life: names, stories, spirit-things. Public rituals in Ambonwari are very well prepared. It is arranged in advance who will enact which role, and the actions of all participants in a ritual are discussed and rehearsed prior to its commencement. In the way that any practice has its sequential order of acts, ritual, the procedures of which are not embodied, has to be structured properly in order to achieve its ends. Thus ritual follows and reflects previous scenarios, plots, and patterns of village life. The emphasis on correct procedure can be seen in the continual attempts to confirm the recollections of those who have been through the rituals before. But ritual is also that kind of practice which enables people to invent and make the rules. Not only do people expect rules in ritual (Lewis 1980: 11; Parkin 1992: 14–15), that is, some kind of proper behaviour and sequencing if it is to be effective, but the very idea of 'the rule' is the ultimate product of ritual. Understanding (*wambung*) and way of doing things (*kay*) are interrelated; *kay* produces *wambung* and vice versa. Ritual and rules are intertwined, inseparable, each being the function of the other. Both rituals and rules, as well as ideas about them, can be seen as antidotes to existential and epistemological anxiety, as antidotes to the fear of disorder and the loss of identity.[2]

The focus of ritual is *kay*: either its formation (a new beginning or transformation) or its presentation. In either case, what matters is that embodied *kay* is not something that one simply possesses (like knowledge) but something that one *is* (Bourdieu 1990: 73). Such rituals as initiation and first menstruation are concerned with the formation of being and with the transition from the state of not-yet-being to being. The *kay* of the initiates is in a process of becoming. But such a process always depends on the presence and capacity of those who have already embodied such transformations—those who have already 'become'. For them, the ritual seniors, the ritual is an occasion for the representation of

[2] In other words, one might say that 'ritual is one kind of custom. Custom . . . implies behaviour guided by rules whose claim to validity is based upon the general acknowledgement of those who practise them that these rules are theirs and proper to them, ancient and to be observed because things should be done in that way, as they were done before by their forebears: tradition justifies their practice' (Lewis 1980: 11). On the other hand, we should not forget that practice also justifies their tradition. Children embody many practices before they begin to think of them in terms of their symbolic meanings or in terms of tradition. Habitual practices, then, as much as initiation rituals, over time both construct and justify the idea that they are meaningful (Toren, pers. comm.).

their *kay*. For children, this means that they 'have to construct meaning anew and through the social relations in which it is embedded, [and] they have also inevitably to submit to the meanings that others have made before them' (Toren 1990: 230). Other rituals and ceremonies, such as the all-night dance, focus purely on representation of *kay* and perhaps its celebration or commemoration. Thus we might say that ritual as an ensemble of practices effects both the embodiment of new values and identities (in formation, change, and transition) and the reembodiment of old values and identities (in representation and commemoration).

Initiation and first menstruation rituals separate those who have undergone them from those who cannot undergo them in any sense, that is, men from women; and separates those who have undergone them from those who have yet to do so, that is, adults from children. Thus ritual institutes 'a lasting difference between those to whom the rite pertains and those to whom it does not pertain' (Bourdieu 1991: 117). To reiterate, a girl who has gone through a first menstruation ritual is differentiated not only from boys and men, whose initiation ritual embodies those aspects which pertain to males only, but also from those girls who have still to go through the ceremony. In this particular ethnographic case the point of rituals is not to produce men and women but to produce Ambonwari men and women. Before these rituals a boy or girl might die or be abducted, in which case such a person is lost before they become a complete Ambonwari. This is highly undesirable, and people often go to some lengths—at least in the case of a boy—to preempt its occurrence by initiating the corpse (Telban 1997).[3]

What distinguishes these two rituals from other ceremonies is not only that they include participation of spirits but, as Ambonwari men emphasize, they address, please, and stabilize *imɨnggan saki*, the main spirit of the village, the snake which lies under the village. People say that while male initiation shows to this spirit the new men, often referred as 'dogs' (warriors), who will protect the place and preserve the knowledge, the first menstruation ritual shows to this spirit a new woman who will anchor the place and secure reproduction and continuity.

First Menstruation

While male initiation is an event that is concerned with the entire set of boys who are eligible to become initiands, and therefore directly involves many households in preparations, the first menstruation is an event directly involving

[3] On 28 Nov. 1990 a male baby, only a month old, died. The post-mortem one-day initiation ritual was performed, in this way giving the boy his identity before he was buried (Telban 1997). Of course, it often happens (as I witnessed and was told by several mothers who had done so) that a baby is simply buried by its mother before others realize what has happened.

only the girl's own and her mother's lineages. Nevertheless, most if not all villagers participate at some stage of the ritual, and those linked to her *konggong* have a particular interest.[4]

The first menstruation ritual seems to be disappearing faster than other rituals, but I was able to talk to young women who were among the last of those initiated into adulthood in the early 1980s. Yet there is evidence of revival: in 1993 a letter from the village informed me that a girl had just been placed in the fenced area inside the house and that the whole ceremony would be performed for the first time in a decade. The abandonment or transformation of rituals can be attributed to social changes induced by the cash economy, by Christianity, by travel and school education, all of which require new, or rather, different values to be embodied in a person belonging to the nation state of Papua New Guinea. As one old man commented, once upon a time the big men directed village life, but today it is the young people's decisions which dominate the village. Nevertheless, because opposition to practices and habits which are at the same time also part of young people's lives would seriously threaten the stability of the village, even young people endorse those activities which they see as providing that stability.[5] Young people also recognize the powers of their ancestors and their spirits, and when they go on a trip they say that the village spirit will follow them and protect them. The identity and powers which Ambonwari young people want to embody combine their own ancestral ones with the ones introduced by recent transformations of their world.

But let us return to the description of the first menstruation ceremony. In general, mothers do not explain menstruation to their daughters, nor do their elder sisters. It is her classificatory sisters who tell her about it. In turn, a girl does not tell her parents when she first menstruates. The most visible sign of her condition is that she behaves as if she were sick: on the morning of her first menstruation she stays 'in bed'. Menstruation is usually referred to as 'the way of the moon' (*sunggwin kay*). It is said that 'she is in the way of the moon' (*sunggwin kay mɨn kaykan*).

As it was explained to me, the whole procedure took the following course. Female members of the family and their classificatory sisters sectioned off a small area at the back of the house with woven mats and cloth. The newly menstruating girl, who was called *pandayma*, was secluded in this temporary room, called *manggay*.[6] Her father then discussed the matter with his brothers

[4] Among the Chambri, girls are initiated only when their mother's brothers wish to demonstrate their wealth by sponsoring a ceremony (Errington and Gewertz 1987: 160 n.).

[5] I do not deny that these changes also become embodied and internalized: they are not just external events, and are important in the construction of Ambonwari life. However, during my stay in the village people themselves emphasized the ancestral *kay*, continually discussing the possibility of organizing a male initiation ritual. They held a small, one-day initiation for one boy, complaining that this would not please their *imɨnggan saki*. At the end of 1993, many months after I had left the village, the proper ritual finally took place when six boys were put into a men's house.

[6] Ambonwari do not (and did not) build special menstrual huts for women outside their living-houses as do Abelam and Arapesh people (Kaberry 1941: 361; Mead 1963: 92; Scaglion 1986: 158).

and the men from his maternal and wife's lineages. They needed lots of food for a period of one to several months while 'the menstruating girl slept' (*pandayma mɨn paykan*).

During that period of seclusion, classificatory sisters instructed the girl about the need for her to sit well. She was not allowed to be seen by men. A special exit was made by removing either a section of floor or wall. If she needed to relieve or wash herself she had to cover herself with a woven hood, called *andaypa*, before leaving the house. Most often she had to use two hoods: one all the way from her head to the ground at her back and another covering her face at the front. The latter had holes for her to see through. The two hoods protected her from the gaze of other people, which in her condition would be equivalent to a physical contact with outside world. In her state of becoming, such 'tactile' gazes could interfere with the transformation of her *kay*. She was allowed to communicate only with those women who looked after her. The Ambonwari say that she had to forsake her childish practices and take on the identity of an adult woman. This means that she was instructed about the women's *kay* (habits), most importantly, about the importance of marriage and anchoring her household. Once in the confinement of *manggay*, it was important for her to remove the woven hoods. She could influence other people's activities and daily processes such as growing plants (they would not thrive) in the gardens or hunting. Such an act, people say, would 'cover' all the animals in the forest and hunters would be unable to find any food.[7]

A *pandayma* was instructed to put on several pads.[8] During her first days of seclusion an older female relative took the girl's pads with her first menstruation blood and buried them. A female relative planted a betel nut, banana, or coconut in the soil so that the palm's roots would cover the pads. This practice supposedly prevented a girl from becoming pregnant. The girl herself was not even aware of this. She was shown the palm growing out of her menstrual blood only later, when she was about to get married. She then pulled it out of the ground, thereby making her pregnancy possible.

Each day after sunset, the *pandayma* came out of her room and sat on a mat in the middle of the house. She had to keep her legs crossed and hold her hands in her lap. Her right hand had to lie on the left one with both palms turned up. She was forbidden to move the right hand, which later in her life was going to be used for catching fish. When women gave her betel nut or cigarettes she was allowed to use only her left hand. By sitting still and by holding her right hand steady a girl embodied two of the most important aspects of her future life: firmly anchoring the household and the path between her own and her future husband's place, and acquiring the necessary stillness in her right hand for

[7] Women wear such hoods during mourning as well, and are required to uncover their heads if men go hunting.

[8] In the past women used *kumunggrum*, pads made of the tender young leaflets of the sago palm. After these leaflets were beaten, the fibres were fastened at front and back to a rope around the belly.

catching fish.[9] In her state of becoming, her stillness did not frighten away the game which her menfolk were hunting. The instructions of the older women constructed her *wambung* ('insideness', Heart), that is, understanding, feeling, and way of thinking appropriate to an adult Ambonwari woman.

Every night the *pandayma* had to wait until everyone in the village was asleep before she was allowed to lie down. If not, people say, the whole village would sleep 'on top of her' and she would not only remain thin but would not embody the knowledge and practices of an Ambonwari woman. When all the men inside the house had gone to bed, the *pandayma* and her mother retreated into the mosquito net near the *manggay*. Throughout these weeks and often months the girl stayed inside the fenced room in the house, keeping her breasts uncovered so that they could grow. The embodiment of womanhood included putting on weight and getting larger and fuller breasts.

At dawn, when the village was still asleep, the women of the household and the classificatory female relatives who took care of the *pandayma* brought water from the creek and they all washed together behind the house. All the participating women then covered their faces and bodies with white clay (male members of the household also covered parts of their bodies with clay). One could say that this sign of mourning denoted the sorrow caused by 'cutting off' the bond between a girl and her parents, and the anticipation of her departure from the clan area after her future marriage. One could also say that all the participants in this sensitive period of transition from being an extension of her parents to being an individual and conscious Ambonwari woman (i.e. being herself) have to protect, because of their temporary connection with spirits and those issues which pertain to cosmogony, their own as well as other people's skins. The ontological change of her being, if not carefully guided, could destroy not only her life but the whole Ambonwari village and its cosmology. After the young woman had washed and smeared the clay over her skin, she returned through the back entrance of the house and sat in the *manggay*, the whole day long. The women brought her materials for weaving baskets and mats. Small children of the household were also allowed to keep her company. She was not supposed to be left alone at any time; there was always a *kwanma* ('watch woman') or *sɨnggandarma* ('the one who sits with her') nearby. If her marriage had already been arranged, her future husband helped all the men from her lineage with hunting, fishing, and cutting sago palms for sago grubs. He tried to show himself to be a good, generous man, able to find food. After the young woman's lineage had collected an impressive amount of meat and smoked it, they went to cut and wash sago.

[9] During male initiation boys also have to acquire stillness in their right hands. In their case the purpose was to confirm the requisite skills for hunting and fighting with spears. Michael Jackson, in his discussion of initiation ritual among the Kuranko of Sierra Leone, writes that 'the bodily practices mediate a personal realization of social values . . . Bodily and moral domains are fused, and, as Kuranko say, maturity is a matter of common sense which is achieved when inner thoughts are consistent with spoken words and external actions' (1989: 131–2).

The day before the final celebratory rite the girl's mother removed all the mats and fences except one, and everyone went to sleep earlier than usual. In the morning the girl sat on the mat in the middle of the house for the last time. The women from her maternal lineage brought her sago pudding to strengthen her. The girl was then allowed to talk to other people beginning with her *warɨmbɨnma* (dancing partner), who also brought sago pudding to the house. Women from all the houses of the village began to arrive with their contributions of sago pudding topped with meat or fish, while they received similar food from the house in return. The whole village was finally able to see the new woman.

The young woman's maternal lineage placed a stool painted with red clay on the mat in the middle of the house and dressed her in a new skirt (a grass skirt in the past). While she sat on the stool, a practice which is otherwise solely the prerogative of men, her *away* (mother's brother) cut her hair as short as possible, leaving only a small tuft of hair ('knot of hair', *wapɨs mamɨnɨng*) at the front of her head. Her real or classificatory father's father (*wasamari*) and all those who took care of her during her months of seclusion decorated her with red and yellow clay, and with shells and necklaces. They showed her all the food they had received. While all the women and small children came into the house, talking and joking, the men, including the girl's father, went to the men's house.

All the women of the young woman's lineage and clan then dressed as men to represent, it was said, the woman's 'fathers' (*awkum bun anunggum*). They put *Cordyline* leaves over their buttocks and flying fox skins to cover their genitals; they covered their heads with plumes made of cassowary feathers and held cassowary bone daggers in their hands.

Simultaneously, the women of the girl's maternal lineage dressed as 'mothers', and acted as 'women' (*awkum bun awkum*). Those dressed as 'fathers' began to tease and joke with those who represented 'mothers'. With their feet they 'hooked' the 'mothers' between the legs (cf. Bateson 1958). They joked: 'Is this the vulva which this nice girl came out of?' The women from her maternal lineage presented the girl with a fishing spear, lime container, male and female paddles, and a mallet for pounding sago. The girl's lineage would later pay for these presents with shells and money.

The women called the men to fetch the food which would later be distributed to the whole village, and to take the girl's stool to the men's house. The men would make incisions with a bamboo knife (a razor blade) on the chest and arms of the girl's future husband and all those who were prepared (or—in the case of his *warɨmbar*—obliged) to share his pain. At the same time the men would cut the skin on the forehead, above the breasts, or on the back of the newly initiated woman. She would stay painted and decorated for two days.

In the meantime a man who knew the slit-drum 'language' beat the drum signals identifying every men's house in the village, beginning with the Bird of

Paradise clan down-river. Thus the clans were called starting with the one at the 'snake's head' and ending with those at its 'tail'. All the men gathered at the men's house whence the signals came. The young woman's clan distributed the food and betel nuts to the other clans; they would later redistribute them between individual lineages and households.

After two days the young woman went to the stream early in the morning and washed. People say that 'she wanted to throw away dirt', *usɨk pasararin.* This was the last day she would be covered with a hood. That same day she followed her brothers and sisters to the sago forest to collect sago grubs. But first she had to collect some firewood and tie it with a vine into a bundle. Then she had to make a bundle of greens. The men cut down a sago palm and the women washed the flour. After they had gathered the sago into their baskets the whole group moved towards the wild sago palms which were cut a month or two before.

While collecting sago grubs the young woman's small tuft of hair was removed from her head. If she had already decided on her future husband and if he was among the company, he would look for a sago pupa in its woven pouch. *Wuniam*, a transitional form between a grub and a beetle, was also in the process of 'becoming', like the young woman. She was then supposed to face the sun while everyone else turned their backs on it. Therefore, we may notice, having had intercourse with the moon she entered into a similar kind of relationship with the sun too. They say that the sun should help a new woman in her future marriage: she would have to have her eyes wide open, to see the needs of her husband (a similar instruction about caring for their wives is given to young men before they marry). While the woman faced the sun her future husband approached her and pressed and smeared the *wuniam* on her forehead. He then took a young sprout of sago palm which is also in transition—ripening to maturity—and beat her on the forehead until the sprout burst. In this way he put his mark or taboo on her (*yan awsar*). She still belonged to her parents and was still wearing their decorations, but after the next time she washed she would call him *mayn*, 'husband'. When she returned to the village her parents would notice the dark greasy mark on her forehead and know what had happened.

The whole company washed in a stream, covered their bodies with white clay, and paddled back to the village. The young woman carried everything back to the village: a basket full of sago flour and sago grubs was suspended from her forehead and hanging down her back; on top of it was a bundle of firewood, a bundle of greens, and a basket of young sago sprouts. She also carried a bundle of sago palm stems which would be used for torches. She had had to find and bring back all those things which were most important for every household, showing at the same time her worthiness as a future wife. Her hood was folded. All day long the whole party had been forbidden to eat, to chew betel nut, or to smoke. If they ate or chewed something, people commented that they would become deaf, that is, they would lose understanding (Heart). For the young woman this would mean that she would not hear her future husband's demands and would be thought useless for neglecting her work. I should emphasize that similar instructions concerning his own behaviour are given to a young man during his initiation and before he gets married.

In the village her *arkay* (father's mother) had already put the ashes from her father's fireplace, symbolizing the hearth of the household, at the front of the house. Before the young woman entered the house she removed the ashes with her foot showing that from then on she would manage her own hearth in some other household. At the door her *arkay* beat her gently with a broom, acknowledging in this way her womanhood. She put down everything she had carried (symbolizing not only her last work in this household but her capability as a new wife) and checked the small basket prepared for her by her *arkay*. In the basket was baked sago (*sipi*) and leaves of tobacco. Because she wanted to 'cut' her past she was not allowed to eat the baked sago.[10] She took the fried sago pancake to the nearest house, and the mother of that household beat her again with a broom. Everyone present repeated: 'Take from her that baked sago.' In this way she ended her transitional period along with her childhood past. She removed or 'threw away the dirt' (*wusɨk mɨn pasan*) and 'cut her [childish] hands off' (*yanggri mɨn kran*). The young woman returned to her parents' house empty-handed and, for the first time since the seclusion began, ate salted sago pudding with others. Salt in the pudding, as I mentioned in the first chapter, represents the close of her period of transition and reincorporation into the daily stream of social life.

The young woman continued to live an ordinary life. Her paternal relatives would try to find a pig, a cassowary, a possum, or a few large fish and use this meat for their final obligation: the removal of the *manggay*, or what was left of the separate section inside the house where the girl had spent most of her seclusion. They gave sago pudding with this meat to those women who had made the *manggay* and kept her company.

It sometimes happened that a girl got married even before her first menstruation. In such a case the ritual was modified. Upon betrothal, the girl moved to her husband's house together with one of her younger sisters. There she stayed and slept with her sister and her husband's sisters. When they observed that she had begun her first menstruation they built the *manggay* and the 'watch woman', chosen from among her own or her husband's sisters, was appointed to look after her. The young woman washed every second day at the back of the

[10] The literal translation of the Ambonwari expression *kra-* 'to cut' (also 'to beat', 'to fight') expresses the actual bodily change more powerfully than the Tok Pisin expression *wasim* ('to wash'; *sɨpa-* in Karawari) which Ambonwari also use sometimes to speak of the final act of transition. The old way of doing things, her childhood *kay*, has to be cut off, while new habits, those which pertain to an adult woman, have to become embodied. As Lewis (1980: 57) notes, Gnau also use the verb root meaning 'to cut across' or 'to cut off' when referring to puberty rites, while 'washing' represents 'the way of doing it'.

house. Whenever she returned she marked with clay one of the horizontal house posts inside her *manggay*. She painted two 'marks of clay' (*andin simbang*): the white line representing herself and the red line representing her husband (white being the colour of mourning, grief, worry, isolation; red being the colour of happiness). She and her husband covered themselves with white clay. She stayed secluded in the *manggay*, often for several months.

During the first day of his wife's first menstruation a husband did not eat anything except a fried sago pancake made by his *arkay* (FM). He was not supposed to smoke (people said that he might get white hair and suddenly become an old man). If his wife wanted to smoke he had to place the cigarette between the points of a pronged spear and pass it to her. They were not allowed to touch, and the wife was not supposed even to look at him because, as I have noted above, in her state of transition she could affect his *kay*. In the case of her young husband it would mean the rapid transition from a young to an old man. Any contact between them is likened to sexual intercourse. The Ambonwari prefer not to leave a young husband and wife alone under such circumstances. Also, all those who had had sexual intercourse the same or the previous day were too dangerous to come close or give food to the girl in *manggay*, as she could get sick and die. On the other hand, men are afraid of getting sick or losing strength as a consequence of contacts with menstruating women. They believe that their skin might become 'heavy' and they will become short of breath so that they will be unable to hunt or fight. They are afraid of becoming soft, watery, or, as they joke, 'like over-ripe bananas'.[11]

On the day of cutting the young woman's hair her 'mother's brothers' also cut her husband's. They decorated her with two bird of paradise feathers. They called her 'female flute' (*sanggurma*) and her husband, also decorated, was called 'male flute' (*sanggur*). Men, while sitting in a men's house, sometimes used the names of individual flutes to refer to the couple. This association of the couple with the paired sacred flutes obviously suggests the importance of the couple to those concerned, and perhaps implies their satisfactory transition to complete Ambonwari personhood, that is, a being of a specific gender complemented by one of the opposite gender. At this time the real marriage was performed. The woman's brothers brought axes, clay hearths, bowls, and so forth to their sister's husband's house while the husband and his lineage prepared shells and money. Then his father's mother (*arkay*) taught the young wife how to make a new fireplace by using a new clay hearth. The first time she moved to her husband's house she did not get these things and was not permitted to cook. While they waited for her first menstruation, his mother looked after both of them. Now, after she put the first ashes into the new clay hearth, her sisters could leave her. The couple began their adult married life.

[11] The same is said of those with many children. In the past the fiercest fighters were single young men 'uncontaminated' by copulation and menstruation.

The themes in this ritual are patent and little commentary is needed: the seclusion of the girl, the threat her liminal condition poses to others, and the rites and celebrations which mark her 'reaggregation', are all paradigmatic of rites of passage. The attention paid to her future capacities, the stress upon the need for her 'to change her ways', as we say, and the symbolism of becoming all bespeak a concern with moulding of the girl's *kay*. The involvement of different categories of kin, which is an integral part of the formation of Ambonwari womanhood, shows that this is a formation of a social as well as an individual being, so that the formation of her *kay* is simultaneously a formation and representation of the way of the village (*iminggan kay*). Indeed, the distinction between the individual and social is peculiarly inappropriate in relation to *kay*. As always, the individual *kay* is a reflection and a product of *iminggan kay* and vice versa. 'Rites take place because and only because they find their *raison d'être* in the conditions of existence and the dispositions of agents who cannot afford the luxury of logical speculation, mystical effusions, or metaphysical anxiety' (Bourdieu 1977: 115; see also 1990: 96). Throughout her childhood a young woman has already constructed her understanding (*wambung*) about several aspects of her womanhood in a community of which she is a product and a producer. It is because her understanding and behaviour are already embodied throughout her childhood that she can attach symbolic meanings to a set of procedures and make sense of them. There are, however, new practices to be formed in a new environment (when she marries, for instance), and new understanding of the world constituted (if she becomes a 'mother of the village', for example). That is why initiation means an absolute beginning for her: some understanding has been formed but not yet practised, some practices are performed but not yet understood. *Wambung* and *kay* need time to adjust to novelties.

The All-Night Dance

Yamɨn siria or *yambɨr siria* ('dancing and singing inside the house') is a spectacular and extremely popular event which involves the participation of the whole village—men, women, and children. It can be held for a number of reasons: as a celebration of the restoration of social relations following a dispute; to celebrate the opening of a new house; as celebration for the parents of a boy who is about to enter the men's house.[12] The ceremony is simply called *siria* ('dancing and singing') or *sia* ('dance') and it is performed from dusk to dawn. Old people remember that such a dance used to be performed night after night, for many weeks, prior to initiation. Such dances are also arranged between different villages to settle disputes. Because of the complexity of the

[12] The ceremony for achievement, called *kurang*, also involves dancing and singing. But since the whole ceremony very closely resembles the *naven* described by Bateson (1958) I will not deal with it here. Moreover, the dance associated with *kurang* is an attenuated form of *yamɨn siria*.

performance, it will be best to describe the event in the order in which it unfolds and at the same time to discuss the most important aspects of the dance and their meanings.

It is up to a particular 'father' of the household, his lineage, and his clan to organize the dance. The organizers have to provide a large amount of meat, such as pork, sago grubs, bandicoots, possums, large lizards, large fish, and eels, that will be served on sago pudding to the dancers throughout the night. This is the main obligation of the organizers, who will not dance themselves. The major role on the dancers' side is taken by the maternal clan of the father of the household in whose house the dance will take place.

Regardless of which lineage stages the dance, the main dancers represent all the clans of the village and dance in a formation which represents the hierarchical relations amongst them (see also the arrangements of flutes during initiation, discussed in the previous chapter). Consequently, members of the sponsor's maternal clan dancing in the main positions have to borrow the decorations they will wear from the clans who own them. First they have to ask the fathers of the two leading clans, Crocodile-1 and Bird of Paradise, if they will lend them their body decorations to adorn the two principal dancers who dance at the front of the formation.[13] Then they choose a man from their own clan who will, together with his partner, *warɨmbar*, lead the dancers. He and his partner need to be strong men, because they will be unable to rest throughout the night. By dressing in the decoration of the two main clans these two dancers, called 'two at the front' (*kanggiakngri*), represent Kapi and Akɨmbrɨkupan, the two brothers who founded the village. During the dance the one who is decorated with the shells of Kapi (Crocodile-1) stays on the left side, also called the female side; and the one who wears the adornment of Akɨmbrɨkupan (Bird of Paradise) stays on the right, male side (see Plate 15).[14] The dancers also have to borrow those adornments which will decorate the two men at the end of the two parallel lines of dancing men (these two will be followed by women). They are called 'back ones' (*akupi*). The man on the left will wear the decoration of the Cassowary clan, the man on the right the decoration of the Crocodile-2 clan. Again, they will represent the first ancestors of these two clans. Those who borrow these adornments from their 'fathers' have to compensate them with a payment.

Once the sponsors and dancers have agreed to hold the dance, the men begin to beat slit-drums in every men's house in the village. This signal can be heard as far as a day's walk away, so all those who happen to be in their forest camps

[13] The ancestral ornaments and decorations used in the dance are stored in 8 baskets: the Bird of Paradise and Crocodile-1 clans have 1 each, while Eagle-1, Wallaby (with Hornbill), and Cassowary have 2 each. Cassowaries keep one basket all the time for Crocodile-2.

[14] It happened during one such dance that one of the leading dancers had to sit for half the night because of a stomach ache. People commented on this as being his punishment; he had refused to wear the decoration of Crocodile-1 and had used his own from Wallaby instead.

PLATE 15. An all-night dance inside a house: the leading dancers representing the first two ancestors, Kapi of the Crocodile-1 clan (right) and Akɨmbrɨkupan of the Bird of Paradise clan (left)

learn about the forthcoming celebration. Every evening the men beat the slit-drums and small groups gather and practise the words of songs (see below). Except for two or three men there is no one in the village who knows the whole texts. This is understandable because each corpus of songs, of which there are three, comprises many individual songs.

Several other things must be prepared for the dance: different kinds of ochre (see previous chapter), soot, and *Bixa orellana* fruits have to be found and made ready; lizard skins have to be soaked in water overnight and stretched on hand drums; shells have to be polished and the ropes which connect them checked and replaced if necessary; chickens have to be caught and the small white feathers which will decorate the headdresses have to be pulled out of their tails; the bundles in which the feathers of birds of paradise, cassowaries, goura pigeons, cockatoos, and other birds are kept have to be checked and the feathers cleaned.[15]

On the day of the actual performance everyone is excited. If the dance is to be held inside the village those who are not closely related to the main dancing group might prefer to dance only until about midnight and then sleep the rest of the night. But if the dance is being held in some other village, like Imanmeri or Konmei, the whole of Ambonwari participates. Only the very old, the sick,

[15] Ambonwari are not interested in using eagle's feathers, mainly because they feel there is not enough contrast between the brown colour of the feathers and the colour of their own skin.

and those who are abroad will be missing. When going to another village, people have to remember to take mats on which the children will sleep, kerosene lamps, bowls for leftovers, and other personal belongings. People do not paint their bodies in the village but carry their decorations with them. On the way they stop to get clusters of betel nut and young palm shoots which, split into strips, will be later attached to their belts over the buttocks along with red and green *Cordyline* leaves. They stop to decorate themselves near the village in which they are going to dance. Women and children find shade under some large tree, men under another.

The ochres are mixed with water in coconut shells and painted on using dry, chewed betel nut husks. The dancers help each other with painting their faces and their backs. Some are experts and many ask them to give a hand. *Warɨmbar* dancing partners have the same design on their faces. These patterns, many of them resembling the carvings on shields, slit-drums, and posts, have names: *imbian* (pig), *yimbakima* and *ambrɨkukum* (kinds of crab), *akuriyak* (a kind of bird), *kɨndɨm* (design, pattern), *marama* (big face), *mambaynggi* (bananas), *wunduma* (female spirit), *yas* (tears). People paint their bodies mainly with red, using black only around the elbows, knees, and waist. A few who are still mourning the death of someone very close paint their bodies with white ochre (*yaykwan andi*, 'crying clay'). However, they still go to dance.

Women dress in their colourful grass skirts while men wear belts made of shells, with dried flying fox skins at the front and *Cordyline* leaves at the back. Some men complain that rats and mice have chewed their flying fox skins so they have to use *simbis* leaves instead. Both men and women attach the strips made from the young palm shoots to their belts. Small shells are tied below the knees and on the ankles, so that they rattle whenever a dancer moves. Some men stick cassowary bone daggers under their woven armbands which are pushed tight around their biceps, while others use decorative or aromatic leaves. Different kinds of shell hang around the neck or sideways over the shoulder. The two principal dancers at the rear of the formation hang a special kind of string bag decorated with feathers over their backs. These are supposed to mark a line between men and women. Those who do not have plumes can put a headband or possum fur around their heads and stick some feathers in it. Plumed headdresses are put on last. The longest and most beautiful feathers are tied in such a way that they move as if they were attached to a spring.

The leading pair of dancers, representing the two main clans of the village, stand on the path leading to the house where the celebration will be performed. The others slowly line up behind them looking for their partners, *warɨmbar* and *warɨmbɨnma*. The women wait until the men are in two lines and then attach themselves behind: girls first, young women next, old women last of all. Some men try out their hand drums. Others chew betel nut. Some girls move their hips from side to side so that their grass skirts wave in both directions. Some laugh. Someone begins to sing with a voice which chokes on every word and

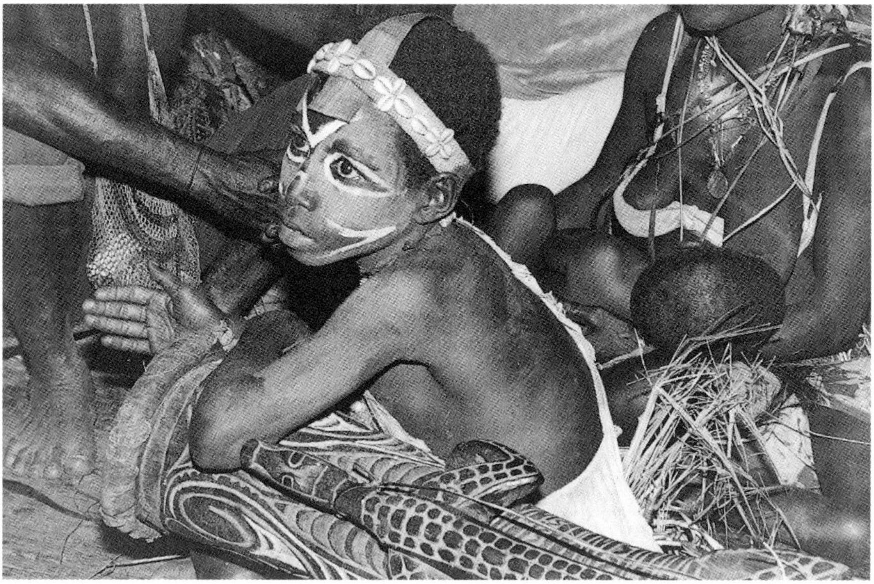

PLATE 16. Boy beating a hand-drum during an all-night dance

draws out the last vowel at the end of the line. The chorus joins in. The beating of hand-drums (see Plate 16) is synchronized into a uniform rhythm and the singing becomes louder. The procession begins to move, following the proper steps, towards the dance house. The dance is a combination of stepping forward with hands hanging down and bobbing on the spot. While dancing through different wards the inhabitants provide betel nut or even food, and thus detain the dancers for a while. The dancers circle three times the house they will dance in. Then they enter and dance counterclockwise. The beating of a slit-drum joins the beating of the hand-drums. Now the singing becomes very loud, and the dance steps are not just steps but a vigorous pounding of the palm bark floor. Will the floor survive the pounding of so many dancers? Somewhere the floor breaks, at another point a large gap appears between two bark planks. The whole house shakes. This keeps the spirits of the house and spirits of the clan awake, and they are pleased to see the decorated people dancing for a whole night in honour of their house.

Some time during the first hour the hosts bring the first portion of food, prepared by mothers and wives of the father of the household and his real and classificatory brothers. One of the big men from the visitors' group calls out the names of the fathers of individual men's houses and distributes the food among them. They divide it between individual lineages and households. The hosts do not eat. Those who are related to the father of the host household as 'brother', or who have married one of his sisters, also refuse to eat. They have been able to eat some food before the dance started and they might even go during the dance to some other house where they may eat in privacy. But inside the dance house it would be shameful not to show empathy with their 'brothers' or 'wife's brothers' who, as hosts, must abstain. People say: 'The food carries the name of those who dance; it carries the name of those who are decorated; it carries the name of ornaments.' The leading and the hindmost male dancers are also not allowed to eat food with the others. They are not even supposed to sit down during the whole night. These four, the most spectacularly decorated men, can accept food only from the fathers of the clans whose decorations they wear. So the one representing Kapi can accept food only from the father of the Crocodile-1 clan, that is, food which comes in the name of the Crocodile-1 clan or in the name of the Yanbonman men's house. The same applies to the other three dancers, who respectively wear the decorations of the Bird of Paradise, Crocodile-2, and Cassowary clans.

Women with small children spread their mats around all sides of the house close to the walls and in the middle, around the slit-drum, so that there is enough space for dancers to move in a circle. Only a few people from the host group stay near the entrance: they bring food, betel nut, and water and repair the broken floor.

The dance goes on for at least two hours without a break. The one who knows the songs well sings the lead lines alone, the others joining in with a characteristic *ay*, *ay*, *ay* at the end when the dancers jerkily move forward. The leading pair of dancers stop and wait for someone to start a new line. Sometimes one of the dancers repeats the previously sung lead line and the chorus repeats the line they have already sung. It often happens in such cases that the order of words has changed. Sometimes young female dancers become very loud. This makes their menfolk angry: 'Who is leading this dance, men or women?' The women giggle. The young men beating the slit-drum relieve each other every hour or so. Drumming is hard work; their skin is shiny with sweat. The dancers smoke and chew betel nut with pepper and lime; some add to this mixture a special aromatic bark, while a few men chew pieces of ginger. Those big men who keep lime in a gourd and have a carved stick use it to rattle while dancing and singing.

At about eleven o'clock, and then off and on throughout the rest of the night, some women of the host group begin to dance in front of those dancers who are 'sister's sons'. They face the dancer, their 'son', and move backwards following the rhythm of the dancers. They wave fishing-nets and fish-traps in front of his face (see Plate 17). As people explain, they are 'catching' him in the fish-traps; they are showing him those nets and traps with which they had caught fish to feed him. They are like the belly in which his mother carried him. They are happy that their 'son' has grown up. Though he is a big man now he is still their 'son'. His classificatory mother's brothers carry him on their shoulders for

PLATE 17. Holding fishing-traps and nets, 'mothers' dance in front of their 'son', who is carried on the shoulders of his 'mother's brother'

several circles (like his mother did when he was a child) while at the same time his 'mothers' are 'catching' him in their nets. The dancer has to pay his 'mothers' and his 'mother's brothers' with one kina or so each for all their 'hard work', for all the food and care they provided for him.[16] But the money which the 'mothers' collect from their 'sons' will not be kept by the hosts. They have their own 'mothers' to pay: that is, the main dancers. At the end of the dance the women of the visiting 'maternal' group usually leave their grass skirts in the house of their hosts, while the men carry the hosts on their shoulders. So at the end of the dance the hosts pay their maternal clan for the grass skirts, and for being carried around on their 'mother's brothers' shoulders.[17]

The hosts bring food around midnight and then again about two o'clock. By

[16] If there is anyone among the dancers who happens to be fully decorated for the first time in his life, he will be surrounded by 'mothers' for quite some time, and as Ambonwari say, he 'will lose money on them.' Only very rarely is the same honour given to a woman.

[17] We might note a similarity with Kiriwina society (despite its being matrilineal) of which Weiner observes that the skirt is 'an embodiment of all that is 'womanness': sexuality, reproduction, and nurture. Further, the transformation of elements of the female self into an object of economic value [skirts in this case] allows women to play a public role with political implications for themselves and for men' (1977: 119). During the *kurang* ceremony the roles of 'mothers' are even more powerfully expressed by men wearing grass skirts, string bags over their heads, and female decoration (see Plate 18). During such a ceremony both men and women of the maternal group can hold the fishing-traps and nets in front of the face of their 'son'. The nature of this celebration of a person's achievement (Bateson 1958: 9) has occasioned some discussion. Handelman, for example, views *naven* as 'ludic' (Handelman 1984).

PLATE 18. *Kurang* ceremony: men from maternal clan dressed as women

this time the majority of spectators and small children are asleep. The chatter dies down and the steps of the dancers, the drumming and singing, are the only sounds echoing through the night. About four o'clock the dancers begin to show the first signs of fatigue. They pause for a short time, some lying on their backs, others sitting with their eyes closed. But there are always those who seek to continue the dance. They call out: *Kwasara*, 'Get up'! And the dancers slowly form two lines again, everyone finding his or her *warɨmbar* or *warɨmbɨnma* partner. The dance goes on. Around five in the morning people begin to sing and dance faster, to a new rhythm. The leading singer begins to sing those lines which are part of *sarkurar siria* ('hold-walk-around dance and sing') or 'round dancing' for short. The slit drum begins to beat faster, the dancers speed up with firm and energetic steps. Everyone sings: 'The two colourful baby cassowaries stood on the other side of the Awinjan forest' (*Awinjan manggwarmɨna gɨma imɨnggwar awan pamɨnggrio sɨrɨmbɨn pamɨnggrio*); then 'first the two went around and jumped over the stream, later the two shook off water and went up, the two colourful baby cassowaries' (*asɨndiambɨn kanggrain gɨma wapar yambrasambɨn sambin gɨma wapar awan pamɨnggrio sɨrɨmbɨn pamɨnggrio*). The dance becomes faster and faster, the lines repeated over and over again without a pause in between, until someone calls a long *oooooooo*. The dance continues as before, but shortly the quick 'round dancing' starts again. 'A young woman from the Awinjan forest, her two shells [her vulva]' (*Awinjangwan wasanma kasanggria kasanggria*). And then again a refrain: 'She wanted to cut grass but the two [her vulva] opened' (*akwi krararin kamariaknggri*, i.e. she began to menstruate). Again the vigorous dance ends, with someone concluding the lines with a prolonged *ooooooo*. The dancing returns to normal. Shortly, the third round of vigorous dancing begins; when the singers recall some fishing event from the past, the tempo again becomes faster and faster. Everyone repeats only a few words from the lines: 'The grasshoppers from the Yaramɨng grassland', and then 'from the Yaramɨng grassland—grasshoppers'. Many of the decorative leaves have already fallen off the dancers; those who were sleeping wake up. The young women become vivacious again. Everyone is sweating; the rhythm is very fast. After a few minutes of such rapid pounding the dance ends with a cry, *yay*, and with a hop on the right leg. Everyone pauses for a second. And again: *yay*, and a hop on the left leg. Everyone pauses again. Then *yay* on the right leg, and *yay* on the left leg. When everyone thinks that the 'round dance' is over, someone starts again with the leading words. Everything begins anew. The decoration of men bounces up and down, the decoration of women swings to the left and right. Many of those who did not dance throughout the night now join the dancers. Someone asks an older man: 'What's with you? Are you sleeping?' (*Mi wariya? Mi paykan?*). And he replies: 'No. I am a big man. You too, you are a big man.' (*Kaya. Ama kupan. Miok mi kupan.*) The big men have no trouble staying awake all night; only children and women can show their weakness by sleeping. The dancers 'enter' the dawn and the day. While the hosts, at this moment, usually say that the dance should end now, since everyone is tired, the dancers become charged with new energy. They dance till seven, eight, even nine in the morning. Sometimes the hosts end their performance by bringing fresh food. The dancers move out of the house, still dancing, and circle the house outside for the last time, in the way they had begun the dance some twelve or fourteen hours before.

Song

Ambonwari call a single meaningful segment of a song, composed of one or two lines, *suruk*, which I translate as 'verse'. Several such verses then form a song, *siria*, which connotes dancing as well as singing.[18] A song, which is composed either of isolated or several thematically related verses, has its own leading

[18] The actual plural of *suruk*, that is, *surukɨnggi* ('verses') is seldom used to refer to a song. Ambonwari would rather say e.g. *awkunma suruk mɨn yapaykan*, which was translated to me as 'a woman joins the verse(s)', that is, 'a woman sings', than use the plural of *suruk*.

verse. Two such verses bracket a song into a meaningful whole arranged around a single place-name. The sequence of place-names then orders the songs.[19] Several songs are thus grouped into a corpus which is again called *siria* or *sia*. *Sia* and *siria* are used interchangeably to refer both to singing and to dancing.

There are four traditional song forms in Ambonwari: (1) *yamɨn siria*, 'songs of a house', comprising three corpora of songs; (2) *mambawn siria,* 'songs of a crocodile'; (3) *kambɨn siria,* 'songs of a shield'; (4) *masungun siria,* 'songs of an enemy'. Each of the last three consists of only one corpus of songs. It is the context of an event that determines which form is going to be sung. During initiation the first three forms must follow one another. Before a boy from a household is brought to the men's house, the house dance is arranged to ease the grief of his parents and to praise the household. Inside the men's house, before initiation begins, men sing the 'songs of a crocodile', composed of only five lines which are repeated endlessly throughout the night while the stress is on the frantic rhythm of a slit-drum. Here the repetition of a single song makes a corpus. On the day the initiates are released from the men's house, the men sing the 'songs of a shield', indicating the emergence of new warriors, or *wiyas* ('dogs'). This is the only song performed in front of the men's house, without dancing. The last form, the 'songs of an enemy', was sung inside the men's house after the Ambonwari had killed someone from another village.

The three corpora of *yamɨn siria* relate to the three main clans of the village, though they are not exclusively owned by them. Every clan in the village has its own songs included within these corpora. The songs are secular and sung by everyone whenever a large public dance takes place in an individual household.[20] The other three song forms, 'of a crocodile', 'of a shield', and 'of an enemy', which are much shorter, are kept in the men's domain and sung only inside or at the front of the men's house. They contain many secret and sacred names of spirits. The use of ancestral words and metaphorical expressions ensures their secrecy.

The 'songs of a shield' were in the past sung also during the concluding ceremony of a mourning period. However, since the Ambonwari purchased—for 50 kina—from Madang a dance called *mandayp* they have used it in prefer-

[19] This structure recalls the Iatmul knotted ropes described by Wassmann (1991).

[20] It took several months for Bob, Johannes, Dominic, Julian, Elias, and myself to translate about 800 lines of 3 corpora of *yamɨn siria*, sung by the person who acts as lead singer. We also translated an additional 800 lines sung by the chorus. The 3 corpora differ in length. If we consider only those lines of the corpora which are sung by a lead singer, we see that one corpus has about 370, a second one 160 plus an additional 77 lines from the first corpus, and the third about 240. (Lewis, 1980: 59, recorded and transcribed a Gnau song from a night-long performance of just over 150 different 'verses'.) Some additional verses are used on special occasions, called 'funny songs'. As the meanings of all these verses were not clear from the songs themselves, we spent a few more months clarifying the meaning of every line. I typed up these 3 corpora of songs for the Ambonwari, so that in future literate young men will be able to take on the role of leading voice. It happened on one occasion that I myself was holding a text in my hands throughout the night while dancing, whispering the lines to the lead singer in front of me. Because of their length and complexity, these songs cannot be discussed in more detail here.

ence to the 'songs of the shield'. *Mandayp* involves a dance around a tall decorated stick thrust into the ground at the front of the main men's house. The dancers circle this stick the whole night, singing songs in Tok Pisin. Another dance ceremony, *aymalo*, purchased from Moim in East Sepik, involves singing and dancing with a huge carved fish. It has been performed a few times in the past as a replacement for the 'songs of a shield'. Although the carved fish is still in the main men's house, the Ambonwari do not perform this ceremony any more.

Young men also compose short love songs, mainly laments for unrequited love or unrealized marriages (cf. Harrison 1982b). These songs are sung by other young boys and girls in the village, but today they are never incorporated into the songs sung during the house dance. I mention this because it seems that in the past anyone who wanted to express something important, interesting, funny, or sad composed a two-line song which was later added to *yamɨn siria*. Thus these songs represent a kind of secular narrative of Ambonwari's past.

Only one of three corpora of *yamɨn siria* is chosen for a particular celebration. Each corpus, like the dance itself, is a kind of journey. The sequential order of songs follows the sequential order of places. The singers begin their journey in Ambonwari and then 'travel' through grasslands and forests, up creeks and rivers, to familiar places and to some which do not exist any more, between familiar camps and those which were long ago overgrown by forest. The songs are filled with place-names, names of creeks and forests, personal names, names of spirits; all these are names with which villagers are quite familiar (cf. Feld 1989; 1990; Lewis 1980: 59–67; Weiner 1991).

Every song is composed of many short and cryptic two-line verses which describe events of the past (not in the mythological past). The first or lead lines are sung by a single person called *siria yaparar*, 'the one who awakes the singing'. He is supported by a prompter, *siriar pam*, 'a stump of dancing and singing', an old man who truly knows all the songs but might not dance. The second lines, which echo the leading meanings, are sung by the rest of the dancers.

Each song begins with a lead verse (*kanapanggɨn siria*, lit. 'head song') so that the singers know where they have moved to and which lines are to be sung next. Thus, for example, in one corpus the singers begin their 'voyage' in Ambonwari, move just across the Konmei Creek to the Araymarin forest, then go up-river to the grassland near Kansimei, then again down-river, stopping at several camps, then across the land to Yimas and then travel down-river to Karawari, Kundiman, and Manjamai, ending their journey in Masandanai. Each of these places is identified by a song composed of thirty verses on the average.

Verses were in the past composed by any individual, usually male, whose thoughts were at that moment preoccupied with a particular event or a particular person involved in that event. A collection of these verses might be called love poems, since they are about a person of the opposite sex whom the composer cared for. The seductive exchange of small gifts, the beginning of love between young women and men, were often encapsulated in such a poem which then became incorporated into one of the major whole-night songs. If a couple married, the poem was abandoned, the verses no longer sung. Those poems which endure are mainly those about relationships which were, for whatever reason, not approved by kin. The couple may remain 'lovers' for the rest of their lives or might simply remember their past relationship. Though composed by men, the words in such poems are often put into the mouths of their girlfriends, expressing their longing for their lovers.

But there were other events too which people found worth remembering in poems. Most events were those of the everyday life with which the singers are familiar. The verses describe a brief scene, 'a frozen, encapsulated moment' (Harrison 1982b: 15). Because of the use of metaphorical language, obscure synonyms, and antonyms, the main protagonists are hardly recognizable to those unfamiliar with the event and unrelated to the composer. Nevertheless the participants, who sing mainly in the present tense, can identify with these familiar events and people from the past. One is surprised to learn from such songs about love affairs between *kasɨnma* and *akimban*, who should maintain a strict avoidance relationship. In these songs we can hear about a successful catch of fish, about a lonely man waiting for his girlfriend in the rain, about seductions of young women by gifts of food, about first menstruations, about longing for a gift of tobacco from a beloved, about forbidden sexual liaisons, about hunting, house-building, decorations, fights, killings, spirits, deaths, and men's houses.

To give one or two examples: 'You are the man who plants tobacco, you should send some tobacco to me; I caught some fish in a creek but I do not have any tobacco leaves with me'. The next line goes: 'I am a nice young woman who asks you for tobacco, my skin is like the skin of a bird of paradise, send me some tobacco'. Another verse begins, 'You, all young girls, do not cry thinking that Imanmeri have killed us. No, we have just caught a large crocodile, we cut it up in the forest and the night fell before we were able to return.' And the answer goes: 'Why did we cry? We thought that Imanmeri had killed you all. But no, you have killed a crocodile.' Another: 'You, my boyfriend from the Crocodile-1 clan, we slept together till dawn, now the rooster is crowing, you have to leave. The father of the household will find you here if you stay longer.' Or a leading voice sings: 'A young woman from the Kamsmarin forest sleeps together with an egg [i.e. she is pregnant].' And the dancers add: 'I am watching you secretly, I got some white clay and spread it all over my skin, like a pig. I am worrying about you [a previous boyfriend of a young woman expects that she will give birth in few days time].' There is only one overtly postcolonial verse in all these songs. It tells about the arrival of the bishop and the baptizing of the whole village: 'You, young men, he washed us together. He fetched water and washed us together with all the women from our house.'

Places and people are often mentioned by name in songs though, as I have indicated, there are many exceptions: for example, a person, because of his or her secret relationship, is referred to by such euphemisms as 'outside husband' and 'outside wife' (i.e. lovers). Other general references are to 'a widow' or 'a widower', to nicknames derived from the names of creeks, grasslands, or forests; or they might refer to 'a girl from the grassland', 'a woman who knows how to catch fish', or 'a man who is not afraid of killing an enemy'.[21] Metaphors and sobriquets used for people are so plentiful that one has to know the Ambonwari context very well—and at the same time understand what they value and laugh about—if the meanings of these songs are to be properly comprehended. In short, we might say, one needs to share their *kay*.

Yamɨn Siria

So far I have concentrated on the content of the dance. But the way the dance is staged connects it to aspects of Ambonwari life that I have repeatedly referred to throughout this book. The most obvious is the dancing in two lines which recalls the two moieties of Sun and Moon and the two brothers Kapi and Akɨmbrɨkupan. Moreover, all the principal clans of the village are represented by conspicuously well-decorated dancers at every *yamɨn siria*, no matter who the sponsors are. This concern with the inter-clan relations and village hierarchy, relating as it does to the order of arrival of ancestors, seems to represent a stress upon precedence or a desire to commemorate the past. Either way this relates to political relations and the distribution of resources. However, the enthusiasm for *yamɨn siria* and the intensely joyous atmosphere that envelops preparations for and staging of the event make it clear that these considerations are not the most salient for the participants.

The songs which the dancers sing during *yamɨn siria* do not mention mythological heroes, they do not sing about searching for a new land, rivers full of fish, forests full of sago, men for women and women for men, for a home where they can settle down. In these songs there are no strange beings, no strange events, no strange practices which are all characteristic of myths. There is no struggle for identity. The dancers 'depart' on their 'journey' to 'meet' people

[21] Many different flirtatious forms of address are used in songs for both women and men. Thus they call women hen (*yakwayma*), white cockatoo (*andɨma*), cassowary (*awanma*), red parrot (*kawinjɨma*), a kind of fish (*wurumayma* and *wungdɨnggɨnma*), nice fat woman (*waykɨnma*), a woman with a red skin (*arɨm awiyanma*), a woman with a dark skin (*arɨm wurukupɨsanma*), good, nice woman (*yapakupanma*), nice flute-woman (*sanggurma*). Men are referred to by such terms as flying fox (*kɨmbima*) or rooster (*yakwan*), or by terms used for different species of fish. Both a young man and a young woman can be called a 'red possum' because of their shiny skin, and a boy can be called a 'lizard' because he skilfully climbs palm trees. A woman who has many children is referred to as 'female crocodile' (*mambayma*), a young girl who urinates a lot in front of everyone, 'turtle' (*yuwanbak*).

who have the same characteristics as they do. The place-names, in contrast to those in *kupambɨn konggong* (ancestral path) are local. Songs do not feature a single hero (as myths do) and the people mentioned in them are 'just human'. With so many local heroes, male and female, songs present a plurality of characters and situations with which the people of a community are familiar and can identify. These 'heroes' experience all sorts of troubles, unexpected turns, funny or frightening situations of daily life. The songs use individuals as mediators between the past and the present. The past is not presented in general terms; it is not historicized in the sense of providing an 'explanation' of Ambonwari life. Rather, it uses particular individuals and 'frozen moments' of their lives to represent the past. But the most striking thing about this presentation is that those who lived it are no different from those who live now. Accordingly, one might say the past is 're-present-ed'.

The circular movement of the dancers, and the narrative plot of the songs constructed around place-names and the order of movements between them, show how the singers make use of space to represent events in their past as is done in ancestral narratives. 'By serially listing place names, a temporal span is automatically invoked, a sense of life *movement* is appealed to' (Weiner 1991: 6).[22] Thus, events and place names from the songs

> do not just encode memory; they stimulate social recall and imagination, and they revive and reveal and foreshadow feelings, again intensified by the specifically bounded nature of performance. Voicing memories through placenames . . . produces emergent feelings of empathy, connectedness—the sense of sameness through difference. (Feld 1989: 262)

When the scattered events commemorated in songs are collated together into a kind of matrix of travel, the 'episodic dimension' acquires a 'configurational dimension' (Ricoeur 1979: 24), that is, a meaningful whole. What seems to be configured as this whole is *imɨnggan kay*, for the spatial and social references of *yamɨn siria* songs are both characterized by their familiarity. In this they contrast with *kupambɨn mariawk* and *kupambɨn konggong*. Another striking feature of *yamɨn siria* is that it means fun. So one might summarize by saying that, although it is concerned with *kay*, it is non-ancestral *imɨnggan kay*.

In dance, a line of song, a footstep, the sound of a drum, the colours of the painted bodies, the shaking of the dancers, the intimate sociality, the heat, the rhythm, the tempo, all evoke in the individual dancer that sentimental attachment to the familiar, that feeling of belonging to a place and to a people with whom the dancer can emotionally identify. During the dance, Ambonwari

[22] Moreover, the place-names constitute 'space out of a blank environment' (Weiner 1991: 32) and become a familiar 'society of named places' (Lévi-Strauss 1966: 168). 'Because of their inseparable connection to specific localities, placenames may be used to summon forth an enormous range of mental and emotional associations of time and space, of history and events, of persons and social activities, of oneself and stages in one's life' (Basso 1988: 103). On the other hand, unfamiliar names are like 'unfamiliar languages' (p. 99).

experience their existence in terms of 'we', sharing each other's lifetime, with all the strength of what Schutz called a 'mutual tuning-in relationship' (1977: 108, 115). This is why dances offer excellent occasions to settle old disputes and to create new relationships between young men and women.

If the Argentinian tango with its 'ethos of melancholy' reflects a painful insecurity about people's identity, their lack of roots, and the bitterness of immigration and homelessness (Taylor 1987), then the Ambonwari *siria* (singing and dancing) reflects the opposite. Both dances display a common foundation. Both are related to familiar places, familiar ways of doing things, familiar people, and familiar time.[23] Ambonwari dance brings together different individuals and relates them to a common collective past, present and future. '[F]or the individual, much of what goes on in the social world is the actions and sufferings of groups, not as an external spectacle played out before the individual's eyes but as a function of his or her participation' (Carr 1986: 154–5). The relationship and the connection to this collective life into which individuals, without choosing it, are 'thrown', first by birth and then by rites of passage, represent the heritage which they can appropriate as their own and which becomes a function of the temporality of their existence.

[23] Looking at their habitual and customary ways of life, one can see that 'nothing is more impenetrable than the familiar' (Ricoeur 1966: 289).

CONCLUSION

In this study I have explored different dimensions of the Ambonwari life-world. My concern has been to concentrate on *kay*, the term most intimately involved in the conceptualization of life as it is lived in Ambonwari. *Kay*, seen as an embodied but not unchangeable process, is already characterized by specific temporality which encompasses ancestral, historical, and personal time. Though I have stressed the bodily nature of *kay* throughout the book, I have sought to make it clear that *kay* is not the Ambonwari term for a physical body. It would be more accurate to comprehend this term as 'being', provided that its essential condition is understood to be an active corporeality which may or may not be instantiated in material flesh (I have compared, for example, the *kay* of living people and that of spirits). *Kay*, as we have seen, refers to the most secular habits of an individual as well as to the most complex collective ceremonies and rituals. *Kay*, being temporal in itself, relates present with past and future. The most inclusive variant of *kay* is *imɨnggan kay*, which represents the collective identity of Ambonwari village. *Imɨnggan kay*, like the *kay* of an individual, is not an unquestioned and unreflexive state but a process that encompasses transformation. Through their *wambung* (Heart), which is an aspect of *kay*, that is, through understanding, people continually confront and (re)produce their own *imɨnggan kay*.

In this conclusion I do not intend to summarize the many points made in the book. Rather, I will address some important questions that arise from it. First, I want to emphasize the importance of *wambung* (Heart, understanding) as an aspect of *kay* which is able, through transforming itself, to transform *kay* as well. This will lead me to consider a second question: since my focus in this book has been upon *kupambɨn kay*, *kupambɨn konggong*, and *kupambɨn mariawk*, that is ancestral aspects of *imɨnggan kay*, the question of postcolonial transformations arises. Third, I want to relate the principle of duality in Ambonwari thought and practice to the concept of *kay* and its temporality. This will lead me to the fourth question, which has lain beneath the surface throughout the book, namely the relation between Bourdieu's concept of habitus and *kay*.

As people explained, the village itself is 'covered' by *arɨng*, a term used both for a mosquito bag and for the skin. The large spirit-snake which lies under the village also represents the village's Heart (understanding), and 'follows' the villagers wherever they go. Ambonwari also suggested that the main men's house represents the eyes and ears of the village. The customary way of the village is then referred to as *kupambɨn kay* (represented in its most condensed form in male initiation) which is both the 'way of the ancestors' and the 'way of the spirit-snake'. The spirit-snake and any ancestor may be referred to as *kupan*, an elder brother or a 'big man'. People keep this spirit-snake, as the

Heart of the village, firm and stable. Together they constitute Ambonwari and *imɨnggan kay*. People say that if the spirit-snake moved away, he would take away the village's Heart, and the whole village would be destroyed.

The 'sedimentation' of the past in *imɨnggan kay* is as much a function of *wambung* as it is of *kay*. For example, if people forgot the symbolism of *yamɨn siria*, if the leading dancers were just leading dancers and not 'embodiments' of the first two ancestors from whose descendants they have to borrow decoration, if people forgot their songs, if there were no hierarchy, no expression of *konggong* in the dance, it would lose its authenticity, its essential 'Ambonwariness' and hence its existential and political importance. It would become simply a dance Ambonwari perform. But perhaps not; maybe it would gain a fresh significance in a new context. In either case *wambung* would transform the dance but it would still be staged and referred to as *imɨnggan kay*. Perhaps, too, it would be represented as perfectly following *kupambɨn kay*. Thus we could say that *wambung* always makes the present *imɨnggan kay*, and through it reconstructs *kupambɨn kay* of which it is itself a product.

Let me recall another example of the relationship between *kay* and *wambung*. In my discussion of the first menstruation ritual I emphasized two of its aspects. First, the necessity for the girl to remain immobile and keep her right hand still, which is a direct intervention in bodily *kay*. This 'mechanical' intervention or 'practical mimesis' is even clearer in the emphasis on the fattening of the girl and the cutting of her skin. Second, there is another, 'unmechanical' intervention that is crucial to the ritual and the bringing into being of a new *kay*: those women who instruct the girl form the future woman's Heart (understanding), so that the whole and its crucial part are in accordance. This, in addition, points to the close relationship between *mariawk* (speech, story) and *wambung* (Heart), both being, as we have seen, related to thought and thinking.

From the perspective of *imɨnggan kay*, perhaps the most important consequence of the transformation of a girl's *kay* and *wambung* is that she becomes capable of forming a path—*konggong* (marriage)—between two groups which modulates the relational aspect of their identity. Being 'sister' to her clan, which being male-centered looks towards the preservation of its unity, she becomes 'mother' to another clan, which also being male-centered similarly looks towards the preservation of its unity. Thus, as I suggested in Chapter 1 and elsewhere, a woman's *kay* is to relate groups identified as male. I will discuss this in relation to Ambonwari cosmology in more detail below.

Imɨnggan kay has its origin in the past, but people's Hearts and their *kay* together sustain, perpetuate, and modify it. In effect, both collective and individual *kay* are the product and producer of *imɨnggan kay* as well as its past, as represented in *kupambɨn kay*. While *kupambɨn kay*, as its source, unifies the village as a whole, *kupambɨn mariawk* (myth) and *kupambɨn konggong* (paths of the ancestors) differentiate and relate groups inside the village. It is the differentiating and relational characteristic of these concepts which is itself one of the grounds of *kupambɨn kay*. Though I have usually translated the word *kupambɨn* as 'ancestral' or 'of the ancestors', I would like to indicate the other meanings discussed in Chapter 2, such as 'of the elders' and 'of the big men'. This means that *kupambɨn* includes all those elderly people who are still alive, not just 'of the past' but also 'of the present'. Therefore, though being part of *imɨnggan kay*, *kupambɨn kay* covers all those practices which pertain to the elders of the village, who possess adequate knowledge, knowledge of the old and of the dead. The village elders connect the past with the present, each having an influence upon the other. *Kupambɨn mariawk* is not just 'the talk of the ancestors' but also 'the talk of the (living) elders', and *kupambɨn konggong* is not just 'the path of the ancestors' but 'the path of the (living) elders'. Myth and ritual, therefore, are not just 'how the ancestors talked or did it', but more importantly, 'how the elders talk and do it' in their now. Such a conceptualization leaves space for changes both in the present and in the past. Without the adjective *kupambɨn*, *kay*, *mariawk*, and *konggong* designate the mundane present to which I now turn.

I am well aware that it might be said that I have concentrated too much on the past, that I have not presented the reality of Ambonwari's current circumstances. As Bateson remarked in the introduction to *Naven*, it is impossible to present adequately the complexity or the whole of a culture, or to stress every dimension of people's lives in exactly the same way as they do themselves. My concern, however, has been mainly with those aspects of Ambonwari life which people themselves acquire and embody while growing up. Given the historical juncture at which I carried out my fieldwork, adults still embodied an *imɨnggan kay* oriented to the ways of the ancestors. Moreover, they did not regard this aspect of themselves as a relic of the past, and they sought to inculcate this 'way' in their own children. The Ambonwari, although they feel remote from roads and city life, do make use of all the things to which they have access as citizens of Papua New Guinea. These things have become part of their practices and thus of their *kay*. Daily activities such as hunting and fishing in which people use guns, purchased nets, lines, hooks, and torches are obvious examples of these changes. People use cotton mosquito nets, kerosene lamps, and outboard motors and know how to repair them. Moreover, the villagers produce rubber, work on tourist boats and at the Karawari Lodge, and try their hand at business by opening trade stores. They attend local school, visit the aid post when they are sick, speak Tok Pisin whenever they need to, buy and eat rice and tinned fish, and make a holiday of Sunday. All these features, characteristic of other Sepik societies and New Guinea at large, are certainly part of Ambonwari *imɨnggan kay*. They also listen to the radio with its news of a world of computers, space travel and war, which they in some sense share but which is not part of their *kay*. Although the Ambonwari modified the overt expression of *kupambɨn kay* following the baptism of the village during the 1950s and 1960s (preeminently by forsaking the men's house and their lineage spirits), they have never been able to sustain the belief that their lives are possible without it. Christianity is accepted, but as I was told: 'God has too many people to look after. You [speaking of my place of origin] have your own *imɨnggan saki* [spirit of the village] and we have ours. If we follow our own way, and maintain good relations with our spirits, our village will stand well.' Thus *kupambɨn kay* orders relationships between groups and individuals and all new aspects of *imɨnggan kay* are filtered through the understanding (Heart) in which past and present are inseparable. Thus, for example, if a villager buys a new outboard motor he will bring it to the men's house, where his *sɨmɨrinyar* will perform *sɨmɨrin kay* (an inauguration ceremony). In this way a motor will receive its own name and become an Ambonwari being.

The insistence on the way of the village is implicated in the failure of trade stores in Ambonwari (Telban 1993). People understand that a business can be successful only if goods are paid for, but the owners are continually obliged to give them either on indefinite loan or for nothing. Thus all businesses in Ambonwari collapse shortly after they begin. If one does not follow the 'good way' of giving things, someone will break into the store and steal the goods. Of course, things given are remembered and require reciprocation, but this may happen very much later and may be replaced by another gift and not in cash. Whenever men who earn wages return to the village they are put under pressure (from my perspective) by others who ask them for money, petrol, purchased goods, and so on. Conversely, the returnees expect others to build them houses, to prepare sago which they can sell in Angoram, to give them large amounts of garden and forest food. All these practices are part of today's *imɨnggan kay*, and are constrained by kin relationships.

In summary, whatever new growths appear within *imɨnggan kay*, they are nourished by those ancestral roots which, according to the village elders, have always been central to the village's collective life. A piece of evidence which I would cite in support of the claim that *kupambɨn kay* remains central to people's cosmology is the reluctance of Ambonwari's most recent settlers, the Eagle-2 and Pig-2 clans, to put their sons through initiation. Initiation and first menstruation rituals are not simply transformations of boys and girls into Ambonwari men and women, but are means of regaining and preserving the relationship with their spirits, most importantly with the spirit-snake. They are means of re-forming the Ambonwari cosmology and the life-world which it sustains. That is why those groups which settled last in the village, and whose attachment to their own land is still very strong, participate in all ceremonies but hesitate to put their boys through initiation ritual. For this would mean that they became Ambonwari, something that they were, during my stay, still not sure that they wanted to become. It seems that they were also apprehensive about whether Ambonwari spirits would accept them, and if so how their own spirits would react to abandonment. The health and well-being of these settlers as individuals and as a group was in question precisely because of the issue concerning which ancestral way they should follow.

In several contexts the principle of unity as duality structures conceptualizations of beings and processes: for example, in rituals of the men's house, the succession of generations, dance, and intergroup relations. I would now like to examine these principles and relate them to *kay*.

In addition to the obvious case of the implicit moieties, Moon and Sun, there are many other situations in which the dual nature of unity is represented as fundamental. The flutes of the men's house are ordered in two lines. The presentation of dancers during *yamɨn siria* also signifies duality. In both contexts the right side is called male and the left side female, and they are represented as brother and sister (flute names, for example) or as elder and younger brother (in *yamɨn siria*).[1] In the ordinary household, all the middle brothers have to leave while the oldest and the youngest are supposed to stay together. In this way the household itself manifests a unity in duality by combining the oldest and the youngest brother in the same unit. Because of the association of older and younger brothers with brother and sister respectively, even this unity suggests that of male and female. Again, the posts within the men's house are carved with the male figure on the front and the female on the back. They too are referred to as brother and sister. The men's house itself is said to have female characteristics, although it bears a male name and contains sacred male spirit-things.

The above are all examples of what we might call static or structural dualism. However, the crucial process underlying agnatic continuity in Ambonwari is also represented dualistically. The generation of fathers is followed by the generation of sons and then, again, by the generation of fathers. A man takes over his paternal grandfather's identity (through his names, his marriage paths, and by calling him 'elder brother'). Although this is not specifically represented as gendered duality, it is a dualism in which men are differentiated and complementary in relation to women.

An apparent anomaly arises in several crucial contexts where the conceptualization of patrilineages and clans shows a concern with absolute unity, which, following Mimica (1988), I shall call 'oneness'. For example, the distinctions between the generations are ignored or irrelevant to the conceptualization of ancestors. They are all, like their most primordial ances-

[1] The question arises of why the male side in these performances is associated with the eldest brother and the female side with the youngest. This seems explicable by the fact that it is the eldest son who is primarily identified with agnatic continuity; he takes his father's place, occupying his house, and assuming his roles. Moreover, in the main myth of origin (see Chapter 5) the oldest brother, Akɨmbrɨkupan, gives his daughter to his younger brother, Kapi, and so from the point of view of the oldest brother's patriline the youngest brother's patriline represents their female side. Though Kapi was the founding ancestor of Ambonwari, he was the younger brother, and his agnatic descendants—despite their precedence in general social and political life—must dance on the left side.

tors, simply called *kupam* (plural of *kupan*, i.e. 'elder brother'). Also, even in the context of the duality of patrilineal generations men call their own sons 'fathers', while they become 'sons'. Thus, through the identification of paternal grandfather and son's son the distinction between adjacent generations can also be reduced to unity. Mimica writes about the same system of identity between father's father, father and son, among Iqwaye:

> What oneself is to another, that other is so to oneself . . . instead of myself, my own son can marry my (classificatory) mother, and as such he can give birth to myself . . . The two are one, and in empirical situation it is the three persons that are bound into oneness. (1988: 88–9)

This is not a conceptualization unique to men, for father's sister and brother's daughter are similarly encompassed by this 'oneness'. Another example of 'oneness' was discussed in Chapter 5, where I argued that Kapi should be understood not as an individual but as a collective entity and identity which unites all men from Crocodile-1 clan. Kapi as such is not only himself but also his son and his father at the same time. Like any other clan founder, he reproduces himself endlessly in every Crocodile-1 member. The generation of 'oneness' as patrilineal identity is produced from the multiplicity of generations and individuals. Accordingly, we seem to be faced with three basic notions: unity, unity in duality, and unity in plurality. But as Marilyn Strathern observes:

> a plurality of individuals as individuals ('many') is equal to their unity ('one') . . . The suppression of internal differentiation occurs, however, in a pluralized context of sorts. This is the plurality that takes the specific form of a differentiated pair or duo. 'Many' and 'one' may be homologous, but neither is to be equated with a pair . . . the dyad is a unity only by virtue of its internal division. (1988: 14)

This suggests that the basic division is between 'oneness' and duality, because unity in plurality is simply a form of 'oneness'. But if we examine these two notions in Ambonwari we see that 'oneness' is itself an aspect of duality, as indeed Strathern goes on to argue: it is so to speak, duality achieved. The patriline is constituted as 'oneness' only by suppressing its 'dividual' or relational nature. The duality of adjacent generations is irreducible in reality and it is generated by relations mediated by women. The paths of the fathers and the paths of the sons are categorically differentiated and the patriline perpetuates itself through this duality. Similarly, the maintenance of this 'oneness' in the future is absolutely predicated on its female members going in marriage to some other 'oneness', and women coming in marriage from elsewhere in order to produce new members. It is only in relation to the past that this double duality between males and females, and between the pluralized self and the other, can be suppressed and 'oneness' thereby constituted. Overall, then, the patrilineage can perpetuate itself only on condition that it divides its 'oneness';

yet the only reason for dividing itself is to maintain its 'oneness'. This simply restates the proposition that the patrilineage is a relational entity; in regard to the present and the future this truth is obvious, but in relation to the past it can be forgotten. In relation to the past *kupambɨn konggong* can be presented as a simple relation between simple units; but this belies their relational nature. Just as in the men's house it is the male aspect of the unity which is designated the 'front', so too in the patrilineage and clans the male element indexes the dualistic unity. The merging of all men from the same clan into a single 'one', the mythical ancestor Kapi, for example, is possible only because the female gives their unity a processual nature which enables this very same unity to reproduce itself. Only brother and sister share generational *kay*, household *kay*, lineage *kay*, clan *kay*, and can cosmologically be merged by collapsing the generational difference. Father, son, and grandson all 'become' one and all the clansmen identify with the mythological ancestor. At some point even the sexual differences disappear, 'making' sister into a younger brother, that is, into 'oneness'.

The dualism of generations and their perpetual alternation transform *imɨnggan kay* (the way of the village) into *kupambɨn kay* (the way of the ancestors). In temporal terms, for an idealized Ambonwari, we could say that his father represents the 'past' and he the 'present', but because his son is also identified with his father, the 'present' is bracketed by the past, which means that the future is identified with the past. So the 'way' that lies before such a man is the 'way' that lies behind him, and *kupambɨn kay* is the way of the future no less than it was the way of the past.[2] This amounts to the annihilation of time itself, or, what amounts to the same thing, a temporality which is an undifferentiated duration, i.e. eternity. Within such a temporal horizon there are no distinctions which make a difference, and fathers can become sons of their own sons and sons fathers of their own fathers. Though this is a metaphysical illusion it is nevertheless put into practice in certain contexts, as we have seen, in the use of kin terms between fathers and sons.

Such a perspective would make *kay* completely self-reproducing and make *kay*, *imɨnggan kay*, and *kupambɨn kay* coextensive. *Wambung* would be unnecessary since *kay* as habit and way would automatically project itself forward in time. As we have seen, however, *kay* and *wambung* are inseparable, and the *kay* of the individual is not automatically a reproduction of *imɨnggan kay*. *Kay* is therefore necessarily confronted by an open future—which is to say a genuine future—so that reproduction of *imɨnggan kay* is an achievement made possible only by the work of *wambung*. Accordingly, if the temporal horizon of *kupambɨn kay* is constituted as eternity the temporal horizon of *kay* is time itself, that is to say, time 'which is intrinsically unitary and unifying' (Gell 1992: 315).

[2] Thus the Ambonwari push Bloch's (1977) notion of the past in the present a stage further, to make the past present in the future as well!

Let me now explicate Ambonwari *kay* in terms of Bourdieu's habitus, for time is an important point of departure for his theory of practice. His 'practice' is the product of habitus which is itself 'the product of the em-bodiment of the immanent regularities and tendencies of the world [;] it contains within itself an anticipation of these tendencies and regularities, that is a nonthetic reference to a future inscribed in the immediacy of the present' (Bourdieu and Wacquant 1992: 138). Habitus, which Bourdieu defines in many of his writings, is a product of collective history; it is a socially constituted system of cognitive and motivating structures; it is a system of durable, transposable dispositions, in which the term 'disposition' expresses the result of an organizing action, a way of being, a habitual state, with emphasis on the body (1977: 76, 82, 85–6, 214 n. 1). Here we see, besides the reference to embodiment, an emphasis on collective history as the origin of habitus. But, 'because it implies a practical reference to the future implied in the past of which it is the product, habitus temporalizes itself in the very act through which it is realized' (Bourdieu and Wacquant 1992: 138). For Bourdieu the lived experience unites present with past and future in a dynamic equilibrium. This conceptualization follows previous phenomenological studies in which past and future have been seen to create the present (and time itself), as in Husserl's 'consciousness' (1964) and Merleau-Ponty's 'body' (1962: 239–40). It is the dynamic character of practical activity which, for Bourdieu, produces time in the very act of producing itself. At one level Bourdieu emphasizes collective history or collective past as being essential for habitus, while on another level habitus produces individual and collective *practices* (and hence history). Habitus itself becomes 'practical', that is, producing itself in a dynamic equilibrium of past, present, and future. In this theory both habitus and practices are related to body, 'knowing' is 'doing', and the body through movements and gestures makes space (and vice versa).

The concept of habitus is thus evoked by the many dimensions of *kay* I have charted in this book, such as its relationship to body, habits, the importance of collective past, and its social origin. However, there is a difference between the two: *kay* itself has Heart as a crucial dimension which reflects upon and, through its own transformation, transforms *kay*; habitus has no such element. *Kay*, not being the function of some unquestioned historical rule, depends on *wambung*. Through their understanding people then recognize, conceptualize, differentiate, and create all past and present, personal, and collective *kay*, as well as constructing and challenging their hierarchical relationships. Bourdieu in his concept of habitus neglects both the process of the personal development of one's understanding throughout childhood and the creativity of one's understanding in adulthood, which in Ambonwari can be witnessed in the vigourous discussions that take place daily in the men's houses. In short, as I have stated earlier, through Heart *kay* is transformed; but since Heart itself is an aspect if *kay*, Heart is the means of *kay*'s self-transformation. Although I emphasized that *wambung* is socially produced, it is nevertheless the means by which existence itself comes into question and cannot therefore be seen purely as habit; this means that *kay*, of which Heart is a part, should not be seen as such either. This interpretation brings me closer both to Gell, who questions Bourdieu's identification of 'knowing' with 'doing' and his treatment of 'cultural knowledge as a set of dispositional propensities of socialized agents' (Gell 1992: 274), and to Connerton, who regards the term 'disposition' as misleading, in that it 'suggests something latent or potential, something which requires a positive stimulus outside ourselves for it to become actively engaged' (1989: 94; see also Toren 1990: 16, 233).

If *imɨnggan kay* represented as *kupambɨn kay* suggests a duality which is also a unity and an alternation that is almost an alternation of identities, the image it evokes is that of the *yamɨn siria* dance, in which a person jumps on the right leg (male), pauses for a while, and then jumps on the left leg (female), though both legs are part of the same body. Under this aspect, *kay* suggests a state of being which, despite its inherent duality, is still a state of being. On the other hand, *kay* as a conceptualization of life itself, both collective and individual, confronts an open future. Its temporal horizon is time itself, which suggests not simply being but becoming. It may be proper, therefore, to see *kay* as a being whose being is becoming.

APPENDIX A

Ways of Finding Meat

TABLE A1. *Fishing methods*

Method	Description	Accessories	Material	Period	Sex	Prohibitions	Time spent	Water depth	Type of fish	Transport
Ambiapi kar kay ('the way with a light')	At night with a torch	Torch (with batteries) *Kunaypi* (torch) *Akwar* (spear) *Wanya* (bush knife)	Torch made of the sago palm leaf stalk	*Yuwayn maraym* (wet period)	M, Ch	No	3–4 hours	20 cm	All (often eel)	On foot
Aria awur kay ('the way of catching fish')	*Akrɨnggi* (hooks)	*Akrɨng* (hook), purchased hooks, baits	String made of pandanus fibres, sago thorns as hooks	All year	F (M)	No	Hooks are left in water overnight; or from canoe	All	All	Canoe, on foot
Aria apɨr kay ('the way of putting fish inside')	Net	Purchased nets	Cotton, nylon	All year	M (F)	No	Check twice a day	Deep	Large, also crocodile	Canoe
Aria apɨr kay	*Aprang* ('place for fishing')	*Sɨmbang* (fishing basket, trap) *Yuway* (fishing-net)	Made of sago	*Yuwayn kangɨm* (dry period) *Kambɨs kangɨm* ('dry period of flowering grass')	F (M) in group	No	A day	100–150 cm. muddy water	All	On foot
Makwi apanar kay ('the way of throwing sap')	Poisoning of a creek	*Makwi* (sap; poisonous sap) *Akwar* (spear) Net	*Derris sp.*	*Yuwayn kangɨm* *Kambɨs kangɨm*	M (F) in group	Yes	A day	Mostly small creeks	Small, rarely big	On foot
Aria awkwiyar kay ('the way of pulling fish out')	*Sɨnggrɨk* (fencing; barricades)	*Sumbo* (fish-trap)	Bamboo	*Yuwayn maraym*	M	Yes	After being built, checked twice a day	20–150 cm.	Large, also turtle	On foot

Aria awkwiyar kay	Fish-traps	*Sɨmɨn sɨmbang* (vine traps) Bait: sago scrapings	*Kanda* vine with thorns	All year	M	No	Check twice a day (a man can make 5 traps in a day)	Clean, deep water	Large	Canoe
Aria awkwiyar kay	Fish-traps	*Parinjanɨng* Bait: *Amɨnɨng* (ant hill)	Made of sago	*Yuwayn maraym*	F	No	Check twice a day (a woman can make 1 trap in a day)	Deep water	Small	Canoe
Aria awkwiyar kay	Fish-traps; also used at *Aprang*	*Wusɨsaprar*	Made of sago	*Yuwayn kangɨm*, *Kambɨs kangɨm*	F	No	A day	100–150 cm. muddy water	Small	On foot
Aria wandɨr kay ('the way of shooting fish')	*Pɨpɨrkwanmay**	*Akwar* (spear)	Palms (black palm, wild betel nut palm, or bamboo)	*Yuwayn maraym*, *Kambɨs maraym* ('water of the flowering grass')	M	No	Variable	Deep water	All	Canoe
Aria wandɨr kay		*Akwar* (spear)	Palms and bamboo	All year	M (F)	No	Variable	All	All	Canoe
Aria wandɨr kay	*Kurar aria* ('fish go around' in groups, on surface)	*Akwar* (spear)	Palms and bamboo	*Yuwayn maraym*	M	No	When a person comes across the fish	Deep water	Small	Canoe
Aria wandɨr kay	Fishing under the water	Stick with a rubber; iron arrow; goggles		*Yuwayn kangɨm*, *Kambɨs kangɨm*	M	No	1–2 hours	Low water	Small	On foot, swimming

TABLE A2. *Hunting methods*

Method	Description	Accessories	Period	Prohibitions	Time spent	Group/ individual	Animal hunted	Occurrence during my stay
Wandɨr kay ('the way of shooting')	*Ambanyas* (fence made of sago leaves near the wild sago palm, cut down and its bark removed)	*Inggum* (spear) *Pambang* (bow) *Pambangwanya* (arrow, 'the knife of a bow') *Kamaynggwa* (spear thrower; mainly used in fights)	All year	Yes	If the pigs do not eat up the sago, a hunter can use the same spot for 2–3 months; edible sago can be used longer than the wild one	Individual	Pig	Some men often, others never
Awkwiyar kay ('the way of pulling out')	*Inggay* ('trap'; either a heavy tree trunk which falls on a pig or a cage with a door which closes when a pig eats the sago)	*Inggum* (spear) *Wandɨkakɨr* (axe)	All year	Yes	The trap is only used once or twice	Individual	Pig	Rarely (only those who know how to make a trap)
Awur kay ('the way of catching')	*Anggɨm* ('rope'; a sling on the ground with the rope tied to a bent tree; new method)	*Anggɨm* (rope or wire)	All year	No		Individual	Pig	Rarely
Kumɨndɨ kay ('the way of finding meat')	Hunting with *wiya* (dog)	*Inggum* (spear)	All year	No	A day or more	Individual or a small group	Pig, bandicoot, rarely cassowary or wallaby	Very often
Kumɨndɨ kay	Hunting with gun	Gun, cartridges	All year	No	A day or several short excursions	Individual	Pig, cassowary, ducks, pigeons, rarely crocodile	Very often (if cartridges available)
Awi pasar kay ('the way of burning')	Burning method by lighting a circle of fire on grassland	*Inggum* (spear); guns and bows with arrows are feared because people are in a circle	*Yambɨn* (a grassland which is never flooded; all year) *Akɨm* (a grassland which is flooded in the rainy period; in dry period only)	No	A day or half a day	Large group	Pig, bandicoot, lizard	3 times

Imbias warasɨr kay ('the way of chasing pigs')	Two groups of men, on each side of an island; one group calls loudly and beats on tree trunks while the other waits with spears	*Inggum* (spear)	*Yuwayn maraym* (wet period)	No	1–2 hours	Group	Pig	None
Awsar awkwiyar kay ('the way of pulling out bandicoots')	Trap; the heavy sago bark falls on bandicoot	Only a trap	All year (better in dry period; better when there is no moon)	No	1 hour to make a trap	Individual	Bandicoot	Often
With a torch (new method)	Bandicoot is blinded by the light of torch	*Wanya* (bush knife) *Wandɨkakɨr* (axe)	*Yuwayn kangɨm* (dry period)	No	3–4 hours at night	Individual	Bandicoot	Often
Andɨkrangɨndɨ kay ('the way of harkening')	At night a hunter listens for a possum moving through crowns of sago palms; when he spots the possum he waves with firewood or torch and coughs. The possum stays on the palm until dawn, when the hunter cuts down the palm and kills the possum	*Akwar* (fishing spear) *Wanya* (bush knife) *Wandɨkakɨr* (axe) *Awurmasɨnɨng* (firewood with fire)	*Yuwayn kangɨm*	No	Whole night	Individual	Possum	Rarely
Kwandɨr kay ('the way of a long neck')	In the morning a hunter walks around and looks for possums in treetops	*Akwar* (fishing spear) *Wanya* (bush knife) *Wandɨkakɨr* (axe)	All year	No	Morning	Individual	Possum	Rarely

TABLE A3. *Other meat-obtaining methods*

Animal	Period	Procedure	Location	Sex of gatherer	Transport	Importance
Wun (sago grub)	*Yuwayn kangɨm* (dry period) *Kambɨs kangɨm* ('dry period of flowering grass')	A man cuts down a wild sago palm (*wayak*) and makes incisions on a trunk; after a month or two sago grubs are ready to be collected	Sago forest	M, F, Ch	On foot	Very important
Anyamun (large grub)	Any time	When a wild sago (*wayak*) is dry, its crown falls off the plam. The grubs are in the trunk; a man cuts the palm and collects the grubs	Sago forest	M	On foot	Important
Large python	Any time, but more during *yuwayn maraym* (wet period)	*Akwar* (fishing-spear)	Rain forest, grasslands, sago forest	M	On foot	Less important
Frog	*Yuwayn maraym* *Kambɨs maraym* ('wet period of flowering grass')	*Akwar* (fishing-spear) or with hands	Grasslands	M, F, Ch	On foot, canoe	Not important
Mɨkɨs (large lizard)	Any time	*Akwar* (fishing-spear)	Sago forest or near water	M	On foot, canoe	Important
Sukwan (small lizard)	Any time	*Akwar* (fishing-spear) or with hands	Coconut palms	Ch	On foot, climbing	Not important
Kɨmbima (flying fox)	Any time	*Pambang* (bow) *Sambrim* (arrow with blunt head)	Village, forest	M	On foot, canoe	Rarely caught, but prized meat
Yarau (wild fowl)	*Yuwayn kangɨm* *Kambɨs kangɨm*	*Pambang* (bow) *Pambonar* (four-pronged arrow) *Aringgɨnɨm* (arrow) or gun	Rain forest	M (Eggs) M, F, Ch	On foot	Rarely caught but eggs important

Pɨmbɨngga (crowned pigeon)	Any time	*Pambang* (bow) *Pambonar* (four-pronged arrow) *Aringgɨnɨm* (arrow) or gun	Forest at river bank	M	On foot, canoe	Rarely caught but goura meat very prized
Mɨnggɨmbɨnma (pigeon)	Any time	Gun	Rainforest	M	On foot	Rarely caught, prized meat
Wild ducks	Any time, more during *yuwayn maraym*	With a gun or with *akwar* (fishing-spear) and a torch during the rain; ducks come to the light	Grasslands	M	Canoe	Occasionally caught, important
Sɨpɨr (kokomo)	Any time	Gun	Rainforest	M	On foot	Not important
Large birds	Any time	Sap from breadfruit tree is smeared on the branch of a tree where the birds usually sit; the birds stick to the branch. Also caught with fishing hooks	Grasslands	Boys	On foot, canoe	Not important
Ming (birds)	Any time	A sling, with the hand taken from nests	Any place	Boys	On foot, canoe	Less important
Grasshoppers	Any time	After a wild sago palm is cut down grasshoppers are caught with hands between the leaves	Sago forest	F, Ch, M	Canoe	Less important
Praying mantis	*Yuwayn maraym*	With hands	Grasslands	F, Ch, M	Canoe	Less important
Yanjar (flying yellow insects)	*Yuwayn kangɨm*	With hands or with *sɨmbang* (fishing-basket)	On water	F	Canoe	Less important
Apɨk (a type of beetle)	*Yuwayn maraym*	With hands (its acid smarts)	On new leaves near water	M, F, Ch	Canoe	Not important
Yuwanɨnggɨnma (black insect)	*Yuwayn maraym*	With hands	Rainforest	F	On foot	Not important
Yambɨnɨnggɨnma (large grub)	Any time	Gathered from dry or fallen trees	Rainforest	M, Ch	On foot	Not important
Sɨndɨng (house of flies)	Any time	Inside tree trunks; people drink yellow juice of squeezed nest of *kwarmaman* flies	Rainforest	M, F, Ch	On foot	Not important but prized

APPENDIX B

The Stone Breast

Mapmban tokples *amanan*, stori *amanan, njay awkum bɨnang sinja amɨngori.*
Other vernacular mine, story mine, breast women on them we-it drink.

Wasan yan sayk ambambɨn. Kupan sɨmbɨn. Mapmban mayarmasɨnma mɨna ambarar.
Son she held gave birth. Big became. Other other woman she gave birth.

Kupan sɨmbɨn mba nggri warɨmbawarɨm. Warɨmbawarɨm, orait na *nggri*
Big became so they two partners. Partners, all right and they two

warɨmbawarɨm kɨmbu isa san kamayk gra prinying asɨm kɨn parimbɨn
partners they themselves called this sago mother she-them placed

angakɨna imbrum bɨna. *Ambɨn kɨn san sangwakɨna:*
gave leaf in it. Ate he-it he looked:

'*Hey, warɨmbara yandia waria asay sipinya wapaysimbɨn sɨnanma angaria*?'
'Hey, partner these what mother sago put on-fried she-you gave?'

'*Ee, waria sɨnga simbɨn angakria. Asay mɨnang krambɨn imama awur njay*.'
'Eee, what she-me fried gave. Mother from her cut it took breast.'

'*Ako? Mi, asay mɨnma marɨmbio?*'
'Is this so? You, mother she probably died?'

'*O, kaya. Mɨnda asay mɨnma san. Panma pɨn angor njay ngandɨkɨnma.*
'O, no. That mother she lives. Herself she takes breast with-woman.

Prinyɨngbɨn ingama angar. Inja kambra mɨn wamɨnggɨk.'
Cooked in leaves she-me gave. This nothing it went inside.'

Asɨm mɨnang pɨn yayk:
Mother to her he-her said:

'*Asay saman makuriyara imbrumbri prinja krapapɨkɨmbɨn sɨka aurumbri. Prinja*
'Mother here come leaves-two these cut and count I-them got-two. These

andɨka awsambanbri manggɨnɨnggiok awkwiyambɨn awsara. Njay krambɨn
I should put down crumbled sago too take put. Breast cut

imban awun sayn inga prinyingbɨn angakɨr. Prinyinggɨna.'
I-it take again you-me cook in leaves will give. Cooked in leaves.'

Asɨm pɨn yaykɨna bɨni:
Mother she-him said so:

'*Ooo, mi inja njay wasa krambɨn in akura. Jamba minyang kwayangɨndayki.*
'Ooo, you this breast why cut you-it will take. This on you pressed.

Source: Bob Kanjik Anjapi (Sapɨndam, Crocodile-1). Additional sources: Elias Wapun Kandanggway (Yimasɨndam, Wallaby) and others. The translation was made with the assistance of Bob Kanjik Anjapi and Julian Kapyamari Yanggan.

Mi amɨng amɨng inja njay mɨndɨn kupan sɨkɨn. Njay krambɨn sɨnga akuri
You drink drink this breast so big became. Breast cut you-me took

mba ama mba kwasɨkayn ama marikan.'
enough I enough now I die.'

'*Eee, kaya. Mi kambra mariawk mi sukwan. Warɨmbar asɨm mɨnang krambɨn*
'Eee, no. You nothing talk you talk. Partner mother from her cut

inma awur njay sayn prinyinggɨna prinyingbɨn inma angara. Pan kundɨng
he-it got breast again cooked in leaves cooked in leaves she-it gave. Very tasty

ngandɨki sɨka san ambɨnanggwaraki. Wasɨsaki sɨnga angari. Pan kundɨng ngandɨki.'
has me-it tried to eat. Little he-me gave. Very tasty has.'

'*Mi kambra mariawk mi sukwan, kambra mariawk.*'
'You nothing talk you talk, nothing talk.'

'*Kaya, kaya. Sangga mi ama imban krambɨn awun.*'
'No, no. Keep clear you I I-you cut take.'

'Orait. *Mba mi nggok. Kanggɨnɨng amban sɨnggan* mekim
'All right. Enough you wait. Something I-you become make

gut na *kaymbania.*'
good.'

Kanggɨnɨng bɨnang stretim *mɨna kaymbɨn, awsambɨn.*
Something it put in order, put.

Sandanggɨna yan sandak, pɨn yaykɨna:
At entrance she sat, she-him said:

'*Mba anja sura!*'
'Enough you-me kill!'

Sayn mba krasɨmbɨngbɨn in akuna. Asɨm panma mɨnang ambanma
Again enough cut around he-it took. Mother herself from her gave birth

i dai pinis. *Mɨn marɨk.* Em i dai pinis.
died. She died. She died.

Pɨn sɨmɨnggɨk:
He said:

'*Eey, mɨnda asay wasa yaka sura? Yanda warɨmbar mba mɨndɨkɨn waria*
'Eey, that mother why I-her killed? This partner enough same what

yaka may sɨkra.'
I-him will do.'

Em nau. Katim mambu pinis, mambu bilong shave. Kaman.
So now. Cut bamboo, bamboo for shave. Come on.

Orait, *mamyang yan kamiak:*
All right, sister he-her called:

'*Apra. Makuriyara. Wapɨs kanapang kɨnga akrupɨmbɨn pasan.*'
'Come. Come down. Hair head you-me remove throw.'

Orait, em i tok isi long em. Em tok:
All right he spoke softly to her. He said:

'*Wasɨsak wapɨs mamɨnɨng anja sɨndakrɨng! Pan wasɨsak*. Nametna. *Saman*
'Little hair tuft you-me leave it! Very small. In the middle. Here

anja sɨndakrɨng!'
you-me leave it!'

'*Warisayn kɨmban sɨndakɨkra?*'
'Why I-you will leave?'

'*Anja sɨndakrɨng!* Lusim hap gras long namel bilong het.'
'You-me leave it! Leave half hair in middle of head.'

Em i tok:
She said:

'Bilong wanem bai mi lusim?'
'Why will I leave?'

'*Anja sɨndakrɨng!*'
'You-me leave it!'

Orait, *kɨn sɨndakɨkɨna. Andi yambun*, waitpela graun, *wakɨn andi yambun*
All right, she-it left it. Earth took, white earth, snake earth took

kɨn prikwakɨsak. Maray awsawsa. Prikwakɨsambɨn andin sapak mɨnamban awsandar
he-it rubbed. Water put. Rubbed earth clod same put on top

wapaymbɨn. Liklik *wapɨs mamɨnɨng kɨndanɨng sɨn sɨndakɨkɨnak. Mba pan kaya*. I no
went up. Little hair tuft very small she left. So very not. No

mo kamap. Het nating. *Kambra kanapang. Kambra kanapang* nau
more come up. Head nothing. Nothing head. Nothing head now

warwaringgar krapapɨkɨmbɨn. Warwaringgar sɨmbim samandɨn apmbɨn pɨn parik
feathers cut around. Feathers held this way put in he laid

andinya. Ipunggiambɨn mɨn apasɨk. Em nau, poro bilong em nau. *Yan*
in earth. Jumped he came out. So, partner his now. He

andandɨkamiak:
listened-called:

'Poro, *ey. Eey* mi amamas long yu, *pan* amamas na *kamban sanggwasan*.'
'Partner, ey. Eey I happy about you, very happy I-you see-live.'

'*Mi ama mɨnda wariasimbɨn yanma yuwasara?*'
'You I this how you-them put?'

'*Wapɨs* rausim *mɨna kaysapmbɨn. Ingama kriapisapɨr mamyang. Kriapisapmbɨn sayn*
'Hair removed. She-me shaved sister. Shaved again

ingama kwasɨpɨr. Em i bowim.'
she-me drilled. She drilled.'

'*Wariana inyanma* bowim *mɨna kara?*'
'What with she drilled?'

'*Kɨngama* bowim *mɨna kar* tit bilong pik*na*, *sɨsɨnggɨna imbian mɨnak*.'
'She-me drilled tooth of pig, with tooth pig its.'

'*Mba* sel*ok yananma kwasɨpɨr?*'
'So shell also she-it drilled?'

'Em i brukim sel, em i bowim olgeta.'
'She broke shell, she drilled all.'

'Ey poro, no gut mi dai. *Ama amana marikɨr*.'
'Ey partner, not good I die. I I will die.'

'*Kaya kaya*. Em i no *marpora*.'
'No no. It is not death spot.'

'*Amangok mɨndɨn andɨkama sɨmba*.'
'I also so should do.'

Em i tok:
He said:

'*Miok mɨndɨn ma sɨra* yu ken i go wokim olsem. *Yananga minyang kanapang*
'You too this will do you can go make so. Who on you head

makriapɨkanma? Asay sɨnanma kriapɨkanbɨn sayn kananma kwasɨpmbanɨng.
shave-woman? Mother she shaved again she must drill it.

Kwasɨpmbɨn yambandanggɨna, orait, *sayn warwaringgar* as bilong kangar
Drilled in this (hole), all right, again feathers bottom of feather

samandɨn mawusɨrar. Samandɨn sayn mba win *kanma*
in this way went down. In this way again enough wind must

blowim *mɨna kaypiambania. Yandia sanggwaria amangok* win *yanga san*
blow. These look me too wind it-me

blowim *mɨna kaykan. Mi anja sanggwara! Wara maman yan*
blows. You you-me look! Not bad it

blowim *mɨna kaykan*. Nais.'
blows. Nice.'

Em i giamanim em. *Mɨn* dinau man:
He lied to him. He revenge man:

'Bilong wanem mi kilim mama? Bai mi mekim wanem long em? *Asay mɨnang*
'Why I killed mother? Will I make what to him? Mother to her

yaka sungangandukun ama waria yaka sɨkra?'
I-her killed-gave I what I-him do?'

Em nau. Em i gat dispela tinktink. *Mba wambung mɨndana pasa yan may kayk:*
So. He had this thought. Enough thought it only it he is:

'*Sapɨka sukura bio. Panatna kan sumbunanggwamba*.'
'I not him kill yes or no? Himself must try-kill.'

Em nau. *Asay mɨnang pɨn yayk:*
So. Mother to her he said:

'*Asay anja* brukim *mɨn karɨng saman* tit bilong pik*na. Sɨsɨnggɨna imbian*
'Mother you-me break it drill here tooth of pig. With tooth pig

mɨnaknggɨna anja bowim *mɨna karɨng. Saman anja kwasɨprar* sel.'
its you-me drill it. Here you-me drill shell.'

Asay pɨn yaykɨna:
Mother she-him talk:

'*Hoo. Yarmasɨnar wara kwandapria mɨndɨn sɨka sɨkɨr. Mi, asay ama panma*
'Hoo. Man not rubbish so I-him will do. You, mother I myself

minyang ambanma wasa kamban sukwana?'
to you gave birth why I-you kill?

'*Kaya. Wara inga sukwan. Warɨmbarok yanda sanggwarar yanda sɨkɨtna aka san.*
'No. Not you-me kill. Partner also this look this on stool sit.

Iman sariana apasɨmbɨn sandambɨn sɨkɨr awsambɨn mɨna san. Win *yana*
Men's house at front came out sat stool put he sit. Wind it-him

san blowim *mɨna kaykana. Mɨnda miok mba inga sanggwan bɨni* poroporo
hold blew. This you-also enough you-me see so partner

mba nggrika san mban.'
so two of them sit one.'

Sandambɨn, asay pɨn yaykɨna bɨni:
Sat, mother she-him said so:

'*Mba kamban sukwan kwasɨkayn.*'
'Enough I-you kill now,'

'*Kaya.*'
'No.'

Orait. *Kɨn kwasɨpɨkɨna saman. Imbian mɨnak sɨsɨnggɨna kwasɨpasamba.* Sel
All right. She drilled here. Pig its with tooth drilled. Shell

brukim *mɨna kaymbɨn.* Blut tasol istap. Kru. Orait. *Warwaringgar sɨmbim* em
broke. Blood only is. Brain. All right. Feathers tied she

i pasim pinis long wanpela maypas. Orait. *Sɨmbim sarisimbɨn*, fopela. *Saman*
tied in one bundle. All right. Tied arranged, four. Here

yan wusɨk. Yan wusɨkɨna. Em i mekim olsem. *Sɨmbɨnɨngbɨn yan*
she-them put down. She-them put down. She made so. Shook head he

sanggwak.
looked.

'Nau yu lukim. Mi no i dai. *Ama wara ama marikan.*'
'Now you see. I not died. I not I died.'

Asɨm pɨn yaykɨna bɨni:
Mother she-him said so:

'Giaman. *Mba mi marikan kwasɨkayn.*'
'Lies. Enough you die now.'

Pɨn yaykɨna:
He-her said:

'*Mi saman anja awyaynggaymbɨn anja sara ama amama kurapɨkan*
'You here you-me watch you-me stay I I go up

imana. Mi go ausait nau.'
men's house. I go outside now.'

Apasimbɨn win *yan san* blowim *mɨna kaykuraykɨna konggonggɨna. Konggong* bilong
Came out wind it-it blew-walked around on path. Path for

painim haus boi. Orait mɨnma kurapɨk mɨn apasɨk imana bɨni:
find men's house. All right he walked he went out men's house so:

'*Ama wamɨnggɨrbio*'.
'I will go in, or not'.

Ai bilong haus boi yet *pimbɨn kapukpaymbɨn kɨn sandak. Mba mɨn marik.*
Eye of men's house yet fell noise-crashed he sat. Enough he died.

Dai pinis. *Asɨm pɨn yaykɨna bɨni:*
Died. Mother she said so:

'*Em, mba kamban sur. Asay panma ama minyang ambanma, ama kamban*
'So, enough I-you killed. Mother herself I to you gave birth, I I-you

ambandukun ama kamban sur.'
gave birth I I-you killed.'

Em nau. *Asɨm yan sanggwasambinsiak. Yan sanggwasambinsiak.*
So. Mother she-him looked-around-danced. She-him looked-around-danced.

Yamɨn paymbɨn yamba apuk. Olgeta bung long haus. *Yambu sanggwayayk*
In house lay down put in. All gathered in house. They looked-cried

mban kaprapmbɨn. Wi imbu yayayimunggiakɨk mɨnda ngandɨk. Panatna asɨm
one gathered. Night they cried-stood awake this with. Herself mother

sɨn suk. Dinau man. *Sayn waria. Mɨn asɨm mɨnang*
she-him killed. Revenge man. Again what. He mother on her

bekim *mɨn* na *mɨn kayk.* Em tu, em i giaman man. *Yambu wusɨk. Kim*
took revenge. He too, he liar. They put in. Sago bark

urambɨn. Sɨmasum mamba aykrak. Opim skin bilong saksak *wusɨr sar. Sɨpasɨmbɨn*
decay. Sago they cut. Open bark of sago put in stayed. Went in

hap hap, opim olgeta pinis, rausim insaid olgeta, rausim *mɨna kaysapmbɨn,*
half half, opened all removed inside all, removed,

urambɨn pasapmbɨn mba imbu payok yamɨn apmbɨn. Yambu
decayed threw all enough they carry on shoulders in house put. They-him

wusɨk. Wusimbɨn mba mɨn urakɨk. Man kilim mama pastaim *mɨn kayk.*
put in. Put in enough he decayed. Man killed mother first he stayed.

Bus*na mɨn sɨpasɨk.* Bus*na sɨpasɨmbɨn pɨmbɨnggaws mbunam sarakiar arɨm pɨn*
In bush he went in. In bush went in goura pigeons theirs sat branch he

sanggwak. Arɨmbas yandiamba guria olgeta *saman maya sarakianan.*
saw. Branches these goura pigeons all here sat.

Mɨndɨr sɨkɨr yandamba sɨmba apɨkan. Dispela diwai guria i kam pulap
Faeces seat this they put in. This tree goura pigeons came full

long en. Slip tulait long en. Orait, apinun.
to it. Sleep dawn at it. All right, afternoon.

'*Sɨnakmonana andɨkama kuriyamba yaka kaprakɨr. Pambang yakama sɨndakɨr*
'Afternoon I can come I-them will shoot. Bow I-it left

andɨka may sania saman. Mi go nau.'
could sit-they here. I go now.'

Orait, *mɨn mɨn kurayapasɨkiakɨk imɨnggan. Imɨnggan kurayapasɨmbɨn*
All right he he walk around-went out to place. To village walk around-went out

wasɨsangganar yan kamap na *kaykiakɨk.*
little-young man he came to.

'*Apra!*'
'Come!'

Em i singautim wanpela yangpela man. Manki bilong haus boi.
He called one young man. Boy for/from men's house.

Wasasanggɨnar mɨnang pɨn yaykiakɨk bɨni:
Little-young man to him he said so:

'*Apra. Kapa makurapɨkɨr. Pɨmbɨnggaws yuwan sɨmbu sarakiakɨkan. Yakama*
'Come. We-two will go up. Goura pigeons tree they sat. I-they

wandar sanggwarɨmbɨn. Nggok kayama sandamba. Sɨma piamɨnggɨkanbɨn
tree saw. Wait they must sit. They sleep

yambanma san kapramba sakɨr. Ama wapaymbɨn ama pambangɨna. Mi isinar mi
I-you shoot will be. I will go up I with bow. You down you

maimɨnggɨr.'
will stand.'

Wundumbunar bilong em, *wundumbunar* bilong poro bilong em *yan*
Spirit of a dead man his, spirit of partner his he

andandɨkiakɨk. Bi kamban andɨr. Dinau man.
heard. So I-you heard. Revenge man.

'*Kamban andɨr.*'
'I-you heard.'

Tudak liklik nau, manki ya, *mariawk mɨn sɨn sɨngan sɨsɨkiakɨk mɨn* i go pinis long
Dark little now, boy ya, talk he became talk he went to

susa bilong em. Em i weit, weit, *sara, sara.*
sister his. He waited, waited, stayed, stayed.

'*Eey, mɨnda sapɨma kuriyana mɨnda.*'
'Eey, this he not came this.'

Orait, em nau. *Wundumbunar* bilong poro bilong em, *wundumbunar*
All right, so. Spirit of dead man of partner his spirit

warɨmbar mɨnan pɨn yaykiakɨkɨna warɨmbar:
partner his he talked partner:

'*Eey, mi mi san?*'
'Eey, you you are here?'

Em i no kolim em *warɨmbar. Mɨn wara pɨn yaykiakɨkɨna bɨni warɨmbar,*
He did not call him partner. He not he-him talked so partner,

kaya.
no.

'*(Ingana may aykapɨkɨr.) Mi san?*'
'(Not he knows.) You are here?'

'*Eey, mi makuriyar?*'
'Eey, you came?'

'*Yau.*'
'Yes.'

'*Bayo, mi kapa.*'
'All right, you us-two.'

Kisim bunara bilong em wantaim hap paiawut, em tok:
Took bow his with half firewood, he said:

'*Bayo kapama kurapɨn.*'
'All right us-two go.'

Em tupela i go pinis. *Wundumbunar* i go *wasasakɨn* makim lek bilong manki.
Two went. Spirit went little boy marked leg of boy.

Mɨnamban sɨkɨn. Tupela i go. Tupela i go, go. *Nggri makurakiakɨpɨk.*
Same became. Two went. Two went, went. Them-two went up.

Mba, mba, mba. I go klostu, *wasayn kuraywambɨn yuwan panatna pɨkambɨn*
So, so, so. Came near, near walked-went in tree itself at bottom

yuwan pɨnma yaykiakɨkɨna bɨni:
tree he-him said so:

'*Yandia mba sania.*'
'These enough they sit.'

Mbu kɨma sɨsarakiakɨk. Wundumbunar pɨnma yaykiakɨkɨna bɨni:
They cooed-sat. Spirit he said so:

'*Bayo, wapara!* Yu go antap, *wapara!*'
'All right, climb up! You go on top, climb up!'

'*Mi imɨnda!*'
'You stand!

'*Mi wapara!*'
'You climb up!'

Yarmasɨnar wapaymbɨn mba yanma kaprakiakɨk. Yanma kaprakiakɨk: yau,
Man climbed up so he-them shot. He-them shoot: yes,

yau, yau. Em i holim tasol. *Kambra sarisari yanma san imɨnggiakɨkɨna. Mɨn*
yes, yes. He held only. Nothing held-held he stood. He

yarmasɨnar pɨnma sɨmɨnggiakɨk maun wapaykɨn bɨni:
man he talked above climbed so:

'Olgeta guria *sɨka kapran pambanggɨna. Andi kaya sambuka krarkɨkani.*
'All pigeons I-them shot with bow. Ground not they not hit.

Pairap bilong graun mi no harim.'
Noise of ground I not heard.'

Olsem mi Tok Pisin gen [Bob's comment]. *Andi sambuka krarkɨkani.* Orait,
So I Tok Pisin again. Ground they not hit. All right,

pinisim pinis nau, i no olsem em pinisim olgeta, sampela ranawe. *Mapmbaya*
finished now, not that he finished all, some ran away. Others

kwasasandaymbɨn i go pinis. *Mayama mɨnggaykiakɨk maya kwasasandaymbɨn*
got up went. They ran away they got up

i go pinis. *Mayama mɨnggaykiakɨk. Mandia sinma kaprakiakɨkɨnia kambra*
left. They ran away. These he-them shot nothing

wundumbunar mɨnang mayama anggɨkiakɨk saman.
spirit to him they went down in here.

Kamandiambɨn yanma may imɨnggiakɨk. Mba kaya. Mba kaya.
Opened mouth he stood. Enough no. Enough.

'*Aringgɨnbas pasaria! Pambangok pasarar.*'
'Arrows throw! Bow also throw!'

'*Ho, ama pambang wasa yaka pasakra. Panar yaka pambaysɨpasɨkɨr.*'
'Ho, I bow why I-it will throw. Myself I-it will carry-come down.'

'*Pasarar! Karka pasarar pambang!*'
'Throw it! Must throw bow!'

Em nau. Em i trumai bunara bilong em i kam daun. *Wundumbunar* i kisim.
Now. He threw bow his came down. Spirit got.

Yanma san awkwiyan sarikiakɨkɨna.
He-it got held.

'*Nggok andɨka sɨpasɨmba.*'
'Wait I should go down.'

'*Miok ipunggiapiara! Kamban awkwiyan sarikɨr.*'
'You too jump! I-you get will hold.'

'*Hoo, pasi ama. Ama wara wasɨsakɨn bɨni inga awkwiyan sarikɨr.*'
'Hoo, no matter I. I not little boy so you-me get will hold.'

'O.K. *Sɨpasɨpiara! Sɨpasɨmbɨn ipunggiara! Bayo, karka ipunggiara!*'
'O.K. Come down! Come down jump/fall! Come on, must jump!'

Mɨnma ipunggiakiakɨk bɨni kanga sarimba. Nogat, ya. I go pinis long bel
He jumped/felt so he-him must hold. No way. Went in belly

bilong devel, bilong poro bilong em. *Samɨnggɨna mɨnma akurpiapɨkiakɨkɨna.*
of spirit, of partner his. In belly he jumped/detached.

Yanma sɨnggan kapɨpɨngwariakiakɨk. Kamap long ples nau. Hap skin bilong
He did turned around. Came to place now. Half bark of

saksak em i stink long en. *Kimbɨna yanma sɨnggan akurpiapɨkiakɨk.* Em nau ol
sago he decay in it. Sago bark he did jumped in. So, they

i lukautim, painim. Tulait, moning ol i lukautim istap istap. *Mba yama*
looked, find. Dawn, morning they looked stay stay. So they

kamaysak, yama kamaysak. Wandar kuraysanggwasapmbɨn manda bɨnang
found-stayed, they found-stayed. Tree went around-looked this at it

pɨmbɨnggaws sɨnma kaprakiakɨkɨna. Ooo. Mɨnda manki liklik yangpela boi *mɨn*
goura pigeons he-them shot. Ooo. This boy little young boy he

pɨnma sɨmɨnggɨk bɨni:
he-them said so:

'*Ooo, amanang amban mɨn awsakianan pɨmbɨnggaws arɨm bɨnang kaprararin.*'
'Ooo, on me plan he put goura pigeons branch at it like shoot.'

Mba mɨndɨn masɨpasɨk. Yama kamayk. Ooo, kaya. Yandia mba yamɨnggwas
So at this sat. They found. Ooo, no. These enough legs

paykania. Mak bilong lek. *Yandia mba paykania.*
lie/sleep. Footprints. These enough sleep.

'*Ipa sanggwara. Apia yanda kim bɨna ambɨkan pɨmbɨnɨng mɨndɨk wasa kɨn*
'You look. We this sago bark it was in noise this kind why it

papaykianana. Wara isipela pairap *apia andandɨkianan.*'
made noise. Not easy noise we heard.

Em nau, ol lusim. Ol i lusim nau. *Wurupmbɨn awsambɨn bɨni:*
So, they loosen. They loosened. Loosened put so:

'*Ooo, mba yandamba ambɨkan. Wundumbunar yanma pambaypaykiakɨpianan sumbun.*'
'Ooo, so this is in. Spirit he-him carried-lay down killed.'

Orait, *ama mba yandamba* tok *amanan, panar* tokples *amanan,* stori
All right, I enough this talk mine, itself language mine, story

tumbuna, susu bilong tumbuna meri, nem bilong em Mambinja.
ancestors, breast of ancestor woman, name hers Mambinja.

Toktok bilong em i pinis olsem.
Talk her ended so.

PLATE 19. The stone breast

Translation[1]

Another story, which I am going to tell in my language, is about the woman's breast. Breast from which we drink. A woman gave birth to a boy. He became a big man. Another woman gave birth to another boy. He became a big man too. The two were *poroporo* ('partners'). All right. The two used to ask each other for sago whenever a mother cooked it in leaves. They ate half each. Once one of them looked at the sago of his partner:

'Hey, my partner, what is this that your mother has fried with sago and given to you?'

'Ee, it's nothing, she fried it and gave it to me. I cut off her breast and took it.'

'I think you lie. I think your mother died.'

'No. My mother is still alive. She herself took her breast. After she cooked it in leaves she gave it to me. You cannot even notice that her breast 'went inside' [disappeared].' The astonished partner said to his mother:

[1] The translation of individual words should not be taken to be the only possible ones. For example, I translate words for different types of sago as 'sago' without indicating the individual type. The verbal suffix *-bɨn* represents a completed action or an action that is going on at the same time as another action. It is usually followed by some other verb. I have translated the former verbs by using the same tense as the following verbs. The careful reader will also notice some inconsistency in the translation of, among other things, the inner operators (aspect and modality) and the prefixes to the verbs. When Ambonwari use Tok Pisin verbs they invariably do so in the way in which they compound their own verbs. Thus, after a verb in Tok Pisin, they first add additional *na* ('and' in Tok Pisin) and end the phrase with the verb *kay* (to be, to stay) in a relevant tense: for example, marit na *kayk*, 'married and stayed/was' (i.e. 'married'). I do not translate the verb *kay* when it is attached to Tok Pisin verbs, only when it is compounded with another verb in Karawari.

'Mother, come here. I have cut and brought these two leaves. I will put these leaves down, and I will get and put crumbled sago on them. I will cut off your breast so that you can cook it in leaves with sago and give it to me.' The mother said to him:

'Oooo, why would you cut off my breast? I fed you with this breast. You were clinging to me holding it, and you grew up. You drank and drank from this breast and so you became big. If you cut off my breast, I would die at the same moment.'

'Eee, no. You talk nonsense. My partner has cut his mother's breast off and took it. She cooked it in leaves with sago and gave it to him. Very tasty. I also tried it. He gave me a little. It is very tasty, sago with a breast.'

'You talk nonsense. You talk nonsense.'

'No. No. Move! I will cut it and take it.'

'All right. You wait. I will just put everything in order first.' She put everything in order. She sat near the entrance. She said:

'All right, come and kill me!' So he cut it [the breast] in a circle and took it. His mother, she who gave him birth, died. She died. He said:

'Eeey, why did I kill my mother? This partner, he is not the same one as before. What am I going to do to him?' He cut bamboo, the one people use for shaving. Then, he called his sister:

'Come. Shave the hair from my head and throw it away.' He whispered to her:

'Leave a small tuft. Leave a small one, here in the middle.'

'Why should I leave it?'

'Just leave it.' She left a tuft in the middle of his head. He got white clay and rubbed it on his head. He applied water. When he finished rubbing, he squeezed the clay and put it on the top of his head so that a tuft was covered. He cut feathers, put them together and thrust them into the clay on his head. He jumped out of the house. He saw his partner and said to him:

'Eey, I am happy to see you, very happy to see you.'

'In what manner did you put these feathers?'

'I removed all my hair. My sister shaved me. After she shaved my hair, she drilled my skull.'

'With what did she drill it?'

'She drilled it with a pig's tooth.'

'So she drilled your skull?'

'She pierced my skull. She drilled it all the way through.'

'Eey, my partner. If I did so, I would die.'

'No. No. It is not a lethal spot.' [A man who had killed his mother was taking his revenge by deceiving his partner and implicitly suggesting that he should drill his skull. He succeded.]

'Me too, I should make it so.'

'You too. Make it like this. You can go now and make it so. Who will shave your hair? If your mother shaves it, she has to make a hole in your head. After making a hole, put a crest of feathers in it. In this way the wind will blow your feathers. Like these on my head. Can you see how the wind blows them. Look at me. It is not bad. Very nice.' He lied to him. He was seeking revenge.

'Why did I kill my mother? I killed my mother because of his trickery. What will I do to him?' That was what he thought about. That was the only thing on his mind.

'Am I going to kill him or not?' He thought he must make him kill himself. He [the envious partner who also wanted nice feathers in his skull] said to his mother:

'Mother, pierce my head with a pig's tooth. Here. With a pig's tooth. Drill the hole. Here, make a hole in the skull.' His mother said:

'Hooo. A man is not something from afar that I will do something like this to him. I gave birth to you. Why should I kill you?'

'No. You will not kill me. My partner, see him. He sits on a stool at the front of the men's house. He came out and sat. The wind blows his feathers. You will see us, two partners sitting along each other in one place.' He sat. His mother said:

'So, I will kill you now.'

She made a hole with a pig's tooth. She broke the skull. Only blood was on the skull. She fastened four feathers into a crest. She pushed them inside the skull. Her son tried to shake his head.

'You see. I did not die.' His mother replied:

'Lies. Now you are dying.' He said to her:

'Sit here and watch me. I am going to the men's house. I am going out now.' He went outside. Wind blew the feathers as he walked. He walked on the path that goes to the men's house. He walked and came near the men's house.

'I will go inside, I think.' At the front of the men's house he fell on the ground. Screaming he lay down. So, he died. His mother said to him:

'So I killed you. I, your real mother who gave birth to you, I killed you.'

The mother cried and danced around him. She looked at him and danced [she was mourning]. They brought him inside the house. Everyone came. They looked at him and cried. They gathered on one spot. They cried and stayed awake till dawn, with this man. His mother had killed him. It was revenge; not for any other reason. The one who killed his mother paid back. He also lied.

They cut a sago palm. They opened the bark, pressed down its edges, removed the pith and threw it away. They placed him on the bark inside the house. He stank in the bark of a sago palm. He decayed.[2]

The man who had killed his mother was there too. He went into the forest and found the branches where crowned goura pigeons used to sit at night. Under the place where they sat the faeces had grown high. He knew that many pigeons came to this tree to sleep on it till dawn. He thought:

'I should come in the afternoon and shoot them with arrows. I should sit here and wait for them. But I left my bow behind.' So he left. All right, he walked and came to the outside of the village. A young boy was there.

'Come.' He called the young boy, a young boy ready to go to the men's house. He said to him:

[2] In the past, the Ambonwari used to place their dead into the sago trunk from which they removed the pith. They put this bark inside a house and made holes in the bark so that the fluids of a decomposing corpse could flow freely under a house. They built fences of sago leaves around the body. People buried the bones, together with the sago bark, under the house. The skull was washed, cleaned, and decorated. Ears, nose, and face were filled with clay, shells replaced eyes, vines were tied to the skull. These skulls, together with the skulls of enemies killed, were kept in their housholds. When men went hunting or when there was sickness in a household, people sought advice from the ghosts embodied in these skulls. The sudden movement of skulls for whatever reason was understood as the answer of spirits. Only recently the Ambonwari began placing their dead inside old canoes and burying them at the back of their houses, under the men's house floor, or at the forest cemetery accross the Maramun Creek.

'Come. We two will go. The goura pigeons sit on the tree in the afternoon. I saw the place where they sit. Leave them to sit first. When they are asleep I will shoot them and throw them down to you. I will go up the tree with a bow and you will stand under the tree.' The spirit of the man, the spirit of his partner had heard. He heard him talking thus. Vengeful man thought:

'I have already heard you.'

At twilight, the boy whom he asked to come with him went to his sister's place [not telling anyone]. The man waited, waited . . . sat, sat.

'Eey, the boy did not come.' The spirit of the dead man said to his partner:

'Eey, are you here?' He did not call him partner, no. He thought:

'It would not be good if he recognized me. Are you here?'

'Eey, you came.'

'Yes.'

'All right, let's go.' He took his bow and a piece of firewood and said:

'Come on, we are going.' The two of them left. The spirit of the dead man appeared as the little boy. He copied the boy's feet. The two of them left. They went, went. The two of them walked and walked. So, and so, and so. Near the tree trunk he said to the boy:

'There, they sit.' They [the pigeons] cooed and sat. The spirit said to him:

'Come on, climb!

'You stand here!

'You go up!

The man climbed up and shot them. He shot them. Yes! Yes! Yes! [they fell down]. The spirit stood there pretending to catch them. The man on the top of the tree said:

'I shot all the pigeons. But they did not hit the ground. I did not hear them hit the ground.' They did not hit the ground. They fell inside [the mouth and belly] of the spirit. Here [the narrator points at his belly]. He stood with his mouth opened.

'Throw the arrows down! Throw the bow too!'

'Ho, why should I throw the bow down? I will bring it down myself.'

'Throw it! You must throw the bow!' So he threw his bow down. The spirit took it. He took and held his bow.

'Wait. I shall climb down.'

'Why don't you jump down as well. I will catch you.'

'Ho, me? No need. I am not a little boy that you could catch me.'

'All right. Come down. Jump! Come on, you must jump!' He jumped expecting that the boy would catch him. He went straight into the belly of the spirit. The spirit returned to the village and jumped into the bark where the body was decaying. He had the man in his belly now.

They searched for him everywhere. The boy, a 'real' young boy, said:

'Ooo, he told me about his plan of shooting the pigeons.'

They all went into the forest. They searched for him at the place where he went to shoot pigeons. Ooo, nothing. They found the footprints. Everyone tried to find the missing man. At dawn they still looked for him. Someone said:

'Listen! The man we put in the sago bark, why does he make such noise? It is not the faint noise that we heard at night.' So they loosened the bark. They loosened and opened it.

'Ooo, this man is inside. The spirit of a dead man killed him and brought him.'

All right. Enough of this talk of mine. This is a story of the ancestors about the breast of a woman. Her name was Mambinja. Her story is now finished.

Variants

One of the variants of this story involves Kapi as the one who cut off the breast of Mambinja [Mambinja in other myths is a sister of Mambi and Sapɨn, from whom Kapi received his knowledge and powers]. It is said that the event occured while Kapi was still in Aibom. Kapi's partner gave him sago with bananas cooked in leaves. Kapi had never tasted or even seen bananas before. After he had eaten them his partner told him that what was inside was one of his mother's breasts. Kapi asked one of the women there if she would give him one of her breasts. She refused. Instead, Kapi asked Mambinja to give him a breast. Kapi took a bamboo knife and cut off her breast. She died. He buried her under the hill as she had requested. She returned to life as another woman who gave her breast to Kapi and told him to carry it with him. She told him that it would become a 'stone of the sea'. Kapi wrapped it in banana leaves. During his further travels it turned to stone.

This is the stone breast that is kept in the village between two halves of coconut shell, one of which has a hole so that the breast can peep outside (see Plate 19).

In other variants of the story the two partners had been tricking one another for some time. Thus, A cut a tree with a stone axe but told B that he had used a wooden one. When B tried his hands became sore. He took revenge by telling A that he had caught a fish simply by beating the water with his hands (though he had used a bow and arrow). A tried this in a creek full of *nil pis* (a fish with spines) and damaged his hands. A then performed the trick with bananas as mentioned above. B took revenge by telling his partner about the hole in his skull and how to put feathers in. The stories invariably end with the revenge of the spirit of the dead man, A. People say that they both rotted in the sago trunk. The first sago grubs grew from their bodies and that is, people say, why the Ambonwari have so many grubs.[3]

[3] Wassmann (1991: 85–6) recorded a similar myth among Iatmul. In the final part it represents the myth of origin of 2 crocodiles called Palingawi and Mavakngawi, carved by one of the two friends. These crocodiles killed the other friend and some other men. Their bodies were laid down in Paleagwi, near Chambri, where the new earth was created from their blood. Obviously, the myth's origin can be traced to the period when the Ambonwari lived in Aibom, that is, before they left for Karawari. It seems too that only Karawari people retained the practice of carving the crocodiles.

REFERENCES

Ackerman, C. (1976). Omaha and 'Omaha'. *American Ethnologist* 3(4): 555–72.

Adam, B. (1990). *Time and Social Theory*. Cambridge: Polity Press.

Aitken, T. C. (1990). Tapets: Drumsignals of the Sawos. In Lutkehaus et al. (1990: 546–7).

Alonso, A. M. (1988). The Effects of Truth: Re-presentations of the Past and the Imagining of Community. *Journal of Historical Sociology* 1(1): 33–57.

Appadurai, A. (1981). The Past as a Scarce Resource. *Man* 16(2): 201–19.

van Baal, J. (1966). *Dema*. The Hague: Martinus Nijhoff.

Bach, R. (1970). *Jonathan Livingston Seagull*. London: Turnstone Press.

Barnes, J. (1971). Time Flies Like an Arrow. *Man* 6(4): 537–52.

Barnes, R. H. (1974). *Kédang: A Study of the Collective Thought of an Eastern Indonesian People*. Oxford: Clarendon Press.

——(1976). Dispersed Alliance and the Prohibition of Marriage: Reconsideration of McKinley's Explanation of Crow–Omaha Terminologies. *Man* 11(3): 384–99.

Basso, K. H. (1988). 'Speaking with Names': Language and Landscape among the Western Apache. *Cultural Anthropology* 3(2): 99–130.

Bateson, G. (1932). Social Structure of the Iatmül People of the Sepik River. *Oceania* 2(3): 245–91; 2(4): 401–53.

——(1958 [1936]). *Naven*. London: Wildwood House.

Bergson, H. (1912). *Time and Free Will: An Essay on the Immediate Data of Consciousness*. London: George Allen.

Biersack, A. (1990). Histories in the Making: Paiela and Historical Anthropology. *History and Anthropology* 5: 63–85.

——(1991). Prisoners of Time: Millenarian Praxis in a Melanesian Valley. In A. Biersack (ed.), *Clio in Oceania: Toward a Historical Anthropology*. Washington, DC, Smithsonian Institution, 231–95.

Bloch, M. (1977). The Past and the Present in the Present. *Man* 12(2): 278–92.

Blong, R. J. (1981). Time of Darkness: Legends and Volcanic Eruptions in Papua New Guinea. In D. Denoon and R. Lacey (eds.), *Oral Tradition in Melanesia*. Port Moresby: University of Papua New Guinea and Institute of Papua New Guinea Studies, 141–50.

——(1982). *The Time of Darkness: Local Legends and Volcanic Reality in Papua New Guinea*. Canberra: Australian National University Press.

Bohannan, P. (1953). Concepts of Time among the Tiv of Nigeria. *Southwestern Journal of Anthropology* 9(3): 251–62.

Bourdieu, P. (1977). *Outline of a Theory of Practice*. Cambridge: Cambridge University Press.

——(1990). *The Logic of Practice*. Cambridge: Polity Press.

——(1991). *Language and Symbolic Power*, ed. J. B. Thompson. Cambridge: Polity Press.

——and Wacquant, L. J. D. (1992). *An Invitation to Reflexive Sociology*. Cambridge: Polity Press.

Bowden, R. (1983). Kwoma Terminology and Marriage Alliance: The 'Omaha' Problem Revisited. *Man* 18(4): 745–65.

——(1984). Art and Gender Ideology in the Sepik. *Man* 19(3): 445–58.

——(1990). The Architecture and Art of Kwoma Ceremonial Houses. In Lutkehaus et al. (1990: 480–90).

Brison, K. J. (1992). *Just Talk: Gossip, Meetings, and Power in a Papua New Guinea Village*. Berkeley, Calif.: University of California Press.

Bühler, A. (1961). Kultkrokodile vom Korewori (Sepik District, Territorium Neuguinea). *Zeitschrift für Ethnologie* 86: 183–207.

Burman, R. (1981). Time and Socioeconomic Change on Simbo, Solomon Islands. *Man* 16(2): 251–67.

Burridge, K. (1969). *Tangu Traditions*. Oxford: Clarendon Press.

Carr, D. (1986). *Time, Narrative, and History*. Bloomington, Ind.: Indiana University Press.

Cassirer, E. (1955). *The Philosophy of Symbolic Forms*, ii: *Mythical Thought*. New Haven, Conn.: Yale University Press (originally published in German in 1925).

——(1972 [1944]). *An Essay on Man: An Introduction to a Philosophy of Human Culture*. New Haven, Conn.: Yale University Press.

Clifford, J. (1982). *Person and Myth: Maurice Leenhardt in the Melanesian World*. Berkeley, Calif.: University of California Press.

Connerton, P. (1989). *How Societies Remember*. Cambridge: Cambridge University Press.

Copley, B. T. (1954). Village Population Register. Patrol Report No. 6/54–5. District of Sepik.

Craig, B. (1992). National Cultural Property in Papua New Guinea: Implications for Policy and Action. *Journal of Anthropological Society of South Australia* 30(2): 72–139.

Crapanzano, V. (1979). Preface to Leenhardt (1979: pp vii–xxix).

Durkheim, E. (1961 [1912]). *The Elementary Forms of the Religious Life*. New York: Collier.

Eliade, M. (1972 [1954]). *The Myth of the Eternal Return*. Princeton, NJ: Bollingen.

Errington, F. (1974). Indigenous Ideas of Order, Time, and Transition in a New Guinea Cargo Movement. *American Anthropologist* 1(2): 255–67.

——and Gewertz, D. B. (1987). *Cultural Alternatives and a Feminist Anthropology: An Analysis of Culturally Constructed Gender Interests in Papua New Guinea*. Cambridge: Cambridge University Press.

Evans-Pritchard, E. E. (1974 [1940]). *The Nuer*. Oxford: Oxford University Press.

Fabian, J. (1983). *Time and the Other: How Anthropology Makes Its Objects*. New York: Columbia University Press.

Fajans, J. (1985). The Person in Social Context: The Social Character of Baining 'Psychology'. In G. M. White and J. Kirkpatrick (eds.), *Person, Self, and Experience: Exploring Pacific Ethnopsychologies*. Berkeley, Calif.: University of California Press, 367–97.

Farriss, N. M. (1987). Remembering the Future, Anticipating the Past: History, Time, and Cosmology among the Maya of Yucatan. *Comparative Study of Society and History* 29(3): 566–93.

Feld, S. (1989). Wept Thoughts: The Voicing of Kaluli Memories. *Oral Tradition*: 241–66.

FELD, S. (1990). *Sound and Sentiment: Birds, Weeping, Poetics, and Song in Kaluli Expression*, 2nd edn. Philadelphia: University of Philadelphia Press.

FOLEY, W. A. (1986). *The Papuan Languages of New Guinea*. Cambridge: Cambridge University Press.

——(1991). *The Yimas Language of New Guinea*. Stanford, Calif.: Stanford University Press.

FORGE, A. (1962). Paint: A Magical Substance. *Pallete* 9: 9–16.

——(1966). Art and Environment in the Sepik. *Proceedings of the Royal Anthropological Institute* 1965: 23–31.

——(1971). Marriage and Exchange in the Sepik: Comments on Francis Korn's Analysis of Iatmul Society. In R. Needham (ed.), *Rethinking Kinship and Marriage*. ASA Monograph No. 2. London: Tavistock, 133–44.

——(1972). The Golden Fleece. *Man* 7(4): 527–40.

——(1973). Introduction to Forge (ed.), *Primitive Art and Society*. London: Oxford University Press, pp. xiii–xxii.

——(1979). The Problem of Meaning in Art. In S. M. Mead (ed.), *Exploring Visual Art of Oceania*. Honolulu: University Press of Hawaii, 278–86.

FOUCAULT, M. (1979). *Discipline and Punish: The Birth of the Prison*. New York: Vintage.

FRASER, J. T. (1987). *Time: The Familiar Stranger*. Amherst, Mass.: University of Massachusetts Press.

FROMM, E. (1995 [1976]). *To Have or to Be?* London: Abacus.

GARDNER, D. (1983). Performativity in Ritual: The Mianmin Case. *Man* 18: 346–60.

GELL, A. (1975). *Metamorphosis of the Cassowaries: Umeda Society, Language and Ritual*. London: Athlone Press.

——(1979). The Umeda Language-Poem. *Canberra Anthropology* 2(1): 44–62.

——(1992). *The Anthropology of Time: Cultural Constructions of Temporal Maps and Images*. Oxford: Berg.

GELLNER, E. (1964). *Thought and Change*. Chicago: University of Chicago Press.

VAN GENNEP, A. (1960 [1908]). *The Rites of Passage*. London: Routledge & Kegan Paul.

GEWERTZ, D. B. (1977a). The Politics of Affinal Exchange: Chambri as a Client Market. *Ethnology* 16(3): 285–98.

——(1977b). 'On Whom Depends the Action of the Elements': Debating among Chambri People of Papua New Guinea. *Journal of the Polynesian Society* 86(3): 339–53.

——(1983). *Sepik River Societies: A Historical Ethnography of the Chambri and Their Neighbours*. New Haven, Conn.: Yale University Press.

——and ERRINGTON, F. K. (1991). *Twisted Histories, Altered Contexts: Representing the Chambri in a World System*. Cambridge: Cambridge University Press.

GILLISON, G. (1993). *Between Culture and Fantasy: A New Guinea Highlands Mythology*. Chicago: University of Chicago Press.

HABERLAND, E. (1975). Kannibalismus und Kultkrokodile am Mittleren Korowori (Sepik-District, Neuguinea). *Abhandlungen und Berichte des Staatlichen Museums für Völkerkunde, Dresden* 34: 541–53.

HANDELMAN, D. (1984 [1979]). Is Naven Ludic? Paradox and the Communication of Identity. In B. Kapferer (ed.), *The Power of Ritual: Transition, Transformation and Transcendence in Ritual Practice*. *Social Analysis* (special issue) 1: 177–91.

HARRISON, S. (1982a). Yams and the Symbolic Representation of Time in a Sepik River Village. *Oceania* 53(2): 141–62.

——(1982b). *Laments for Foiled Marriages: Love Songs from a Sepik River Village.* Boroko: Institute of Papua New Guinea Studies.

——(1984). New Guinea Highland Social Structure in a Lowland Totemic Mythology. *Man* 19(3): 389–403.

——(1985). Names, Ghosts and Alliance in Two Sepik River Societies. *Oceania* 56(2): 138–46.

——(1990a). *Stealing People's Names: History and Politics in a Sepik River Cosmology.* Cambridge: Cambridge University Press.

——(1990b). Concepts of the Person in Avatip Religious Thought. In Lutkehaus et al. (1990: 351–63).

——(1993). *The Mask of War: Violence, Ritual and the Self in Melanesia.* Manchester: Manchester University Press.

HEELAS, P. (1986). Emotion Talk across Cultures. In R. Harré (ed.), *The Social Construction of Emotions,* Oxford: Basil Blackwell, 234–66.

HEIDEGGER, M. (1971). *Poetry, Language, Thought.* New York: Harper & Row.

——(1987 [1959]). *An Introduction to Metaphysics.* New Haven, Conn.: Yale University Press.

——(1988). *The Basic Problems of Phenomenology.* Bloomington, Ind.: Indiana University Press.

HILL, J. D. (ed.) (1988). *Rethinking History and Myth: Indigenous South American Perspectives on the Past.* Urbana, Ill.: University of Illinois Press.

——and WRIGHT, R. M. (1988). Time, Narrative and Ritual: Historical Interpretations from an Amazonian Society. In Hill (1988: 78–105).

HOSKINS, J. (1993). *The Play of Time: Kodi Perspectives on Calendars, History and Exchange.* Berkeley, Calif.: University of California Press.

HOWE, L. E. A. (1981). The Social Determination of Knowledge: Maurice Bloch and Balinese Time. *Man* 16(2): 220–34.

HUSSERL, E. (1964). *The Phenomenology of Internal Time-Consciousness.* Bloomington, Ind.: Indiana University Press.

JACKSON, M. (1983). Thinking Through the Body: An Essay on Understanding Metaphor. *Social Analysis* 14: 127–49.

——(1989). *Paths Toward a Clearing: Radical Empiricism and Ethnographic Inquiry.* Bloomington, Ind.: Indiana University Press.

KABERRY, P. M. (1941). The Abelam Tribe, Sepik District, New Guinea: A Preliminary Report. *Oceania* 11(3): 233–59, 345–67.

——(1967). The Plasticity of New Guinea Kinship. In M. Freedman (ed.), *Social Organization: Essays Presented to Raymond Firth.* Chicago: Aldine, 105–123.

KAPFERER, B. (1984 [1979]). Introduction: Ritual Process and the Transformation. In Kapferer (ed.), *The Power of Ritual: Transition, Transformation and Transcendence in Ritual Practice. Social Analysis* (special issue) 1: 3–19.

KLEINMAN, A. (1986). *Social Origins of Distress and Disease: Depression, Neurasthenia, and Pain in Modern China.* New Haven, Conn.: Yale University Press.

KORN, F. (1971). A Question of Preferences: The Iatmul Case. In R. Needham (ed.), *Rethinking Kinship and Marriage.* ASA Monograph No. 2. London: Tavistock, 99–132.

Kulick, D. (1992). *Language Shift and Cultural Reproduction: Socialization, Self, and Syncretism in a Papua New Guinean Village*. Cambridge: Cambridge University Press.

Lacey, R. (1990). Temps Perdu et Temps Vécu: Cross-cultural Nuances in the Experience of Time among the Enga. *Pacific Studies* 13(2): 77–102.

Lawrence, P., and Meggitt, M. J. (eds.) (1965). *Gods, Ghosts, and Men in Melanesia: Some Religions of Australian New Guinea and the New Hebrides*. Melbourne: Oxford University Press.

Lea, D. A. M. (1964). Abelam Land and Sustenance. Ph.D. dissertation, Australian National University.

Leach, E. (1965 [1954]). *Political Systems of Highlands Burma*. Boston: Beacon Press.

——(1966). Two Essays Concerning the Symbolic Representation of Time. In *Rethinking Anthropology*. London: Athlone Press, 124–36.

Leenhardt, M. (1979 [1947]). *Do Kamo: Person and Myth in the Melanesian World*. Chicago: University of Chicago Press.

LeRoy, J. (ed.) (1985a). *Kewa Tales*. Vancouver: University of British Columbia Press.

——(1985b). *Fabricated World: An Interpretation of Kewa Tales*. Vancouver: University of British Columbia Press.

Lévi-Strauss, C. (1965). The Future of Kinship Studies. *Proceedings of the Royal Anthropological Institute of Great Britain and Ireland* 1965: 13–22.

——(1966). *The Savage Mind*. Chicago: University of Chicago Press.

——(1968). *Structural Anthropology*. London: Allen Lane The Penguin Press.

——(1969). *The Elementary Structures of Kinship*. Boston: Beacon Press.

——(1992). *The View from Afar*. Chicago: University of Chicago Press.

Levy, R. I. (1973). *Tahitians: Mind and Experience in the Society Islands*. Chicago: University of Chicago Press.

——(1984). Emotion, Knowing, and Culture. In R. A. Schweder and R. A. LeVine (eds.), *Culture Theory: Essays on Mind, Self, and Emotion*. Cambridge: Cambridge University Press, 214–37.

Lewis, G. (1975). *Knowledge and Illness in a Sepik Society: A Study of the Gnau, New Guinea*. London: Athlone Press.

——(1980). *Day of Shining Red: An Essay on Understanding Ritual*. Cambridge: Cambridge University Press.

——(1990). Gestures of Support. In Lutkehaus et al. (1990: 255–65).

Lounsbury, F. G. (1964). The Formal Analysis of Crow- and Omaha-type Kinship Terminologies. In W. H. Goodenough (ed.), *Explorations in Cultural Anthropology*. New York: McGraw-Hill, 351–93.

Lutkehaus, N., Kaufmann, C., Mitchell, W. E., Newton, D., Osmundsen, L., and Schuster, M. (eds.) (1990). *Sepik Heritage: Tradition and Change in Papua New Guinea*. Bathurst: Crawford House Press.

Lutz, C. (1988). *Unnatural Emotions: Everyday Sentiments on a Micronesian Atoll and Their Challenge to Western Theory*. Chicago: University of Chicago Press.

Mai, P. (1981). The 'Time of Darkness' or *Yuu Kuia*. In D. Denoon and R. Lacey (eds.), *Oral Tradition in Melanesia*. Port Moresby: University of Papua New Guinea and Institute of Papua New Guinea Studies, 125–40.

Malinowski, B. (1922). *Argonauts of the Western Pacific*. London: Routledge & Kegan Paul.

——(1927). Lunar and Seasonal Calendar in the Trobriands. *Journal of the Royal Anthropological Institute* 57: 203–15.

MALTZ, D. N. (1968). Primitive Time-reckoning as a Symbolic System. *Cornell Journal of Social Relations* 3: 85–111.

MAY, R. J. (1984). *Kaikai Aniani: A Guide to Bush Foods, Markets, and Culinary Arts of Papua New Guinea.* Bathurst: Robert Brown.

MCDOWELL, N. (1976). Kinship and Exchange: The *Kamain* Relationship in a Yuat River Village. *Oceania* 42(1): 36–48.

——(1977). The Meaning of 'Rope' in a Yuat River Village. *Ethnology* 16: 175–83.

——(1978). Flexibility of Sister Exchange in Bun. *Oceania* 48(3): 207–31.

——(1980). It's Not Who You Are But How You Give That Counts: The Role of Exchange in a Melanesian Society. *American Ethnologist* 7(1): 58–70.

——(1985). Past and Future: The Nature of Episodic Time in Bun. In D. Gewertz and E. Schieffelin (eds.), *History and Ethnohistory in Papua New Guinea. Oceania* Monograph No. 28: 26–39.

——(1990). Intergenerational Exchange in Diachronic Context: A Melanesian Example. *Anthropos* 85(4–6): 393–401.

——(1991). *The Mundugumor: From the Field Notes of Margaret Mead and Reo Fortune.* Washington, DC: Smithsonian Institution.

MCKINLEY, R. (1971a). A Critique of the Reflectionist Theory of Kinship Terminology: The Crow/Omaha case. *Man* 6(2): 228–47.

——(1971b). Why Do Crow and Omaha Kinship Terminologies Exist? A Sociology of Knowledge Interpretation. *Man* 6(3): 408–26.

MEAD, M. (1963 [1935]). *Sex and Temperament in Three Primitive Societies.* New York: Morrow Quill.

——(1970 [1938–40]). *The Mountain Arapesh,* ii: *Arts and Supernaturalism.* New York: Natural History Press.

MERLEAU-PONTY, M. (1962). *Phenomenology of Perception.* London: Routledge & Kegan Paul.

MIHALIC, F. (1971). *The Jacaranda Dictionary and Grammar of Melanesian Pidgin.* Milton, Queensland: Jacaranda Press.

MIMICA, J. (1988). *Intimations of Infinity: The Mythopoeia of the Iqwaye Counting System and Number.* Oxford: Berg.

——(1991). The Incest Passions: An Outline of the Logic of Iqwaye Social Organization. *Oceania* 62 (1 and 2): 34–58, 81–113.

MULLER, J.-C. (1980). Straight Sister-Exchange and the Transition from Elementary to Complex Structures. *American Ethnologist* 7(3): 518–28.

MUNN, N. (1986). *The Fame of Gawa: A Symbolic Study of Value Transformation in a Massim (Papua New Guinea) Society.* Cambridge: Cambridge University Press.

——(1992). The Cultural Anthropology of Time: A Critical Essay. *Annual Review of Anthropology* 21: 93–123.

MYERS, F. R. (1979). Emotions and the Self: A Theory of Personhood and Political Order among Pintupi Aborigines. *Ethos* 7(4): 343–70.

NEEDHAM, R. (1971). Remarks on the Analysis of Kinship and Marriage. In Needham (ed.), *Rethinking Kinship and Marriage.* London: Tavistock, 1–34.

OHNUKI-TIERNEY, E. (1973). Sakhalin Ainu Time Reckoning. *Man* 8(2): 285–99.

PANOFF, M. (1969). The Notion of Time among the Maenge People of New Britain. *Ethnology* 8(2): 153–66.

PARKIN, D. (1992). Ritual as Spatial Direction and Bodily Division. In D. de Coppet (ed.), *Understanding Rituals*. London: Routledge, 11–25.

READ, K. E. (1955). Morality and the Concept of the Person among Gahuku-Gama. *Oceania* 25(4): 233–82.

RICOEUR, P. (1966). *Freedom and Nature: The Voluntary and the Involuntary*. Evanston, Ill.: Northwestern University Press.

——(1979). The Human Experience of Time and Narrative. *Research in Phenomenology* 9: 17–34.

——(1980). Narrative Time. *Critical Inquiry* 7(1): 169–90.

——(1984). *Time and Narrative*, i. Chicago: University of Chicago Press.

——(1991). Narrative Identity. In D. Wood (ed.), *Narrative and Interpretation*. London: Routledge, 188–99.

——(1992). *Oneself as Another*. Chicago: University of Chicago Press.

RIGBY, P. (1983). Time and Historical Consciousness: The Case of Ilparakuyo Maasai. *Comparative Studies in Society and History* 25(3): 428–56.

ROSALDO, R. (1980). *Ilongot Headhunting 1883–1874: A Study in Society and History*. Stanford, Calif.: Stanford University Press.

ROSCOE, P. B. (1989). The Pig and the Long Yam: The Expansion of a Sepik Cultural Complex. *Ethnology* 28(3): 219–31.

SAHLINS, M. (1985). *Islands of History*. Chicago: University of Chicago Press.

SARTRE, J. P. (1948). *The Emotions: Outline of a Theory*. New York: Citadel Press.

——(1956). *Being and Nothingness*. New York: Philosophical Library.

——(1992). *Truth and Existence*. Chicago: University of Chicago Press.

SCAGLION, R. (1986). Sexual Segregation and Ritual Pollution in Abelam Society. In T. L. Whitehead and M. E. Conaway (eds.), *Self, Sex, and Gender in Cross-Cultural Fieldwork*. Urbana, Ill.: University of Illinois Press, 151–63.

SCHIEFFELIN, E. L. (1976). *The Sorrow of the Lonely and the Burning of the Dancers*. New York: St. Martin's Press.

——and GEWERTZ, D. B. (1985). Introduction. In Gewertz and Schieffelin (eds.), *History and Ethnohistory in Papua New Guinea*. *Oceania* Monograph No. 28, 1–6.

SCHINDLBECK, M. (1980). *Sago bei den Sawos (Mittelsepik, Papua New Guinea): Untersuchungen über die Bedeutung von Sago in Wirtschaft, Sozialordnung und Religion*. Basle: Museum für Volkerkunde.

SCHUSTER, M. (1990). Aspects of the Aibom Concept of History. In Lutkehaus et al. (1990: 7–19).

SCHUTZ, A. (1972). *The Phenomenology of the Social World*. London: Heinemann Educational.

——(1977). Making Music Together: A Study in Social Relationship. In J. L. Dolgin, D. S. Kemnitzer, and D. M. Schneider (eds.), *Symbolic Anthropology: A Reader in the Study of Symbols and Meanings*. New York: Columbia University Press, 106–19.

SCHWIMMER, E. (1973). *Exchange in the Social Structure of the Orokaiva*. London: C. Hurst.

SOLOMON, R. C. (1980). Emotions and Choice. In A. O. Rorty (ed.), *Explaining Emotions*. Berkeley, Calif.: University of California Press, 251–81.

STANEK, M. (1983). *Sozialordnung und Mythik in Palimbei*. Basel: Ethnologisches Seminar der Universität und Museum für Völkerkunde.

——(1990). Social Structure and the Iatmul. In Lutkehaus et al. (1990: 266–73).

STEINER, G. (1992 [1975]). *After Babel: Aspects of Language and Translation*, 2nd edn. Oxford: Oxford University Press.

STEPHEN, M. (1989). Dreaming and the Hidden Self: Mekeo Definitions of Consciousness. In G. Herdt and M. Stephen (eds.), *The Religious Imagination in New Guinea*. New Brunswick, NJ: Rutgers University Press, 160–86.

STRATHERN, M. (1968). *Popokl*: The Question of Morality. *Mankind* 6(2): 553–62.

——(1988). *The Gender of the Gift: Problems with Women and Problems with Society in Melanesia*. Berkeley, Calif.: University of California Press.

TAYLOR, J. (1987). Tango. *Cultural Anthropology* 2(4): 481–93.

TELBAN, B. (1988a). People, Illness, and Plants: Ethnomedicine in the Highlands Fringe of New Guinea. M.Sc. thesis, University of Zagreb.

——(1988b). The Role of Medical Ethnobotany in Ethnomedicine: A New Guinea Example. *Journal of Ethnobiology* 8(2): 149–69.

——(1993). Having Heart: Caring and Resentment in Ambonwari, Papua New Guinea. *Bulletin of the Slovene Ethnographic Museum* 54(3): 158–77.

——(1997). Being and 'Non-being' in Ambonwari (Papua New Guinea) Ritual. *Oceania* 67(4): 308–25.

THORNTON, R. J. (1980). *Space, Time, and Culture among the Iraqw of Tanzania*. New York: Academic Press.

TISCHNER, H. (1965). *Das Kultkrokodil vom Korewori*. Hamburg: Hamburgisches Museum für Völkerkunde und Vorgeschichte im Selbstverlag.

TITIEV, M. (1956). The Importance of Space in Primitive Kinship. *American Anthropologist* 58(5): 854–65.

TOREN, C. (1988). Making the Present, Revealing the Past: The Mutability and Continuity of Tradition as Process. *Man* (NS) 23: 696–717.

——(1990). *Making Sense of Hierarchy: Cognition as Social Process in Fiji*. London: Athlone Press.

——(1993). Making History: The Significance of Childhood Cognition for a Comparative Anthropology of Mind. *Man* (NS) 28: 461–78.

TOWNSEND, P. K. (1974). Sago Production in a New Guinea Community. *Human Ecology* 2: 217–36.

TURNER, T. (1988). Ethno-ethnohistory: Myth and History in Native South-American Representations of Contact with Western Society. In Hill (1988: 235–81).

TURNER, V. (1980). Social Dramas and Stories About Them. *Critical Inquiry* 7(1): 141–68.

TUZIN, D. F. (1972). Yam Symbolism in the Sepik: An Interpretative Account. *Southwestern Journal of Anthropology* 28(3): 230–54.

——(1976). *The Ilahita Arapesh: Dimensions of Unity*. Berkeley, Calif.: University of California Press.

——(1980). *The Voice of the Tambaran: Truth and Illusion in Ilahita Arapesh Religion*. Berkeley, Calif.: University of California Press.

——(1992). Sago Subsistence and Symbolism among the Ilahita Arapesh. *Ethnology* 31(2): 103–14.

WAGNER, R. (1972). *Habu: The Innovation of Meaning in Daribi Religion.* Chicago: University of Chicago Press.

——(1978). *Lethal Speech: Daribi Myth as Symbolic Obviation.* Ithaca, NY: Cornell University Press.

——(1986). *Symbols That Stand for Themselves.* Chicago: University of Chicago Press.

WASSMANN, J. (1990). The Nyaura Concepts of Space and Time. In Lutkehaus et al. (1990: 23–35).

——(1991). *The Song to the Flying Fox: The Public and Esoteric Knowledge of the Important Men of Kandɨngei about Totemic Songs, Names and Knotted Cords (Middle Sepik, Papua New Guinea).* Boroko: National Research Institute.

WEINER, A. B. (1977). *Women of Value, Men of Renown: New Perspectives in Trobriand exchange.* St Lucia: University of Queensland Press.

——(1982). Sexuality among the Anthropologists: Reproduction among the Informants. In F. J. P. Poole and G. H. Herdt (eds.), *Sexual Antagonism, Gender, and Social Change in Papua New Guinea. Social Analysis* (special issue) 12: 52–65.

WEINER, J. (1988). *The Heart of the Pearl Shell: The Mythological Dimension of Foi Sociality.* Berkeley, Calif.: University of California Press.

——(1991). *The Empty Place: Poetry, Space, and Being Among the Foi of Papua New Guinea.* Bloomington, Ind.: Indiana University Press.

WHITING, J. W. M. (1941). *Becoming a Kwoma.* New Haven, Conn.: Yale University Press.

——and REED, S. W. (1938–9). Kwoma Culture: Report on Field Work in the Mandated Territory of New Guinea. *Oceania* 9: 170–216.

WILLIAMS, F. E. (1932). Trading Voyages from the Gulf of Papua. *Oceania* 3: 139–66.

WILLIAMSON, M. H. (1980). Omaha Terminology and Unilateral Marriage on the Sepik. *American Ethnologist* 7(3): 530–48.

YOUNG, M. W. (1971). *Fighting with Food: Leadership, Values and Social Control in a Massim Society.* Cambridge: Cambridge University Press.

——(1974). Private Sanctions and Public Ideology: Some Aspects of Self-Help in Kalauna, Goodenough Island. In A. L. Epstein (ed.), *Contention and Dispute: Aspects of Law and Social Control in Melanesia.* Canberra: Australian National University Press, 40–66.

——(1983). *Magicians of Manumanua: Living Myth in Kalauna.* Berkeley, Calif.: University of California Press.

ZEMP, H., and KAUFMANN, C. (1969). Pour une transcription automatique des 'langages tambourinés' Mélanésiens: un exemple kwoma, Nouvelle-Guinée. *L'Homme* 9(2): 38–88.

GLOSSARY OF KEY KARAWARI TERMS

Some terms already specified in tables and figures, such as temporals and kin terms, are excluded from this list.

Akɨmbrɨkupan the ancestor of the Bird of Paradise clan, Kapi's brother
andi land, ground, earth, ochre
anduk apianak our mouth, our language
anggɨndar-kwanar (*kwanma*) person's spirit (watchman, watchwoman)
arɨm skin
arkɨsan hot-tempered
awkunma female, woman
iman men's house
imɨngga village
imɨnggan anay the father of the village
imɨnggan kay the way of the village
imɨnggan saki the spirit of the village
imɨnggan yandang the mother of the village
kambra nothing
kanapanggɨnar head son, first-born son
kanggɨnɨng something
Kapi the ancestor of the Crocodile-1 clan, the village founder
karis sago pudding
karisikɨn stingy; strong, hard, firm
kay way, habit, manner; ritual; custom; law; being; canoe
konggong path; marriage
kunaypa a splinter of sago leaf's stem, debt (in adoption)
kupambɨn kay the way of the ancestors; custom, tradition
kupambɨn konggong the path of the ancestors; ancestral marriage
kupambɨn mariawk the story of the ancestors, myth
kupan big; elder brother
kurang *naven* ceremony, a celebration of someone's notable achievement
maman bad
mambara resentment
manggay temporary room for a menstruating girl
mar dead
mari sick
mariawk speech, discourse, narrative, story, talk; thought
ngandɨk with (being with, meaning 'to have')
pan very
pandayma menstruating girl
pan kupambi very ancestral, prohibition, taboo
panmari male, man
pɨsɨnɨm period, season, limited duration, span of time
saki male bush-spirit

sakima female bush-spirit
sambukwarinyar tail son, last-born son
sɨ- to do, to become, to feel
sɨmari sun
sɨmɨrinyar ritual inaugurator
sia dance
sipi baked sago
siria dancing and singing
sukunang resentment
sunggwi moon
sunggwin kay the way of the moon, menstruation
wambung insideness, Heart, understanding
warɨmbar exchange partner, dancing partner
warɨmbarar generous
wi name; night; dream
wunduma spirit of a dead woman
wundumbunar spirit of a dead man
yam house
yamɨn siria dancing and singing inside the house
yamɨnɨng house-insideness, lineage, clan
yapakupan good
yaprisipasɨkɨn quiet
yay blood

INDEX